Annual Conference and Expo of the Association of Avian Veterinarians (AAV 2017)

Soaring to New Heights

Held at ExoticsCon 2017

Washington, DC, USA
29 July – 2 August 2017

ISBN: 978-1-5108-4949-5

Printed from e-media with permission by:

Curran Associates, Inc.
57 Morehouse Lane
Red Hook, NY 12571

Some format issues inherent in the e-media version may also appear in this print version.

Printed by Curran Associates, Inc. (2018)

For permission requests, please contact Association of Avian Veterinarians
at the address below.

Association of Avian Veterinarians
PO Box 9
Teaneck, NJ 07666
USA

Phone: 720-458-4111
Fax: 720-398-3496

office@aav.org

Additional copies of this publication are available from:

Curran Associates, Inc.
57 Morehouse Lane
Red Hook, NY 12571 USA
Phone: 845-758-0400
Fax: 845-758-2633
Email: curran@proceedings.com
Web: www.proceedings.com

2017 PROCEEDINGS

SOARING TO NEW HEIGHTS

Association of Avian Veterinarians Conference 2017

JULY 29 - AUGUST 2, 2017 (EXHIBITS 30-2) | Washington, DC

Association of Avian Veterinarians

Advancing & Promoting Avian Medicine and Stewardship

AAVCONFERENCE.ORG

2017 PROCEEDINGS

SOARING TO NEW HEIGHTS

Association of Avian Veterinarians Conference 2017
JULY 29 - AUGUST 2, 2017 (EXHIBITS 30-2) | Washington, DC

Association of Avian Veterinarians
Advancing & Promoting Avian Medicine and Stewardship

AAVCONFERENCE.ORG

Introduction

We invite you to enjoy the benefits provided by these Proceedings of the 2017 Association of Avian Veterinarians (AAV) Annual Conference & Expo. The 2017 Annual Conference & Expo provides access to the leading avian researchers and practitioners world-wide sharing their experience and knowledge in person and through these proceedings.

AAV recognizes and thanks the following committees and staff for their time and hard work in assuring a quality program and publication:

AAV Conference Committee

Kenneth Welle, DVM, DABVP (Avian Practice), Conference Chair and Immediate Past-President
Geoff Olsen, DVM, DABVP (Avian Practice), Education Co-chair
Anna Osofsky, DVM, DABVP (Avian Practice), Education Co-chair
Michelle Hawkins, VMD, DABVP (Avian Practice), President
Lauren V. Powers, DVM, DABVP (Avian Practice), DABVP (Exotic Companion Mammal Practice), President-Elect and Veterinary Review Coordinator
Steven Metz, DVM, Practical Workshop Coordinator
Robert Groskin, DVM, Executive Director
Meghan R. Golden, MBA, Conference Director

Veterinary Manuscript Reviewers

Heather Barron, DVM, DABVP (Avian Practice); Greg Burkett, DVM, DABVP (Avian Practice); Sue Chen, DVM, DABVP (Avian Practice); Robert Dalhausen, DVM, MS, PhD (candidate); Laurel Degernes, DVM, MPH, DABVP (Avian Practice); Scott Echols, DVM, DABVP (Avian Practice); Thomas Edling, DVM, MSpVM; Jesse Fallon, DVM, MS; Scott Ford, DVM, DABVP (Avian Practice); Jennifer Graham, DVM, DAVBP (Avian Practice), DAVBP (Exotic Companion Mammal Practice), Dipl ACZM; Cheryl Greenacre, DVM, DABVP (Avian Practice), DABVP (Exotic Companion Mammal Practice); Vanessa Grunkemeyer, DVM, DABVP (Avian Practice); Michelle Hawkins, VMD, DABVP (Avian Practice); Jerry LaBonde, MS, DVM; Angela Lennox, DVM, DABVP (Avian Practice) DABVP (Exotic Companion Mammal Practice); Geoff Olsen, DVM, DABVP (Avian Practice); Anna Osofsky, DVM, DABVP (Avian Practice); Christal Pollock, DVM, DABVP (Avian Practice); David Sanchez-Migallon Guzman, LV, MS, DipECZM (Avian), DipECZM (Small Mammal), Dipl ACZM; Nico Schoemaker, DVM, PhD, DipECZM (Small Mammal), DipECZM (Avian), DABVP (Avian Practice); Marcy Souza, DVM, MPH, DABVP (Avian Practice), Dipl ACVPM; Brian Speer, DVM, DABVP (Avian Practice), DipECZM (Avian); Anneliese Strunk, DVM, DABVP (Avian Practice); Thomas N. Tully, Jr, DVM, MS, DABVP (Avian Practice), DipECZM (Avian); Yvonne van Zeeland, DVM, MVR, PhD, DipECZM (Avian), DipECZM (Small Mammal), CPBC; Kenneth Welle, DVM, DABVP (Avian Practice); Bianca Zaffarano, DVM; Ashley Zehnder, DVM, DABVP (Avian Practice)

Technical Editor

Charlotte Kirk Baer, MS

Conference Office Staff — North American Veterinary Community

Peter Scott, MBA, APR, Chief Operations Officer; Meghan R. Golden, MBA, Vice President of Industry Services & International Relations; Joe Sorrentino, MBA, CMP, Vice President of Operations; Mandy Taylor, MBA, APR, Director, Industry Services

Proceedings Layout and Design — BrightLizard, Inc.

Elizabeth Smith

Disclaimer

The material appearing in this publication comes exclusively from the authors and contributors identified in each manuscript. The techniques and procedures discussed reflect the personal knowledge and experience of the authors and contributors, and demonstrate their views of the methods that may be used for these medical procedures. The procedures demonstrated do not incorporate all known techniques, are not exclusive, and other techniques and technology may also be available. Any questions or requests for additional information concerning any of the manuscripts should be addressed directly to the authors.

AAV does not research, review, or otherwise verify any of the information contained in this publication. Opinions expressed in this publication are those of the authors and contributors and not necessarily those of AAV. AAV is not responsible for errors or for opinions expressed in this publication.

AAV expressly disclaims any warranties or guarantees, expressed or implied, and shall not be liable for damages of any kind in connection with the material, information, techniques, or procedures set forth in this publication.

Manuscripts contained in these conference proceedings vary in length and format based on author intentions for possible future publication. Authors are given the option to publish their complete manuscript herein as the final, official publication of the manuscript. Authors who wish to retain the right to publish a complete manuscript in another professional publication at a later date have limited their submission herein to an abstract only.

For permission to reproduce or order copies of this publication, contact the AAV office, www.aav.org, PO Box 9, Teaneck, New Jersey 07666, USA, e-mail pubs@aav.org.

For information on AAV educational conferences, visit www.AAVConference.org. For information on membership and other association programs visit www.aav.org.

Table of Contents

Section 1: Foundations to Avian Medicine

Section 2: Aviculture

Section 3: Poultry

Section 4: Sunday Master Class

Section 5: Plenary Session

Section 6: Research

Section 7: Clinical Session

Section 8: Tuesday Master Classes

Section 9: Wednesday Master Classes

Section 10: Wildlife

Section 11: Posters

Author Index

Klaus Hansen – page 379
Don J. Harris, DVM – page 3
Peter J. Havel, DVM, PhD – page 355
Michelle G. Hawkins, VMD, DABVP (Avian Practice) – page 101, 125
J. Jill Heatley, DVM, MS, DABVP (Avian Practice), DABVP (Reptile and Amphibian Practice), Dipl ACZM – page 133, 267, 339, 351
Melody Hennigh, RVT, KPA CTP – page 11
Johanna Heseltine DVM, MS, DACVIM – page 351
Jan Hooimeijer, DVM, CPBC – page 85
Sharman Hoppes DVM, DABVP (Avian Practice) – page 23, 49, 137
Ed Hsu, PhD – page 103
Caleb Hudson, DVM, Dipl ACVS–SA – page 149
Minh Huynh, DV, MRCVS, DipECZM (Avian) – page 105, 129
Mark Jackwood, PhD – page 75
Cathy A. Johnson–Delaney, DVM – page 285
Rick Jordan – page 49
Heather K. Knych, DVM, PhD – page 125, 355
Douglas Kondro, BSc – page 143
Rüdiger Korbel, Prof Dr med vet, DipECZM (Avian) – page 107, 123, 201
Fred Kromm – page 103
Pankaj Kumar, DVM – page 147
Stephanie Lamb, DVM, DABVP (Avian Practice) – page 145, 151, 361
Angela Lennox, DVM, DABVP (Avian Practice) – page 385
Marcie Logsdon, DVM – page 229
Kate Lyngle–Cowand – page 103
Melodiya Nyela Fampo Magno, DVM – page 343, 353
Christoph Mans, Dr med vet, Dipl ACZM – page 111
Anna Martel–Arquette, DVM – page 111
Anna McRee, DVM – page 113
Todd J. McWhorter, BSc, MSc, PhD – page 161
Samer Merchant – page 103
Brian Mihura, MS – page 267
Florent Modesto, DV, IPSAV (Zoological Medicine) – page 129
Jennilee Morrison, DVM – page 127, 375
Kristina Moser, Dr med vet – page 107
Bianca Murphy, DVM – page 141
Maureen Murray, DVM, DABVP (Avian Practice) – page 341
Javier G. Nevarez, DVM, PhD, Dipl ACZM, DipECZM (Herpetology) – page 113
Geoff Olsen, DVM, DABVP (Avian Practice) – page 141
Susan E. Orosz, PhD, DVM, DABVP (Avian Practice), DipECZM (Avian) – page 285
Anna Osofsky, DVM, DABVP (Avian Practice) – page 187
Seth Oster, DVM – page 345
Janet Gordon Palm, DVM, CVCP – page 367
Terri Parrott, DVM – page 171
Peter J. Pascoe, BVSc, DACVA – page 125
Maria E Passarelli, VMD – page 149
Seema Patel, BS – page 121
Joanne R. Paul–Murphy, DVM Dipl ACZM, Dipl ACAW – page 355

Juan C. Pavez, DVM, DACVA – page 125
Irene Pepperberg, PhD, SB, MA – page 85
Lesley Phillips, DVM – page 139
Julia B. Ponder, DVM, MPH – page 99, 337
Bruno H. Pypendop, PhD – page 125
Cameron M. Ratliff, DVM – page 339
Drury Reavill, DVM, DABVP (Avian Practice) DABVP (Reptile and Amphibian Practice), DACVP – page 385
Rachel Rech, DVM – page 339
Bianca Reisenbichler, med vet – page 107
Greg Rich, DVM – page 33
Jessica A. Robertson, DVM – page 355
April Romagnano, PhD, DVM, DABVP (Avian Practice) – page 139, 385
Michael Romero, MS, PhD – page 113
Karen E. Russell, DVM, PhD, DACVP – page 351
Sharon Schmalz, BS – page 267, 339, 351
Annika Schneeganss, med vet – page 107
Nico Schoemaker, DVM, PhD, DipECZM (Small Mammal), DipECZM (Avian), DABVP (Avian Practice) – page 237, 247, 255
David E. Scott, DVM – page 349
Leslie C. Sharkey, DVM, PhD, DACVP (Clinical Pathology) – page 99
Bruce A. Singbeil, DVM – page 75, 79
Victor Daniel Mateuta Sirbu, DVM – page 343
Amberly Sokoloff, VMD – page 101
Brian L. Speer, DVM, DABVP (Avian Practice), DipECZM (Avian) – page 11, 101, 141
Anneliese Strunk, DVM, DABVP (Avian Practice) – page 187
Boston Clark Terry – page 103
Linnea Tracy, BS – page 155
Jennifer Trail, CVT – page 121
Thomas N. Tully, Jr, DVM, MS, DABVP (Avian Practice), DipECZM (Avian) – page 113, 177
Katherine K. Turpen, DVM – page 121
Malou van der Sluis, BSc – page 97
Olivia van Steyvoort, DV – page 105
Yvonne van Zeeland, DVM, MVR, PhD, DipECZM (Avian), DipECZM (Small Mammal), CPBC – page 97, 237, 247, 255
Marike Visser, DVM – page 119, 131
Michiko Wallace, CVT – page 103
Kenneth R. Welle, DVM, DABVP (Avian Practice) – page 121
Valerie J. Wiebe, PharmD – page 355
Crystal Wilcox, LVT – page 103
Jackie M. Williams, DVM, MS, Dipl ACVR – page 111
Andrea Winkel-Blair, DVM, MPH – page 133
Louden Wright, DVM – page 147
Shangzhe Xie, BSc, BVMS, MVS – page 161
Graham Zoller, DV, IPSAV (Zoological Medicine) – page 105, 129

Section 1

Foundations to Avian Medicine

Thomas N. Tully, Jr, DVM, MS, DABVP (Avian Practice), DipECZM (Avian);
Paul Welch, DVM
Moderators

Avian Hematologic Testing

Don J. Harris, DVM

Session #2093

Affiliation: From the Avian & Exotic Animal Medical Center, 6380 S. Dixie Hwy. Miami, FL 33143, USA.

Introduction

Clinical pathology holds the key to unravelling much of the mystery surrounding the avian patient. Although many avian diseases present with identical clinical signs, laboratory data can often distinguish between infectious and metabolic disease, bacterial versus fungal, renal versus hepatic, etc. In a clinical situation, an appropriately broad selection of laboratory tests offers the best odds of quickly determining the nature of the patient's problem.

Complete Blood Count

The complete blood count (CBC) is one of the most important components of an avian diagnostic panel. For analytical purposes the CBC may be divided into components describing: the volume and character of the red blood cells; the numbers, percentages, and characteristics of the white blood cells; the concentration of solids in the plasma; the relative number of thrombocytes; and the presence or absence of blood-borne parasites. While many other tests provide information not demonstrated by the CBC, no other single test provides such a broad range of information. Differences exist between avian and mammalian blood but, once these differences are recognized, the similarity in functions of the various components becomes evident.

The primary differences between mammalian and avian blood are that:

- normal mature avian red blood cells are nucleated, and regenerative anemia is therefore demonstrated by polychromasia among the red cells in a stained smear; (basophilia of cytoplasm, increased nucleus to cytoplasmic ratio, and more spherical shape to the nucleus and cell).
- white blood cell types parallel mammalian cells except that the heterophil is present in avian blood instead of the neutrophil; and
- thrombocytes are present in avian blood instead of platelets.

Unlike dogs and cats, cell counts of birds may vary widely among members of a given species. To determine normal cell count for an individual, baseline data must be collected during periods of apparent good health. Reference values for various species have been published, but these tables should only be used as rough guidelines. Published ranges will typically be wide, therefore subtle patient variations may not be apparent.

Beyond these differences, the functions of the various cellular components in avian blood are roughly comparable to those of mammals. Infections, non-infectious inflammation, necrosis, neoplasia, etc may cause a leucocytosis.

A moderate heterophilia often indicates the presence of bacterial infections or cellular necrosis, and extremely high heterophil counts often accompany pansystemic illness such as chlamydiosis, aspergillosis, or tuberculosis. These changes are usually characterized by varying degrees of toxic changes in the white cells. Subtle-to-moderate heterophilias, without toxic changes in the white cells, may reflect stress leucograms.

An overwhelming bacterial infection, sepsis, or a severe viral infection may result in a leucopenia with a heteropenia or occasionally a lymphopenia. The leucopenia may be due to decreased production or increased consumption of the cell line. Increased consumption is evidenced by the presence of immature and toxic cells, findings not present with decreased production.

In some avian species, the relative lymphocyte count may be higher than in others. An absolute lymphocytosis may suggest a viral infection or certain stages of chlamydiosis.

A lymphopenia may occur in severe viral infections, such as avian circovirus in young African grey parrots. (This presentation, in fact, usually demonstrates a pancytopenia.)

A monocytosis implies the presence of chronic infection, granulomatous disease, or extensive necrosis in which a large amount of phagocytosis is occurring. Classic examples of this include chronic forms of aspergillosis, tuberculosis, and chlamydiosis.

Eosinophil functions have not been clearly defined. Intestinal parasitism may produce an eosinophilia, but not consistently. Peripheral eosinophilia does not appear to occur in allergic conditions.

Basophils are uncommon findings in normal avian haemograms. Conditions that cause their appearance include respiratory infections, resolution of tissue damage, parasitism, and some chlamydial infections.

Electrophoresis

The fractionation of plasma proteins via protein electrophoresis (EPH, SPE) is analogous to the separation and identification of white blood cells in the differential cell count. Just as different families of white cells are separately quantitated, so the relative percentages of the plasma proteins are measured. The technique of performing an electrophoresis does not produce absolute values of each protein fraction; rather, it reveals the percentage of each as part of the premeasured total protein. The absolute values must be calculated after the total protein has been determined through another method. The electrophoresis then yields information regarding a variety of physiologic and immunologic states of the patient. The primary categorization of avian plasma proteins includes pre-albumin, albumin and globulin components. Globulins are then divided (and sometimes subdivided) into alpha, beta, and gamma fractions.

One aspect of protein determinations that should always be observed is the albumin:globulin (A:G) ratio. More important than the patient's total plasma protein are the relative quantities of pre-albumin, albumin, and globulin. The ratio is calculated using the formula (pre-albumin + albumin)/globulins, and the normal A:G ratio ranges from 1.6-4.5. The importance of this ratio is illustrated by the following example. Snowflake and Peaches each have total plasma proteins of 4.0. At first glance, according to published data, each patient's proteins appear to be normal. Snowflake's albumin is 3.0 and his globulin is 1.0, resulting in an A:G ratio of 3.0, normal. Peaches, however, has an albumin of 1.0 with a globulin of 3.0. Peaches' A:G ratio is 0.33, grossly abnormal. Peaches' albumin is very low and her globulins are very high–an indication of a potentially serious condition. Peaches is losing or failing to produce albumin, while at the same time some portion of the globulins is being produced at an accelerated rate.

Pre-albumin and albumin fractions

The significance of the pre-albumin fraction of the serum protein in birds is uncertain. It may function as a transport protein, similar to albumin. There does not appear to be a comparable component in mammalian blood. In avian samples, it may comprise as much as 40 per cent of the total serum protein. In some species, it appears that low pre-albumin values may have the same significance as low albumin.

The albumin fraction typically comprises 45%–70% of avian serum protein in species that have high pre-albumin values, and tends to be lower in species with low pre-albumin values. Albumin functions primarily as an osmotic pressure regulator and a transport protein, as it does in mammalian species.

Globulin fraction

The globulin fraction has alpha (α), beta (β), and gamma (γ) components, and high-resolution electrophoresis will divide the globulins into the protein components listed under the α, β, and γ subgroups discussed below. Each of the 3 primary globulin fractions contains proteins active in different physiologic and pathophysiologic conditions.

Alpha globulins: Alpha globulins consist of 2 principal fractions: α1 and α2. Contained within this group of globulins are acute phase inflammatory proteins such as α-lipoprotein, α1-antitrypsin, α2-macroglobin, and haptoglobin. The α2-macroglobin sometimes migrates into the β range. One condition associated with elevated α globulin levels in birds is parasitism. Other consistent correlations have yet to be identified. Elevations in alpha globulins are somewhat uncommon.

Beta globulins: Beta globulins constitute other acute phase inflammatory proteins, including β2-macro-globulin, fibronectin, transferrin, and β-lipoprotein. In some species, namely the African grey parrot, the β component of the EPH consists of 2 primary components: β1 and β2. Elevated beta globulin levels may be indicative of chronic liver or kidney disease, or chronic inflammatory diseases such as aspergillosis or chlamydiosis. The most common reason for elevated beta globulins in birds, which is attributable to the transferrin component, is egg production. A significantly elevated beta globulin level, combined with a 1.5- to 2-fold increase in the blood calcium level, in birds of unknown sex, is almost 100% suggestive that the bird is an ovulating female.

Gamma globulins: In mammals, gamma globulins appear as 2 primary fractions: γ1 and γ2. In avian species, only one fraction is demonstrated. The primary components of the gamma globulins are antibodies, complement, and complement degradation products. Elevated gamma globulins are a common finding in birds suffering from acute *Chlamydia* infection.

Serum Biochemistries

Alanine aminotransferase

Alanine aminotransferase (ALT, SGPT) is an enzyme found in the cells of many avian tissues. In other animals, elevations have been shown to be associated with hepatocellular disruption, but no such association has been consistently demonstrated in birds. Little clinical significance can therefore be applied to ALT values in avian patients.

Albumin

The function of albumin is discussed above. It should be noted here that an accurate albumin determination is best calculated through electrophoresis. Because of the linear range of most laboratory instrumentation, currently utilized chemistry assays often do not provide accurate avian albumin measurements.

Amylase

In birds, the pancreas, liver, and small intestine produce amylase. Elevations have been associated with acute pancreatitis and enteritis. Because more than one source of amylase exists, an elevation is not in itself diagnostic.

Alkaline phosphatase

Alkaline phosphatase (AP, SAP) is found in bone, kidneys, intestine, and liver. The hepatic fraction composes only a very small proportion of the total reported in routine testing. Because changes in the hepatic fraction have little and inconsistent influence on the overall value, no correlation can be made between liver disease and AP levels. The inconsistent elevations from variable sources cause AP to be of almost no value in avian diagnostics. Disruption of bone probably causes elevations of AP more than other sources.

Aspartate aminotransferase

The intracellular enzyme most useful for diagnosing hepatocellular disruption in avian species is aspartate aminotransferase (AST, SGOT). Although present in liver, skeletal muscle, kidney, heart, and brain, elevations are frequently associated with liver disease or muscle damage. Whenever an elevation in AST is detected, the creatine kinase (CK) level should be reviewed. An elevated AST without a concurrent elevation in CK is highly suggestive of hepatocellular disruption. It should be emphasized that this does not confirm liver disease; nor does a normal AST positively rule out liver disease. As with all diagnostics, the AST provides evidence towards a diagnosis but does not in itself determine the diagnosis. Also, the AST in no way indicates the functional capacity of the liver. The bile acids test, discussed shortly, is more appropriately used to evaluate hepatic function.

Bile acids

Bile acids are produced by the liver to aid in the digestion of fats. After excretion into the intestinal tract, bile acids are reabsorbed and returned to the liver via the portal circulation. The liver then extracts the bile acids from the blood for recycling. Elevation of bile acids in the general circulation implies decreased ability of the liver to extract the bile acids from the portal circulation, and therefore suggests impaired liver function.

Confusion arises when it is noted that the liver is the organ of bile acid synthesis. It would seem logical that hepatic insufficiency would result in decreased production of bile acids, and therefore in decreased circulating levels. However, hepatic extraction of bile acids from the portal circulation is apparently more dependent on efficient liver function than is the synthesis of bile acids. It is reasonable to presume (and it does appear to happen) that at some point the production of bile acids does diminish and values fall. As with aspartate aminotransferase, a normal bile acid level does not absolutely rule out hepatic disease.

It is important to distinguish between the information provided by the AST and bile acids. Aspartate aminotransferase is a leakage enzyme, and therefore an inverse indicator of hepatocellular integrity, while the bile acids is an inverse indicator of liver function. One is not necessarily dependent on the other. For example, a patient may exhibit a normal AST but elevated bile acids. This would imply impaired liver function, even though the cells are intact; such conditions include hepatic lipidosis, chronic fibrosis, etc. Conversely, many diseases such as salmonellosis or acute chlamydiosis may cause hepatocellular damage without being extensive enough to impaire overall liver function. Marked elevations in the AST may be observed without concurrent elevations in bile acids. Again, normal values of either or both do not rule out hepatic disease. A totally fibrotic end-stage liver lacks enough functional hepatocytes to produce measurable AST or bile acids. Finally, a significantly elevated AST along with a significantly elevated bile acids presents a worse-case scenario in terms of liver disease.

Bilirubin

Since biliverdin is the major avian bile pigment, bilirubin is uncommonly observed in avian serum samples. In occasional cases of severe liver disease, significant bilirubin levels are present; therefore, hepatic pathology may be suspected in patients demonstrating elevated bilirubin levels.

Calcium

Calcium levels are profoundly influenced by a number of normal as well as pathologic conditions, and great care should be exercised in the interpretation of abnormal findings. Almost all pathologic changes are secondary to conditions not associated with dietary levels. Because of the effectiveness of the parathyroid gland, dietary deficiencies of calcium will rarely cause obvious subnormal blood levels. Blood calcium levels are also directly linked to albumin levels. Hypoalbuminaemia will result in artifactual depression of measured calcium levels. Other causes of lowered blood calcium levels include hypoparathyroidism in African grey parrots, glucocorticoid administration, and insufficient exposure to full spectrum lighting.

Dehydration will sometimes elevate albumin, and therefore blood calcium. Two-fold or greater elevations typically occur with ovulation. Elevated levels of calcium have been associated with vitamin D_3 toxicity, osteolytic bone tumours, renal adenocarcinoma, and dehydration.

Cholesterol

Cholesterol levels in birds may accompany various physiologic or pathologic conditions, but there is inconsistency and a lack of specificity associated with abnormal findings. Generally, elevations are associated with liver disease, hypothyroidism, high fat diets, and starvation, especially in obese birds. Subnormal levels are rarely significant but may be observed with endotoxaemia, aflatoxicosis, spirochaetosis, and low dietary fat. Unfortunately, there are no clear indicators to determine whether or not an abnormal cholesterol level is associated with a particular condition. For example, if an obese bird were to display elevated cholesterol, it would be unclear whether the elevation was a result of hypothyroidism, excessive dietary fat, hepatic lipidosis, or mobilization of body stores during anorexia. Other tests and observations usually provide evidence of these conditions, with or without the support of the cholesterol level. Again, normal values do not rule out the aforementioned conditions.

Creatinine

Creatinine levels in avian serum samples typically fall below a measurable range, and are rarely useful in avian clinical pathology. Also, certain technical factors contribute to a high incidence of artifactual changes. Elevations have been associated with kidney disease, but creatine is not considered to be a reliable indicator of renal function. Many commercial labs include creatine in an avian profile, but its significance should be regarded with suspicion.

Creatine kinase

The primary sources of creatine kinase (CK) include skeletal muscle, cardiac muscle, and nervous tissue, and elevations are associated with significant disruptions of these tissues. The primary usefulness of this enzyme is in distinguishing between hepatic and non-hepatic causes of elevated aspartate aminotransferase (AST). Any elevation in AST should be compared with the patient's CK level. If the CK is normal, it is relatively safe to conclude that the liver is the source of the elevated AST. If the CK is elevated along with the AST, muscle should be considered a possible source of the elevated AST. Other possibilities for the dual elevations would, of course, be concurrent liver and muscle or liver and neurologic disease. Lastly, an elevated CK in the absence of other biochemical pathology is often the result of significant neurologic disease.

Glucose

Pathologic changes in avian blood glucose levels principally involve elevations. Hypoglycemia is extremely rare in birds and, when present, is almost never associated with starvation. The primary cause of hypoglycemia in pet birds is septicemia.

Hyperglycemia occurs commonly due to stress or recent intake and, occasionally, from diabetes mellitus. Because of the frequency with which hyperglycemia is caused by stress, a diagnosis of diabetes mellitus should be considered carefully and only if other evidence supports it. A visibly normal hyperglycemic patient displaying no polydipsia, no polyuria, and no weight loss, etc should not automatically be considered diabetic. Repeat testing, other investigative tests and observation are necessary to confirm a diagnosis.

Lactate dehydrogenase

Lactate dehydrogenase (LDH) is found in skeletal and cardiac muscle, liver, kidney, bone, and erythrocytes. Elevations can be observed with disruption of any of these tissues or in hemolysis, and are therefore extremely non-specific. One benefit of measuring LDH levels may be in following the progress of liver disease, in which LDH levels apparently change more quickly than AST levels; lowering LDH values may imply improvement even though AST levels remain elevated.

Lipase

Serum lipase levels may be elevated in cases of acute pancreatitis. Currently, the only reliable antemortem confirmation of pancreatitis is by pancreatic biopsy.

Phosphorus

Elevated serum phosphorus is frequently observed in advanced renal failure. An elevated phosphorus level resulting from renal disease suggests chronicity and presents a guarded prognosis. Elevations are also observed in hypoparathyroidism and nutritional secondary hyperparathyroidism. Hemolysis may artifactually elevate serum levels. Malabsorption and vitamin D deficiencies may cause lowered blood phosphorus levels.

Potassium

As with sodium, pathologic changes in potassium levels indicate a serious and usually life-threatening clinical situation. Hyperkalemia develops with advanced kidney disease, adrenal disease, muscle damage, and during episodes of acidosis. Hypokalemia may result from loss through diarrhea, and during states of alkalosis.

Sodium

Changes in sodium values usually reflect serious conditions. Elevated levels occur with salt poisoning, water deprivation, and dehydration. Decreased levels occur due to sodium loss in kidney disease or diarrhea.

Total protein

A total serum protein (TP) level must be evaluated in light of its components, albumin, and globulin. The total value is influenced by various factors but, as discussed previously, a normal value does not rule out abnormalities of the individual protein components. Overall, dehydration and immune stimulation may cause a hyperproteinemia. Hypoproteinemia may be caused by overhydration, protein loss in kidney disease, starvation, liver disease, or intestinal disease. As stated earlier, the protein should never be considered normal until the A:G ratio is known to be normal.

Urea

Because of the low level of urea (blood urea nitrogen [BUN]) in avian blood, its usefulness is limited. Also, the avian kidney appears able to excrete most urea as long as the patient's hydration is adequate. Therefore, blood urea may be a better indicator of hydration than renal function.

Uric acid

The blood uric acid level is the primary indicator of renal function in birds. An elevated uric acid (UA) level is a reliable indicator that kidney function is impaired. With many tests, substantial elevations are necessary before there is reason for concern; however, even a subtle elevation in the uric acid warrants suspicion of renal disease. Conversely, serial determinations should be made after adequately hydrating the patient before concluding a diagnosis of renal disease. While it has been argued that dehydration has little effect on uric acid levels in birds, this is not the author's experience. Many patients with profoundly elevated UA levels have reverted to complete normalcy after fundamental rehydration. An artifactual elevation of UA often occurs if blood is collected via nail trim due to contamination of the nail with the patient's droppings.

Summary

Laboratory testing is an essential component of avian medicine. When examining avian patients, both routinely and in the face of illness, it is necessary to utilize an appropriate assortment of tests in order to obtain a reasonably complete profile of the patient. Knowing how to interpret the findings is as important as knowing which tests to utilize. By employing the right tests, and correctly interpreting the meaning behind the findings, avian patients can be diagnosed and treated with maximal accuracy and effectiveness.

High Impact Ways to Apply Behavioral Medicine in a Healthcare Setting

Brian L. Speer, DVM, DABVP (Avian Practice), DipECZM (Avian), and
Melody Hennigh, RVT, KPA CTP

Session #2038

Affiliation: From the Medical Center for Birds, 3805 Main Street, Oakley, CA 94561, USA.

Introduction

There is an increasing level of veterinary interest in in behavior and behavior-associated problems in avian species. In addition, there is improved and broadened awareness of how profoundly behavioral medicine can impact a veterinary practice's ability to provide optimal healthcare services to their patients and for their clients. Unfortunately, our understanding of the proper methods for the identification, description and address of these problems still remains inconsistent for most veterinarians caring for birds in general practice. For the most part, most medical records still tend to reflect the documentation of historical and physical examination findings, test recommendations and results, and treatments for various perceived or confirmed states of disease. Comments regarding behavioral observations remain rarely mentioned in most medical records. Recommendations for behavioral intervention overall are few or far from complete, and are uncommonly followed up and documented as to their outcome. In the absence of recognition and documentation of problems or a proactive discussion of behavior, relatively advanced states of behavioral problems should logically be expected to be seen in practice. As a result, behavioral intervention efforts seem to be comparatively futile, frustrating, and hence, seemingly not realistically practical to pursue in day-to-day practice. Furthermore, a more pervasive implementation of behavioral science into other aspects of practice remains fragmented and oftentimes less than ideal.

This manuscript and presentation explores specific key concepts and topics that veterinarians and their practice team can apply, expand their practice, and succeed with. Low stress handling and restraint, careful assessment of body language, strategic and appropriate use of conscious sedation or anesthesia to facilitate some medical procedures, the regular implementation of target training as a gateway to other trained behaviors and enrichment, and a different view of applying Skinner's paradigms to increase patient and client return visits are explored in their role in practice management concept and strategy development. Each topic presented can have a favorable impact on the success of your avian practice, satisfaction for you and your clients, and most importantly, great favorable impact on the health and welfare of your patients. Video is used to help identify behaviors, body language and behavior-change for attendees.

Low Stress Handling and Restraint

"*By restraining pets in a forceful, crude or unskilled manner, you could be breaking your promise to do no harm. Such handling can make pets behaviorally worse and even lead to aggression and, ultimately, euthanasia.*" This quote by Dr. Sophia Yin applies to how we, as veterinary professionals, interact with animals in all aspects of veterinary medicine.

It is through the interactions we have with our clients and their parrots in the exam room that we make our first impressions, and many lasting impressions. Based on what the client has experienced, clients will anticipate the exam to unfold in a certain way: such as a low-stress exam or the traditional "capture and restraint." The same is true of our patients. Before restraining a patient for any purpose, keep in mind that behavior is individual, meaning each animal behaves differently, and that an animal behaves in response to its environment as well as its individual learning history. This mindset allows us to obtain more information that may be helpful from the pet owner, and prevents us from lumping in "Fluffy" with every other Goffin cockatoo. For example, you may ask:

> How does your bird prefer to exit the carrier?
> Does he like hands?
> Can he fly?

Information offered about the bird's preferences for handling can help save time and allow you to work to reduce stress for the bird. In addition, by observing the bird's behaviors while working with it, with this additional historical information from its stewards, you will be more empowered to travel the low-stress handling and restraint road.

Capture and restraint: terms and actions that need to be left in the past

Many veterinary textbook references, conference proceedings and discussions still describe techniques and methods by which to "capture and restrain" parrots. In most current circles, these methods are increasingly seen as more forceful than necessary. Loosely, "capture and restrain" often includes combinations of techniques that include physically overpowering the bird, sneaking up on or surprising the bird, or other means to quickly get the bird restrained and get the task at hand done. In addition, videos of these methods and techniques, good and bad, live in perpetuity on you-tube and other sources, are repeated and even referenced in current presentations. This, in turn, can result in further confusion and misunderstanding about the controversial views on the topic. In reality, there is enough clear science out there to simply state that force, coercion and capture-oriented methods, more than ever before, are increasingly less appropriate for most of the animals in our stewardship and care. Numerous trainers and veterinary professionals have shown that procedures such as nail trims, medicating, and exams can be performed in a low-stress manner in the same amount of time and effectiveness as the more forceful techniques. Low-stress techniques also do not have the same problems associated with forceful treatment such as increases in learned fear-eliciting stimuli, increase in learned aggression, increased risks to the bird and handlers, increased risk of problems during medical procedures, and increased difficulty interpreting some laboratory diagnostics due to iatrogenic and stress-influenced changes. These old-school methods typically overlook some of the more important ethical considerations and best practices for companion parrots, and remain principally focused on accomplishing the task at-hand alone. When veterinarians and healthcare personnel did not look back and critically evaluate outcomes, and when the intelligence and learning capacity of the birds was discounted, most people were quick but incorrect to pronounce those methods effective. In reality, these "successes" were in fact often quite far from their ideal or intended mark. The adverse effects of our own failure to appreciate learned fear in the development of and shaping of behaviors during in-patient treatment and clinical procedures are immense. The management traps that can occur secondary to unrecognized adverse behavior changes in avian patients are outlined in Table 1 below.

Low-stress handling techniques are time-effective, and can directly enhance the comfort of the avian patient in daily clinical practice. These basic skills are essential not only to avoid the old "capture and restrain" methods of avian practice and their adverse consequences, but to enable veterinarians and technicians to be more ethical and effective in their healthcare delivery for companion birds, and to enable training of medical behaviors in the future.

Table 1. Undesired behavioral changes and their potential management traps if not recognized.

Adverse behavioral changes and effects observed with forceful restraint methods	Clinical management "traps" if behavior is not incorporated in daily in-patient care
Increases in learned fear-eliciting stimuli.	Changes in body language and behavior are missed or attributed to constructed labels for the patient, preventing proactive or early intervention.
Increases in learned aggression.	Changes are missed or attributed to constructed labels for the patient, allowing or enabling justification of increased use of force and other methods in this patient and others in the future.
Increases in risks to bird and handlers when the birds are being examined or treated.	Risks are managed by the use of increasing speed, drugs, or are attributed to constructed labels for the patient, enabling increased force and intrusion in treatment methods.
Increased risk of problems during or after performing medical procedures.	Problems, when encountered, are explained by constructed labels for the patient, its species, its disease process, or the procedure.

The appropriate use of conscious sedation and anesthesia

Where indicated, conscious sedation or general anesthesia should be considered in particularly stressed or fearful patients. Although these maneuvers can be very effective, they should not be uniformly used in all patients, and should not be viewed as a means with which to avoid the need for low-stress handling methods. Midazolam 0.5 mg/kg to 1.0 mg/kg IM, paired with butorphanol 1.0 mg/kg to 2.0 mg/kg IM can produce effective conscious sedation for many, but not all, companion bird species, and midazolam 2.0 mg/kg intranasally (IN) also can be used where appropriate. The advantages that conscious sedation offers include minimal cardiopulmonary effects and the potential for an amnesia effect as is described in humans. Benzodiazepines, including midazolam are not, however, analgesics. Although sometimes referred to as sedated with isoflurane, volatile anesthetics are not sedatives and render the patient entirely unconscious. Even though volatile anesthetics offer a relatively wide margin of safety, they are not benign, and can produce apnea and hypotension, especially at the higher concentrations required when used as sole agents. In addition, the process of mask induction can be quite stressful and fear-evoking. Although the task at hand may be able to be accomplished, the greater goal of minimizing fear and stress, and facilitating learned cooperative interaction between the bird and staff is reduced.

Let the Force of Body Language Guide You: Observe and Communicate

Watch the body language of the bird; listen, and act accordingly. If you offer your hand with a towel draped over it, and the yellow-naped Amazon parrot you are working with fans out its tail, pins its eyes, and crouches down on the perch, stop. Try a different approach. Take your time. By watching body language, you are empowering a bird to say no and are giving it a choice. When a bird is empowered to say "no" with its body language there is less need for it to escalate from a certain type of body language to a grumble, lunge or bite when the latter will work just fine. Body language is a language all in its own, and can give us tons of useful information about how to handle our patients in a low-stress manner.

The list below, compiled by Barbara Heidenreich (www.goodbirdinc.com), and Susan Freidman, PhD (www.behaviorworks.org), provides principles and suggestions to optimize a training experience. These also can be applied to handling and restraint in a minimally aversive manner.

1. Make it your goal to create an environment in which the parrot appears comfortable and relaxed. This can only be inferred from the bird's body language, specifically behaviors involving feather position, eyes, wings, head, legs and feet.

2. Approach unfamiliar parrots calmly and quietly to avoid creating any signs of anxiety, fear responses or aggressive behavior.

3. If a parrot shows any signs of fear, anxiety or aggressive behavior, discontinue the actions that helped generate those behaviors. This may include lowering your hands and/or stepping away from the parrot.

4. If a parrot exhibits aggressive behavior, immediately discontinue actions creating this response.

5. Keep your attention and your eyes focused on the parrot.

6. If you need to direct your attention away from the parrot for more than a few seconds, put it back in the enclosure or carrier if that is a location where the bird feels more relaxed.

7. Be aware of where you place or hold food in the presence of a parrot, as this can cause anxiety.

8. Be aware of how every action that you do influences a parrot's behavior, and adjust your behavior moment by moment to maintain a calm bird.

9. Be aware of moving objects and how they may influence a parrot's behavior.

10. Move crates, carriers, and cages with extreme caution to avoid bumping or jostling the bird. This could result in increased anxiety and a decrease in the effectiveness of your training efforts.

11. Prior to removing a flighted parrot from its enclosure or carrier, evaluate the surroundings for safety and address any potential safety issues (eg, cover large mirrors, pull down shades, shut doors).

12. If a parrot launches into flight, offer your steady, raised hand as a safe place to perch.

13. Be aware of the parrot's body proximity to adjacent objects. Avoid hitting tail, wings, or head on anything if possible.

14. Avoid creating a high level of excitement (eg, head bobbing or crest raising) by talking loudly or using animated actions. This can sometimes result in the presentation of aggressive behavior.

Body language: what it is, and what it is not

When describing any behavior, it is important to describe observable actions. IE: what we see and can be seen easily by others. What body language and observed behavior is not, is labels. These types of labeled descriptors often have different meanings to different people. For example, an aggressive bird may look like a bird slamming its toys around the cage to one person, but to another person aggressive may look like a lunging, biting bird. Another example is of a depressed bird, which needs to be petted constantly to be happy. The terms happy and depressed can have different meanings to different people, yet they are commonly used as explanations for why a bird behaves as it does. For example, what is the difference between an obnoxious or rude bird and a bird that screams constantly when the owner leaves for the day? One seeks to describe the bird's personality (the owner might say "That's just how the bird is!"), while the latter describes an actual behavior, something the bird is doing, that can be measured, observed, repeated, and therefore is subject to behavioral change. A common label that is used in popular practice is the "hormonal" one. This and other similar labels can easily predispose veterinarians to consider using counter-hormonal treatments, surgery, and incomplete behavior change strategies to treat the label, rather than the patient and specific behaviors that are before them. Without taking time to paint a clear picture of a bird's behavior, we can sometimes be a greater hindrance than help for our patients and clients.

Learn to identify fear, and employ counter conditioning and desensitization

Counter conditioning and desensitization are both techniques that can be used to address elicited fear responses. Counter conditioning focuses on pairing a desired or appetitive stimulus with the aversive stimulus (the undesirable stimulus which the animal avoids and causes a fear response) to change the stimulus from aversive to valued. The appetitive stimulus should elicit a stronger response from the animal than the aversive stimulus. For example, if a bird shows elicited fear (eg, increased heart rate, feathers flat on body) to the towel, and if we know that this individual bird enjoys a head rub or a piece of pizza, we can pair the appetitive response from eating the pizza with the towel, overriding the fear response. Slowly, the end result is a bird that doesn't show a fear response when it sees the towel because the towel is paired with the appetitive response of eating pizza.

Desensitization is the process of incrementally exposing the animal to the aversive stimulus without crossing the threshold causing the animal showing any fear response. Desensitization and counter conditioning are focused on altering respondent behavior. In the previous example, if the bird is preening when the towel is brought within 10 feet of it, but then screams and flattens its feathers against its body when the towel is brought within 8 feet, we know that 10 feet is an appropriate distance to start. Small approximations are used to bring the towel closer to the bird without it demonstrating fear or escape behavior, the size of the approximation is dependent on the bird. Desensitization, counter conditioning and positive reinforcement work very well together by combining the pairing of stimuli from counter conditioning at each approximation used with desensitization. In this example, the towel would be brought close to the bird and if no fear response or escape behavior were shown, a reward (pizza or a head scratch) would be delivered. Each successful approximation of decreased distance between the towel and bird is reinforced with a positive stimulus, which makes the appetitive stimulus dependent (contingent) on the behavior.

Always Keep Key Behavioral Concepts in Mind

One of the most fundamental principles of behavior is known as "the law of effect." Simply, this law states that behavior is a function of its consequences. In other words, the frequency of a response is changed by the consequences that follow that response. We apply the law of effect with two basic procedures: reinforcement and punishment. Reinforcement increases the frequency of behavior and punishment decreases the frequency of behavior. Although people often think of reinforcers as rewards, that can be misleading. Rewards generally refer to prizes, trophies or accolades but many reinforcers don't fit that image. For example, if you shout at a bird every time it screams and the screaming continues because of your shouting, shouting is a reinforcer for that bird. Punishment can be an equally confusing term. People often use it to mean retribution, retaliation and revenge. In applied behavior analysis, if a behavioral decrease is not observed, the consequence is not a punisher for that particular bird. The ultimate classification of which consequences function as reinforcers or punishers is highly individual. The proof their classification lies strictly in the assessed future rate of the bird's behavior(s): if a bird continues to behave in a particular way, it is being reinforced regardless of your intentions.

Two of the most important characteristics of effectively delivered consequences are 1) contingency (ie, the dependency or relationship between the behavior and the consequence) and 2) contiguity (ie, the closeness or timing with which the consequence follows the behavior). When a consequence is delivered inconsistently, it is hard for the learner to associate the two events. If the consequence is delivered too far in time after the behavior, this lack of immediacy decreases the effectiveness of the consequence as well. Perhaps it is inconsistent delivery and poor contiguity that accounts for the fallacy that punishment doesn't work with parrots. Parrots clearly understand cause and effect as they navigate hundreds of behavioral choices daily based on experience from which they predict outcomes. Parrots go to food bowls or learn to forage because doing so causes the effect of access to food; parrots step onto human hands because doing so causes the effect of being removed from their cages and experiencing interaction with humans; and parrots scream because doing so causes the effect of favorite people appearing.

Target Training: The Gateway to Many Desirable Behaviors!

We can teach birds many behaviors using target training. With this method, we teach the bird that if it to "targets" (ie, touches with a body part, for example its beak) an object (such as a chop stick, or closed hand), it will receive a favorable outcome (ie, positive reinforcement). Target training is used commonly to train a large number of animal species. Target training can be used to train birds to do anything from playing games with their owner, allowing intramuscular injections, to entering and exiting the carrier. Using small approximations and the individual strength of different reinforcers we can create new behaviors. A reinforcer is any consequence that increases the strength or frequency of the behavior. A very simple example of target training is teaching a bird to take medication directly from a syringe. Initially, the bird is reinforced just for touching the tip of the syringe. The owner holds the syringe up to the bird's beak (antecedent); the bird touches the tip of the syringe with its beak (behavior); the owner feeds millet to the bird (consequence). Assuming that millet is a reinforcer for this individual, we can predict that the bird will continue to touch the syringe to eat millet from the owner. In this situation, small approximations will be planned until the bird will take a drop of medicine from the syringe and eventually take all of the medication from the syringe (even if it is in small quantities at a time). Not only is this behavior practical (and fun) for clients and veterinary staff, it can be a low-stress way to deliver medication to patients. Getting this started in practice is as easy finding out what is reinforcing for the patient.

Aside from medication training, target training is an important exercise in communication and strengthening the bond that can be started with new clients during new bird exams and wellness checks. When we offer the target and the bird does NOT touch the target, the bird is telling us no. This can mean several things: they might be full, sleepy, bored of training at the moment, unsure of what we are asking them, or physically unable to do what we have asked them. It is empowering to birds that when they say no, we listen! Of course, this is not always possible, for example in the circumstance of medical emergencies. Whenever possible, if we can listen and reassess when birds say "no," then we can keep communication open and encourage it to continue. Target training has other practical applications such as teaching birds to go into their carriers, towel training, several medical uses, and is very useful when working with aggressive birds. Target training is the perfect way to start integrating training into your exam rooms and make visits fun for your patients and clients.

Exam Room Setup and Time Management

While there are several different approaches that can be taken to facilitate time for behavior talk with clients, it is important to find what works best for your hospital. As an example, behavior questionnaires for clients are easily available and can be adapted to fit your client needs. Handouts can be a great way to provide clients with information and give homework between visits. Follow up is important, and can be via recheck exams, technician appointments, telephone calls, email, or a combination of these methods.

Exam room setup can be easily tailored to meet the needs of your patients. Items like appropriately sized perches and a room with minimal distractions will help your patients feel comfortable. Any supplies you may need for the exam or test collection should be stored in the exam room to minimize the amount of time the bird is restrained in the towel. A variety of treats and training supplies like target sticks and a clicker can easily be stored in the exam room. Although it can be challenging, storing properly sized items for various sizes and species of birds (perches, scales, towels, treats) and having these items readily available cuts down on restraint time and stress.

Foraging and Enrichment: Talking the Talk and Walking the Walk

Foraging is a necessary form of enrichment, especially for parrots in captivity. Applied optimally, foraging can be a means of enrichment to redirect a bird's attention and behavior, as a part of a desensitization/counterconditioning plan (DS/CC), to avoid the need for use of restraint collars, to avoid the need for some medical treatments to suppress undesired behaviors, and merely as a means with which to enrich the welfare of the birds themselves. Working for food by means of searching through toys or puzzles, and shredding or chewing through paper or cardboard are great ways to keep birds busy. Many birds will need to be taught these natural behaviors, since they are not innate and oftentimes are limited in the lifestyles of many companion birds. This may be the case when some natural chewing behaviors are considered a nuisance by the pet owner and discouraged. Foraging behaviors and shredding can be taught by demonstration in the exam room and continued at home. These food acquisition activities can range from very simple hidden treats in half-covered bowls to complex puzzles and time-consuming boxes to untie or open. By increasing foraging in daily activities, we often can eliminate or lessen undesired behaviors, and quality of life can be enhanced. These enrichments are best viewed as life-long lesson plans, and veterinarians can deliver their highest impact by following through with patients and their stewards, coaching both from the side, and tailoring recommendations to best fit patient needs.

Practice Management Implications of Behavioral Science

Doctor-centered versus client-centered practice style

Most veterinary practices (private and institutional) are predominately doctor-centered in their manners of operation. Although both the doctor-centered and the client-centered practice operate with shared goals of patient care and client service, the client-centered practice offers a clearer vision that is easiest to maintain by the entire practice team. There are key and recognizable differences between these conceptual operational modes. In a doctor-centered practice, the staff typically waits to do the next duty as assigned. As the doctors move from the outpatient consultation rooms to the inpatient treatment area, a sizeable number of the staff tends to follow those doctors. The staff that does not follow often shifts to a lower productivity mode of work when the doctor is not present in their work zone. Also, in the doctor-centered practice, the doctors more often can be found performing the nursing rounds of inpatients, doing the daily treatments themselves, and in some cases, using well qualified nursing staff as animal holders only, so they can do the nursing work themselves. Veterinarians in a doctor-centered practice typically do not effectively delegate to their staff, nor do they encourage or schedule regular client nursing appointments. In summary, in a doctor-centered practice, the staff works to support the needs and goals of the individual doctors, who in turn strive individually to provide patient care and client service as they best can. In essence and summary, doctor-centered practices look at internal quality factors, the doctor's schedule and pride in procedures; they keep the doctor in the spotlight. Client-centered practices seek client satisfaction, low-stress or fear-free services for the patient, and speak of the healthcare plan (rather than the estimate). The differences in overall practice staff (doctor and staff) functions between these two practice types are compared in Table 2.

Table 2. Practice personnel functions and focus in a doctor-centered and client-centered practice philosophy.

Doctor-centered practice	Client-centered practice
Doctors work individually to provide patient care and client service as they best can.	Doctors function as part of a team with their practice staff.
Staff works to support the needs and goals of the individual doctors.	Staff has a much larger role in most forms of client communications.
Focus is on supporting the doctors, who maintain focus primarily on the client.	Constant focus is on the client by the entire practice team.

In the client-centered practice, the staff has a much larger role in most forms of client communications, and they are empowered to do so by the practice ownership and management. When the doctor moves from outpatient consultation rooms to the inpatient treatment area, the outpatient nurses revert to nursing appointments and client telephone outreach activities and never leave their outpatient zone of responsibility. Concurrently, the inpatient nursing staff meets with the doctor, and both review the priorities for the inpatient demands, and detailed planning for in-patient procedures and surgery. The outpatient nursing staff works to keep the doctor on schedule in the consultation rooms and oversee and augment client and patient service. The inpatient nurse keeps the surgeon/inpatient doctor on schedule within their zone of responsibility.

In summary, in a client-centered practice, the doctor(s) work more with their practice staff, as a functional part of a team that has a constant focus on the client, as opposed to a practice team that focuses on supporting the doctor(s), who maintain their focus primarily on the client. The differences between these two operations-management philosophies become more apparent as the practice grows in size, and numbers of clients, patients, doctors and staff. These same differences can dramatically influence the ability of a practice to consistently deliver optimal patient care and client service.

Staff empowerment and avoiding the use of labels in the workplace

Staff empowerment improves observational acuity, improves pain management, improves treatment accuracy and allows for an ethical hierarchy to be applied to treatment and diagnostic decisions as they are made in a given patient as its condition(s) change. All practice employees, from veterinarians through receptionists and front office personnel, nursing staff and kennel personnel, need to be willing to accept this responsibility for ensuring that clients feel good about the treatment modalities being applied, and that they know how to continue to provide optimal care for their birds once they are discharged from the hospital. It is important to avoid telling veterinary personnel what *not to do*, but to tell them what they can do, and to help them learn to help people be successful on their own terms. Essential in this premise is a foundation of respect for the hospital staff, their feelings, thoughts, and opinions.

During daily in-patient rounds with hospital staff, the behavioral aspects of a patient's healthcare plan should be discussed in conjunction with all other aspects, plans set and adjusted, and these plans should be relayed to owners as well. Careful distinction needs to be made between describing a patient's actual behaviorsand constructs. A *construct* is an interpretive assessment of a characteristic of an animal, often used to describe and explain its behaviors. A construct can also function as an interpretive assessment of what one feels the animal feels your actions and behaviors represent to it. A *behavior* is a specifically described action, and would ideally be described in context with antecedents and consequences. The use of constructs in a clinical setting can function as a trap of categorical thinking or interpretation that can very easily lead the best avian veterinarian astray from a more complete and multidisciplinary approach to integrating behavioral science into a more comprehensive inpatient care strategy. The inadvertent, unrecognized or unaltered use of stressful or fear-evoking treatments can lead to significant adverse consequences. These perceived acute and chronic stressful situations can adversely affect the immediate and longer term outcomes of an examination, a specific procedure or surgical event, patient healing capacity, and the long-term quality of life of the bird. In addition, the observed changes to the behavior of the bird can easily adversely affect a client's desire to return for further care or even annual wellness evaluations. The problems associated with time and repetition using these approaches and the "constructed traps" that can block their recognition are outlined in Table 1 above.

The Importance of Client Perception: Common Courtesy

In the business of veterinary healthcare, taking the time to be courteous is an absolute must. A common component of staff courtesy training is the use of pedestal words during client communication. These are brief words or phrases that function to raise the client to a level above the ordinary, and make clients feel that they have been put onto a pedestal. Examples of pedestal words or phrases may include:

- May I? (Asking permission implies authority.)
- I'd like your advice or what do you think? (Asking suggests superior wisdom.)
- I'd sure appreciate it if..." (There is an implication that s/he has the power to refuse or grant a favor.)
- You are so right or provide a high five. (A pat on the back.)
- As you know (This phrase implies an understanding and professional skill.)
- Please (This is always a great lubricator in human relations.)

Whether the doctors and staff of your practice use a caring approach (body language, thoughtful, listening, interested, compassionate), pedestal words, or a combination of techniques, client confidence in your practice will be reinforced and grow. The birds themselves, and your clients, will have a greater chance of becoming less of a target of rigid medical and wellness protocols in your clinical mind's eye. More importantly, individuals (clients and patients) that you come to know at this level become a part of a greater picture of a tailored delivery of wellness healthcare.

Behavioral Science Applied to Practice Management and Client Retention

There is an interesting paradigm in veterinary practice, which is not exclusively limited to birds. In this paradigm the shared goal to enhance and extend quality of life is focused on procedures, tests and treatments in the search for disease. With regard to wellness or return to health, success often is proclaimed by negative test results: the absence of documentable disease, or apparent recovery from physical illness. This approach contrasts starkly with an approach that combines a focused effort to add in positive impact wellness components, such as training, enrichment and educational discussions regarding welfare and empowerment of the stewards of these animals to facilitate favorable changes. Once a practice accepts that wellness is much more than a diagnostic quest for negative test results, the sooner caseload becomes balanced, and the practice begins to truly stand proud and succeed.

By applying the behavioral principles of the law of effect, the business aspects of the use of standardized wellness testing protocols can be evaluated, from the client's perspective of their desire to return to your practice. The general basics of reinforcement and punishment are displayed in Table 3. In other words, the frequency of a client's return visits is influenced by the consequences that followed their earlier visits. The client's desire to return to your practice is dependent on the same two basic procedures: reinforcement and punishment. Reinforcement increases the frequency of behavior and punishment decreases the frequency of behavior. Reinforcement (both positive and negative) refers to an event that occurs in conjunction with a behavior and increases the likelihood that a behavior will occur again. A positive reinforcer is a stimulus that is sought and added to the environment that the client finds pleasant. A negative reinforcer is a stimulus that is avoided and that the client in this case finds unpleasant, leading the client to avoid this aversive stimulus by repeating their visits to your practice. Both positive and negative reinforcers function to increase the frequency of the target behavior, but by very different means. Punishment (both positive and negative) refers to an event that occurs in conjunction with a behavior and decreases the likelihood that a behavior (client return visits) will occur again. A positive punisher in this case would be a stimulus that occurs in conjunction with the client visit, is added to the environment or experience, and that functions to decrease the probability of the client's return. A negative punisher would be a stimulus that occurs in conjunction with the client visit, is removed from the environment, and functions to decrease the probability of the client's return. The proof their classification lies strictly in the assessed rate or predicted future rate of the client's return visit frequency. If your client's return visit rate is not increasing, it is being punished, regardless of your practice management intentions.

Table 3. Skinner's basic paradigms of reinforcement and punishment.

Stimulus	Operant response increases	Operant response decreases
Presentation	Positive reinforcement (R+)	Positive punishment (P+)
Removal	Negative reinforcement (R-)	Negative punishment (P-)

Reading from the left, if a stimulus is being presented to the learner, this is represented as a positive. Then, evaluating the frequency or probable frequency of the behavioral consequence of this presented stimulus, we can assess if the stimulus is functioning as a positive reinforcer (R+), or a positive punisher (P+). If the behavior increases, the introduced stimulus is functioning as a reinforcer, and if the behavior decreases, the introduced stimulus is a punisher. If a stimulus is being removed from the learner as a result of its behavior, it is represented as a negative. Based on the observed frequency or probable frequency of the behavioral consequence of this removed stimulus, we can also assess if the stimulus is functioning as a negative reinforcer (R-), or a negative punisher (P-).

Skinner's paradigms applied to client return visits

Client-centered veterinary health care providers will strive to ensure that both outpatient and inpatient visits result in client-perceived value (ie, positive reinforcement, see Table 4). These differences highlight the important comparison between output of what we do as compared to the outcome (client perception) of those actions. The probability of return client visits to your practice can suffer the consequences of negative punishment (clients lose money, and return visits decrease), positive punishment (clients receive suboptimal care and recognize it, and return visits decrease). Client return rate can be increased through positive reinforcement (clients are pleased with their experience and feel that they take home something of value), and negative reinforcement (tests performed either confirm the absence of problem or identify a problem, enabling treatment with both leading to a desirable outcome of relief). A common deficit in many wellness protocols that are utilized in avian practice is the belief that negative reinforcement can function as a key drive for return visits; the marketing of peace of mind. In this system, it is argued that clients increase their probability of return visits as a result of the use of protocols and procedures aimed at routinely documenting the absence of disease.

Table 4. Reinforcement and punishment schedule applied to client return visits.

Reinforcement or Punishment	Positive (addition of a stimulus as a result of the visit)	Negative (subtraction of a stimulus as a result of the visit)
Reinforcement (increased rate of return to the practice)	Perceived value. Direct benefit to wellness and strength of the human-animal bond.	Peace of mind. Relief from the perception of disease. Removal of visible or suspected disease. Documentation of the absence of disease.
Punishment (decreased rate of return to the practice)	Suboptimal healthcare. Perceived discourtesy. Stress or injury to the birds.	Loss of money. Inconvenience to client.

This type of a management system can be contrasted to those that are focused on positive reinforcement (ie, client return visits are increased by their perception of increased value at the time of the visit). In general, bird owners state that their perceived value of veterinary service is based on things that directly influence the quality of their relationship with their bird at home. Although a sense of relief of the absence of disease does have value, the direct infusion of fun, knowledge, and improved home relationships with companion birds should be viewed as a strong driver for return visits. In this light, some standardized testing protocols can be counterproductive to the development of a growing clientele if the practice philosophy is not client-centered. Positive reinforcement for regular returns in wellness management (clients receive valued and desired service and increase their probability of returns) should be viewed as the ideally structured, most ethical, and strongest driving factor for return visits. This may be clearly contrasted with a practice that emphasizes a negative reinforcement system for encouraging client return (ie, clients receive the good news that their bird is not diseased at the moment and return regularly for this peace of mind).

References

1. vanZeeland YRA, Friedman SG, Bergman L. Behavior. In: Speer BL, ed. *Current Veterinary Therapy in Avian Medicine and Surgery.* St Louis, MO: Elsevier Saunders;2016:177-251.

2. Friedman SG, Edling TM, Cheney CD. Concepts in behavior: section I, the natural science of behavior. In: Harrison G, Lightfoot T, eds. *Clinical Avian Medicine.* Vol. 1. Palm Beach, FL: Spix Publishing;2006:46-59.

3. O'Neill RE, Horner RH, Albin RW, et al. *Functional Assessment and Program Development for Problem Behavior: A Practical Handbook.* 2nd ed. Pacific Grove, CA: Brooks Cole;1997.

4. Catanzaro T, Speer BL. Practice management and risk management. In: Speer BL, ed. *Current Veterinary Therapy in Avian Medicine and Surgery.* St Louis, MO: Elsevier Saunders;2016:749-763.

Geriatric Avian Medicine

Sharman Hoppes DVM, DABVP (Avian Practice)

Session #2086

Affiliation: From the Department of Small Animal Clinical Sciences, College of Veterinary Medicine, Texas A&M University, 4474 TAMU, College Station, TX 77845, USA.

Abstract: Geriatric medicine is the care and treatment of the elderly. In the past, few pet birds lived to be elderly or geriatric due to poor husbandry, nutrition and infectious diseases. As the avian community has gained knowledge in not only husbandry and in nutrition, but also medicine and surgery, the life expectancy of parrots has lengthened. Unfortunately, even with this increased longevity, few parrots outlive their owners. The life spans of many parrots have been documented to be up to 60 and 80 years depending on the species, but due to sedentary life styles, over abundant nutrition and possibly genetics many parrots will develop old-age diseases much earlier than expected. This article will discuss the common diseases seen in the geriatric parrot, the presenting clinical signs, the diagnostic work up and treatment options.

Introduction

The definition of geriatric is: of, or pertaining to the elderly. So geriatric medicine is the care and treatment of the elderly. In the past, few pet birds lived to be elderly or geriatric due to poor husbandry, nutrition, and infectious diseases. As the avian community has gained knowledge in not only husbandry and in nutrition, but also medicine and surgery, the life expectancy of parrots has lengthened. Unfortunately, even with this increased longevity, few parrots outlive their owners. The life spans of many parrots have been documented to be up to 60 and 80 years depending on the species, but due to sedentary life styles, over abundant nutrition and possibly genetics many parrots will develop old-age diseases much earlier than expected.

Cataracts

Cataracts occur in many species of birds, but in parrots the macaw, Amazon and cockatiel are over represented. They occur most commonly in older birds but can be genetic and also occur secondarily to infection or trauma. If the onset is gradual, birds adapt and often the owner may not notice until the bird is completely blind. Cataracts and the resulting blindness result in a less-active bird, often affecting their appetite and their interaction with other birds and people. In some birds, blindness can lead to an increase in biting, screaming and fearful behaviors.

A complete ophthalmic exam should be part of the annual avian exam so that changes in lens opacity can be detected early. Cataract surgery has become a viable option for larger birds (cockatoos and macaws) with cataracts and it has been shown in several species of birds to lead to an improved quality of life. Common postoperative ocular issues include posterior capsular opacity and corneal scarring.[1] Post- operative medications include ophthalmic antibiotics and non-steroidal anti-inflammatory or steroid drops for 2-4 weeks. A bird's overall health and the effects of diminished vision should be evaluated prior to surgical intervention along with a thorough ophthalmic exam including electroretinography to evaluate the retina.[2]

Unfortunately, typical mydriatics used in mammals are not useful in birds due to the skeletal muscle found in their iris (as opposed to smooth muscle in mammals). If surgery is not an option, maintaining the cage with minimal alterations as far as perching and food and water bowls is important to help the bird maneuver in its cage safely.

Osteoarthritis

The prevalent form of arthritis in birds is osteoarthritis (OA). It is a common cause of lameness and chronic pain in older birds. It is more common in psittacines over 15 years of age or in birds who have experienced trauma. It also occurs at a younger age in waterfowl and poultry kept as pets. Bumble foot, ulcerative dermatitis, and callous formation may be secondary effects of arthritis. The pathology of osteoarthritis is caused by damage to chondrocytes. This damage or stress on the cartilage may be caused by trauma, obesity, or normal wear on abnormal joints or abnormal wear on normal joints. This is a progressive disease and unless the cycle can be interrupted further degradation to the cartilage causes continued damage to the joint.

Joint pain is the essential problem associated with OA. The synovium and subchondral bone are the major sources of OA pain, but pain can also arise from periarticular muscles and ligaments. The symptoms will vary depending on the location of the arthritis and the severity of disease. Birds may present lame, there may be a history of decreased activity, not perching, falling off perches or perching on different perches, or a flighted bird that doesn't want to fly or doesn't fly well. Other less obvious signs of arthritis may include swollen joints, decreased range of motion, feather picking, mutilation or even excessive vocalization.

Diagnosis is determined by clinical signs, physical exam and radiographs. Radiographic lesions include narrowing of the joint space, sclerosis of the subchondral bone, misalignment of the joint and osteophyte formation. Commonly affected joints include the phalangeal joints, the tarsus and stifle. The joints of the thoracic limb can also be affected but appear to be so less commonly than the pelvic limb, probably due to many companion birds not being flighted.[3]

A multimodal treatment plan is recommended to give the bird the best possible chance of relief; it also may allow for decreased individual drug doses in the treatment plan. A multimodal approach incorporates both medical and nonmedical modalities to achieve therapeutic goals. Medical treatment combines the use of nonsteroidal anti-inflammatory drugs (NSAIDs), chondroprotectants, and possibly opioids.[4] The most common nonsteroidal anti-inflammatory drug utilized in avian medicine by this author is meloxicam, a COX-2 inhibitor 0.5-1.0 mg/kg PO q24h to q12h (Boehringer Ingelheim Vetmedica, Inc, St. Joseph, MO) with the goal to use the lowest therapeutic dose possible. There are potential side effects of NSAIDs such as renal ischemia so they should be used with caution. Anecdotally, glucosamine (20 mg/kg q12h or 35 mg/kg PO q24h or every other day) has been used successfully in birds or Adequan (100 mg/ml) injections (5 mg/kg IM once weekly for 4 weeks then monthly). Opioids may be needed for acute exacerbations of an arthritic condition or for conditions not responding initially to NSAIDs. Tramadol 5-10 mg/kg PO q6-12h (Janssen Pharmaceuticals, Inc, South Raritan, NJ USA) or butorphanol 0.5-3.0 mg/kg IM q4h (Torbutrol, Intervet Inc, Millsboro DE, USA) depending on the species of bird, may be used for a short period of time until the anti-inflammatory effects of the NSAIDs take effect.

The non-medical component may include husbandry changes, weight management, an exercise plan, a healthy diet rich in omega-3 fatty acids and physical therapy. An exercise plan in the captive avian patient can be challenging in nonflighted birds. In flighted birds, allowing them to fly or encouraging flight is the best form of exercise if it can be done safely. If this is not possible, encouraging climbing and walking in birds, even stepping up a number of times can be exercise for parrots. Foraging for food can be used to encourage exercise, by putting foraging boxes on opposite sides of the cage or enclosure. Providing foraging opportunities can also entertain and distract the bird from its discomfort diminishing feather picking or mutilation behaviors. If the bird is overweight, then weight loss is essential as studies have shown that obesity is a risk factor for OA in many species. Weight management includes transitioning the bird slowly to a healthy pelleted diet with the addition of essential fatty acids. Fatty acids have been shown in many species to be anti-inflammatory and renal protective (an added bonus when using NSAIDs which can be damaging to the kidneys in high doses or in renal compromised or dehydrated patients). Most recommend flax seed oil as the best source of fatty acid supplementation at 0.1 ml/kg PO q24h. Other husbandry changes such as changes in perch diameter, texture, or padded perches can all help birds with weak or painful legs or feet and help to decrease the development of bumblefoot in a lame bird.

Pododermatitis

Pododermatitis or bumblefoot is a relatively common condition of pet birds. It is a general term for any inflammatory or degenerative condition of the avian foot and can range from mild redness to bony changes. Bumblefoot develops most commonly when birds are either housed on inappropriate perches, or secondarily to an injury in one leg which causes the bird to shift its weight to the other (good) leg, creating increased pressure on the plantar surface of the foot resulting in pressure sores and ulceration. Birds most at risk are birds with leg fractures, but also birds with arthritis of a hip, stifle or tarsal joint, obese birds or birds on poor diet (vitamin A deficiency).[5]

Bumble foot is often a progressive disease. A localized hyperemic lesion followed by ulceration, then abscess formation and if left untreated osteomyelitis. Initially the skin on the metatarsal and digital pads becomes flattened, smooth, and shiny appearing. Over time the skin may become proliferative and then ulcerate allowing bacterial access which leads to inflammation and infection. As the condition progresses tendon sheaths become infected, and osteomyelitis arises along with septic arthritis. Birds may present with lameness, lethargy, and anorexia due to the pain, inflammation and infection.

There is a classification system in place based on severity of lesions.[5]

Grade 1: Desquamation of areas on the plantar foot surface, small shiny pink areas.

Grade 2: Smooth, circumscribed areas on the plantar metatarsal pads of one or both feet with subcutaneous tissue almost visible thru the skin, resulting in scab formation and swelling.

Grade 3: Ulceration of the plantar metatarsal pads, caseous abscess, pain. Mild lameness.

Grade 4: Infection of tendon sheaths, cellulitis, flexor tendon rupture. Severe lameness.

Grade 5: Septic arthritis of the tarsometatarsal-phalangeal joints, osteomyelitis.

Diagnosis is based on clinical signs, physical exam, radiographs, and cultures. Affected birds should be examined thoroughly for predisposing injuries or illness. In all stages, correct the husbandry (padded perches or perches covered with AstroTurf) and improve the diet. In parrots conversion to a healthier, preferably formulated diet is critical. Encourage weight loss and exercise in obese birds (eg, walking, climbing, flying). As the disease progressives a bandage may be necessary to relieve the pressure on the lesion. The lesions should be kept clean. Strict sanitation of the facilities and feet is important to prevent bacterial infections. If there is a scab, it should be softened and removed or surgically debrided. Antibiotic use should be based on culture and sensitivity. *Staphylococcus* sp. is most commonly identified; other reported bacteria have included *E. coli*, and *Proteus*. Effective antibiotics have included amoxicillin clavulanate 125 mg/kg PO q8h,(Clavamox Zoetis Inc, Kalamazoo, MI, USA), enrofloxacin 10-15 mg/kg PO q12h (Baytril Bayer healthcare LLC, Shawnee Mission, KS, USA) and marbofloxacin 10 mg/kg PO q24h for 7-14 days (Bayer healthcare LLC, Shawnee Mission, KS, USA).

Pain management is important and includes a combination of anti-inflammatories and/or opioids depending on the severity of disease and certainly following any surgical debridement. The author routinely uses meloxicam

0.5-1.0 mg/kg PO q24h to q12h (Boehringer Ingelheim Vetmedica, Inc) along with tramadol 5-10 mg/kg PO q6-2h (Janssen Pharmaceuticals, Inc, South Raritan, NJ USA) and in severe cases or immediately post-surgery, butorphanol 0.5-3mg/kg IM q4h (Torbutrol Intervet Inc) depending on species.[6] Local anesthetics may be helpful post surgically.

For severe lesions, surgical debridement may be necessary. Preoperative antibiotics should be administered for several days prior to surgery. Antibiotic impregnated beads may be necessary in cases of osteomyelitis. Bandaging to reduce pressure on the affected site is important to promote healing. An assortment of bandaging techniques have been proposed over the years (ie, ball bandages, rigid foot splints with foam); the most important aspect of any bandage is to remove pressure from the affected site.[7]

Metabolic Bone Disease

Nutritional secondary hyperparathyroidism (metabolic bone disease) occurs commonly in older birds on poor diets and housed indoors. Diets deficient in calcium and vitamin D and high in phosphorus (all seed diets) are risk factors. Decreased bone density leads to pathologic fractures if the bird falls or during restraint.

Clinical signs include weakness, ataxia, inability to perch, and pathologic fractures. Reproductively active hens may present with egg binding or cloacal prolapse.

Diagnosis is based on history, typically a sedentary bird on a poor diet and physical exam. The bird may have decreased blood calcium (plasma and ionized), increased phosphorus and decreased vitamin D (25-hydroxycholecalciferol) levels and radiographs may reveal decreased bone density and/or pathologic fractures.

Treatment includes improving the diet (converting to a formulated diet), providing ultraviolet lighting (indoor) or ideally providing time outdoors. Calcium supplementation and modifying the cage to prevent falling or in the case of an existing fracture, strict cage rest to minimize climbing and any potential for falling. Research has shown that most birds benefit from both oral and UV-B delivered vitamin D3: however African grey parrots (*Psittacus erithacus*) have a greater dependence on UV-B light for maintaining adequate serum calcium levels than do *Amazona* sp.[8]

Renal Disease

Birds of any age can develop renal disease, but it is more common in the older bird. Often there is a history of poor diet and husbandry or it may be due to a viral, bacterial or neoplastic etiology. Diets deficient in vitamin A, high in phosphorus or vitamin D or just long term generalized malnutrition from an all seed diet can lead to renal disease. Numerous viruses can infect and affect the kidneys including polyoma virus and psittacine herpes virus.[8] Avian borna virus (ABV) has been found in the kidney and many birds shed ABV in their urine. Polyuria and polydipsia (PU/PD) and elevated uric acids have been reported in these birds.[10] Chronic bacterial infections or septicemia can potentially lead to nephritis and chronic nephritis can lead to chronic renal insufficiency or renal failure.[9]

Clinical signs include weight loss, PU/PD, depression and lameness. Owners often report the bird is less active and quieter. In some cases articular or visceral gout may develop. Visceral gout is the accumulation of uric acid crystals on the internal organs (eg, heart, kidney, spleen) and articular gout is the accumulation of uric acid crystals in the joints and tendons, particularly in the feet and hocks. Uric acid crystals irritate surrounding tissues and cause severe inflammation and pain. The prognosis for gout is grave.

Diagnosis of renal disease includes blood work revealing a persistent hyperuricemia. The CBC may reveal a leukocytosis and anemia. Radiographs may reveal renal enlargement and mineralization. Kidney biopsy may help determine the etiologic process and prognosis.

Treatment for renal disease includes fluid therapy and nutritional support. Allopurinol 10 mg/kg PO q24h (Zyloprim. DSM Pharmaceuticals Inc, Greenville NC USA), colchicine 0.01-0.2.0 mg/kg PO q12-24h (Colcrys. Takeda Pharmaceuticals USS Inc, Deerfield IL USA) and omega-3 fatty acids are used as anti-inflammatory agents. Butorphanol 0.5-3.0 mg/kg IM q4h (Torbutrol Intervet Inc) or tramadol (5-10 mg/kg PO q6-12h (Janssen Pharmaceuticals, Inc) is important if gout is present. Changes in the diet and husbandry may be indicated.[8] Older birds (especially cockatiels and budgerigars) on poor diets should be converted very slowly to a formulated diet. They may develop PU/PD, lethargy and elevated uric acid levels when placed suddenly on a formulated diet. This may be due to reduced renal and hepatic function from a life- long deficiency in vitamin A and certain amino acids.[11]

Hepatic Disease

Hepatic disease occurs commonly in pet birds and can be related to viral disease (ex: polyoma and avian herpesvirus), bacterial disease (eg, Chlamydiosis) and/or hepatic lipidosis. Hepatic lipidosis is most commonly seen in the older pet bird who has been on a poor diet (seed mix or unhealthy tablefoods) and has a history of obesity. Excessive fat intact leads to lipid accumulation in hepatocytes. In chronic liver disease, fibrosis occurs and results in decreased liver function leading to cirrhosis and liver failure.

Birds may present with depression, anorexia, weight loss, overgrown beak and nails, poor feather quality and biliverdinuria. Depending on the initial weight of the bird and duration of disease the bird may present obese (but has lost weight due to illness) or may be very thin.

Diagnosis of hepatic disease includes blood work revealing elevated aspartate aminotransferase (AST), bile acids and a non-regenerative anemia. In hepatic lipidosis, birds may have elevated triglycerides, cholesterol and be lipemic. Depending on the severity of the disease, radiographs or Cat Scan (CT) may reveal hepatomegaly or microhepatica. Ultrasound may reveal abnormalities and may be diagnostic in hepatic lipidosis. Fine needle aspirates of the liver for cytology, culture and PCR for infectious agents can be performed during ultrasound with minimal risk to the patient. Liver biopsy is preferred for confirming liver disease and determining etiology and prognosis.

Treatment includes supportive care (fluids and nutritional support) antibiotics may be needed depending on the etiology along with ursodeoxycholic acid (UDCA) 10-15 mg/kg PO q24h (Impax Generics Hayward, Ca USA), colchicine 0.01-0.2 mg/kg PO q12-24h (Colcrys. Takeda Pharmaceuticals USS Inc), lactulose 150-650 mg/kg PO q8-12h and silymarin (milk thistle) 50-75 mg/kg PO q12h.[6] Long term modifications of the diet may include converting to a formulated diet.

Cardiac Disease

Cardiac disease is being diagnosed more frequently in our geriatric birds. In the past diagnosis was often post mortem. In one study, heart disease was found in 11% of avian necropsies.[11] With advancing diagnostic capabilities (digital imaging, CAT scan and ultrasound) diagnosing cardiovascular disease antemortum is becoming more common. Predisposing factors for cardiac disease in birds include poor diet, obesity and a sedentary life style. Common diseases of the aging cardiovascular system include cardiomyopathy, valvular disease and atherosclerosis. Vegetative lesions of the aortic and mitral valves may occur secondary to chronic infections such as hepatitis, reproductive infections and pododermatitis.[13]

Congested heart failure is associated with fluid retention resulting from sustained inadequate cardiac output. Pulmonary edema predominates with left sided heart disease and hepatomegaly and ascites predominate in right sided heart disease.

Clinical signs include periodic or sustained weakness, syncopy, lethargy, increased respiratory effort, hepatomegaly and ascites. The physical exam may reveal tachycardia, an arrhythmia, heart murmur, ascites and jugular distention or slow respiratory recovery following restraint (3-5 minutes is normal). There may be muffled heart sounds from pericardial effusion or hepatomegaly.[13]

Diagnostics include radiographs, EKG and cardiac ultrasound. Whole body radiographs can reveal an enlarged cardiac silhouette, the presence of tortuous vessels, hepatomegaly or ascites. Radiographic detection of an enlarged cardiac silhouette with muffled heart sounds is suggestive of pericardial effusion. Ultrasound is the standard in non- invasive cardiac evaluation. The probe can be directed cranially through the liver to get a 4-chamber view of the heart. Chamber size, wall thickness, blood flow and valvular defects may be seen. Computed tomography (CT) allows visualization of the heart and the blood vessels and is more sensitive in evaluating vasculature than radiology and its use in birds is growing. The electrocardiogram (ECG) characterizes arrhythmias and disturbances in conduction, but results may be affected by sedation and anesthesia. Indirect blood pressure can be measured using a Doppler flow detector placed over the superficial ulnar artery or cranial tibial artery and a cuff and sphygmomanometer.[13] However, studies have shown consistently that this method does not agree with direct methods and may be of limited clinical use, especially in small avian patients.[14] Results suggested that indirect monitoring of a single bird during a single cuff placement such as during an anesthetic procedure might be useful to detect changes within an event.[14] Direct blood pressure may also be performed in the anesthetized patient.

Treatment includes supportive care and oxygen therapy for CHF. Furosemide 1-2.2 mg/kg PO q12-24h, (Boehringer Ingelheim Vetmedica, Inc, St. Joseph, MO USA), Enalapril, an angiotensin converting enzyme (ACE) inhibitor 1.25-5 mg/kg PO q12h (Boehringer Ingelheim Vetmedica, Inc, St. Joseph) and a positive inotrope such a digoxin 0.01-0.02 mg/kg PO q12h, Novartis Pharmaceuticals Corp., Morris Plains, NJ USA) or Pemobendin 0.1-0.25 mg/kg PO q12h (Vetmedin, Boehringer Ingelheim Vetmedica, Inc, St. Joseph, MO USA) have been used anecdotally to treat heart disease.[12] The long-term prognosis of heart failure in birds is guarded to grave. Consultation with a veterinary cardiologist is recommended when possible.

Atherosclerosis

Atherosclerosis is a chronic inflammatory and degenerative condition of the wall of the arteries. The mineralization and plaque formation that occurs results in narrowing of the lumen and decreased blood flow. These changes in birds are most commonly seen in the aorta, brachiocephalic trunk and pulmonary arteries. Lesions may also occur in the peripheral vasculature, most notably the coronary and carotid arteries, the descending aorta and the celiac and ischiatic arteries. Atherosclerosis is probably the most common cardiovascular disease in middle aged to older pet birds. Reported incidence has been as high as 91% in some species. Risk factors include high fat diet, sedentary life style, age and there may be a genetic predisposition. Females are over represented as are Amazon and African grey parrots.[15]

Clinical signs depend on the site of the lesion but may include sudden death, dyspnea, depression, lethargy, ascites, ataxia, tremors, seizures and paresis. Owners may report their birds act disoriented or are exhibiting extension and rigidity of a leg or having difficulty perching.

Antemortum diagnosis of atherosclerosis is challenging. Diagnostics include radiographs, CT, and cholesterol and triglyceride levels. Radiology or CT may reveal an increased density and size of the aortic arch and or tortuous or mineralized vessels. Many birds will have elevated cholesterol and triglycerides levels and lipemic blood.

Treatment is challenging and includes dietary modifications (formulated diet with vegetables and fruit), weight loss, and omega 3 fatty acids. Isoxuprine 10 mg/kg PO q24h (Bi-Coastal Pharm International LLC, Red Bank NJ USA) a peripheral vasodilator that causes vascular smooth muscle relaxation has been used anecdotally with some success in birds. Statins are a group of lipid lowering drugs used in humans with atherosclerosis. Statins have been used in birds, but there is little information to support their use. A study investigating rosuvastatin (Crestor) (10 mg/kg and 25 mg/kg (AstraZeneca Pharmaceuticals LP, Wilmington, DE USA) resulted in plasma concentrations below the limits of quantitation.[16]

Reproductive Disease

Common reproductive diseases in pet birds include cystic ovaries, salpingitis, egg yolk peritonitis, egg binding, hernias and neoplasia. Clinical signs include depression, lethargy, ascites, dyspnea, straining, cloacal prolapse and paresis. Birds may present with feather destructive behavior or nesting behavior.

Diagnostics include hematology, biochemistry profile, radiographs, CT and ultrasound. Hematology may reveal a leukocytosis and anemia. The biochemistry profile may reveal lipemia, elevations in AST, creatine phosphokinase (CPK), total protein, cholesterol and triglycerides. Imaging may reveal an enlarged ovary or oviduct, hepatomegaly and ascites. Exploratory surgery may confirm the diagnosis.

Treatment options include dietary modifications if needed, antibiotics, anti-inflammatories or analgesics, and GNRH agonists such as leuprolide acetate 700-800 µg/kg IM (Lupron Depot. AbbVie, Chicago, IL, USA) or a deslorelin acetate implants (Suprelorin. Virbac AH, Inc, Fort Worth TX, USA).[17,18] For severe reproductive disease that is unresponsive to medical management, the only option may be surgical. Surgery in an older debilitated bird often carries a guarded prognosis.

Neoplasia

Neoplasia has been documented in birds of all ages but occurs most frequently in middle aged to older birds. Common neoplasia include: xanthomas, lipomas, fibrosarcomas, squamous cell carcinoma, ovarian neoplasia, gastric carcinoma, biliary and cloacal carcinoma related to avian herpes virus, and lymphoma and lymphosarcoma.[19-27] Pituitary adenoma occurs commonly in older budgerigars.[20]

Clinical signs will vary with the organ system affected and may include lethargy, depression, anorexia, weight loss, infertility, ataxia or lameness, inability to fly, head tilt and blindness.

Diagnostics include hematology, biochemistry profiles, radiology, CT, magnet resonance imaging (MRI), ultrasound and biopsies.

Treatment will vary based on the location of the tumor or the type of neoplasia, but may include surgical excision, radiation therapy or chemotherapy. Treatment protocols are mostly anecdotal and extrapolated from treatment of domestic species. Complete surgical excision of a tumor is recommended if possible. If not, then systemic chemotherapy or radiation therapy is recommended.[18-25] Consultation with a veterinary oncologist is recommended when faced with an avian patient with neoplasia.

Well Bird and Preventive Medicine

Geriatric avian health care should include yearly or bi-yearly exams. A review of the diet and husbandry since the last visit should be obtained. Any changes in behavior and activity and vocalization should be noted. A complete physical exam including a thorough ophthalmic exam should be performed. Owners should be encouraged to weigh their birds at home to pick up on any early changes in the weight. A complete blood count, biochemistry profile with bile acids, cholesterol and triglycerides should be obtained. Radiographs or CT may be indicated depending on the findings of the physical exam and blood work.

Although old age is not a disease, many older birds are more at risk for disease processes. They may also have underlying disease issues such as mild renal insufficiency that may not be evident on a biochemical profile. Geriatric birds placed on long term medications for chronic disease such as arthritis, cardiac disease or neoplasia should be monitored closely (every 3-6 months) evaluating both the birds weight and physical exam findings along with blood chemistry and hematologic findings. Diet conversion should be done slowly and birds monitored closely for signs of renal disease.

Many of the geriatric diseases we see in our pet birds are related to a life long history of poor diet and a sedentary lifestyle. Prevention is the key. Encouraging clients to provide healthy diets and to allow their birds to get adequate exercise (eg, flying if possible) and natural sunlight would decrease the incidence of many of these diseases and increase both the quality and longevity of their lives.

References

1. Rainwater KL, Sykes JM, Sapienza JS. Retrospective investigation of cataract management in avian species in a zoologic collection. *J Zoo Wildl Med.* 2015;46(4):858-69.

2. Hendrix DVH, Sims MH. Electroretinography in the Hispaniolan Amazon parrot (*Amazona ventralis*). *J Avian Med Surg.* 2004;18(2):89-94.

3. Martin H, Ritchie B. Orthopedic surgical techniques, In: Ritchie B, Harrison G, Harrison L, eds. *Avian Medicine: Principles and Applications.* Lake Worth, FL: Wingers Publishing Inc; 1994:1137-1153.

4. Bennet A. Orthopedic surgery, In: Altman RB, Clubb SL, Dorrestein GM, Quesenberry K, eds. *Avian Medicine and Surgery.* Philadelphia, PA: Saunders;1997:733-766.

5. Remple JD. A multifaceted approach to the treatment of bumblefoot. *J Exotic Pet Med.* 2006;15(1):46-55.

6. Carpenter JW. *Exotic Animal Formulary.* 4th ed. St. Louis, MO: Elsevier Saunders;2013.

7. Harcourt-Brown N. Orthopedic conditions that affect the pelvic limb. *Vet Clin North Am Exot Anim Pract.* 2002;5(1):49-81.

8. Stanford M. Calcium metabolism. In: Harrison GJ, Lightfoot TL, eds. *Clinical Avian Medicine.* Vol. I. Palm Beach, FL: Spix Publishing;2006:141-52.

9. Echols S. Evaluating and treating the kidneys. In: Harrison GJ, Lightfoot TL, eds. *Clinical Avian Medicine.* Vol. 2. Palm Beach, FL: Spix Publishing;2006:451-492.

10. Heatley J, Villalobos A. Avian bornavirus in the urine of affected birds. *Vet Med Res Rep.* 2012;3:19-23.

11. Lightfoot TL. Geriatric psittacine medicine. *Vet Clin North Am Exot Anim Pract.* 2010;(13):27-45.

12. Oglesbee BL, Oglesbee MJ. Results of postmortem findings in psittacine birds with cardiac disease: 26 cases (1991-1995). *J Am Vet Med Assoc.* 1998;212;1737-1742.

13. Fitzgerald BC, Beaufrere H. Cardiology. In: Speer BL, ed. *Current Therapy in Avian Medicine and Surgery.* St. Louis, MO: Elsevier Saunders;2016:252-328.

14. Johnston MS, Davidowski LA, Rao S, et al. Precision of repeated, Doppler-derived indirect blood pressure measurements in conscience psitttacine birds. *J Avian Med Surg.* 2011;25(2):83-90.

15. Beaufrere H. Avian Atherosclerosis: parrots and beyond. *J Exotic Pet Med.* 2014;22(4):336-347.

16. Beaufrere H, Papich MG, Nevarez J, etal. Pharmacokinetics of orally administered Rosuvastatin in Hispaniolan Amazon parrots (*Amazon ventralis*). *Proc Annu Conf Assoc Avian Vet.* 2014;25.

17. Ottinger MA, Wu J, Pelican K. Neuroendocrine regulation of reproduction in birds and clinical applications of GnRH analogues in birds and mammals. *Semin Avian Exotic Pet Med.* 2002;1(2):71-79.

18. Rae M. Endocrine disease in pet birds. *Semin Avian Exotic Pet Med.* 1995;4(1):32-38.

19. Coleman CW. Lymphoid neoplasia in pet birds: a review. *J Avian Med Surg.* 1995;9(1):3-7.

20. Lightfoot TL. Clinical avian neoplasia and oncology, In: Harrison GJ, Lightfoot TL, eds. *Clinical Avian Medicine.* Vol 2. Palm Beach, FL: Spix Publishing;2006:560-566.

21. Degernes LA. Multicystic biliary adenocarcinoma in a blue-and-gold macaw (*Ara ararauna*). *J Avian Med Surg.* 1998;12(2):100-107.

22. Filippich LJ. Intravenous cisplatin administration in sulphur-crested cockatoos (*Cacatua galerita*): clinical and pathologic observations. *J Avian Med Surg.* 2001;15(1):23-30.

23. Hillyer EV, Moroff S, Hoefer H, Quesenberry KE. Bile duct carcinoma in two out of ten Amazon parrots with cloacal papillomas. *J Assoc Avian Vet.* 2001;5(2):91-95.

24. Kennedy FA, Sattler-Augustin S. Oropharyngeal and cloacal papillomas in two macaws (*Ara* spp.) with neoplasia with hepatic metastasis, *J Avian Med Surg.* 1996;10(2):89-95.

25. Lamberski N, Theon AP. Concurrent irradiation and intratumoral chemotherapy with cisplatin for treatment of a fibrosarcoma in a blue and gold macaw (*Ara ararauna*). *J Avian Med Surg.* 2002;16(3):234-238.

26. MacWhirter P. Use of carboplatin in the treatment of renal adenocarcinoma in a budgerigar. *Exotic DVM.* 2002;4(2):11-12.

27. Zantopp DA. Treatment of bile duct carcinoma in birds with carboplatin. *Exotic DVM.* 2000;2(3):76-78.

The Long and Winding Road: From Diagnostic Testing to Treatment and Beyond

Greg Rich, DVM

Session #2110

Affiliation: From the Avian and Exotic Animal Hospital of Louisiana, 3635 North I-10 Service Road, W., Metairie, LA 70002, USA.

Introduction

It is not just a well-known concept that all members of the avian species hide symptoms of illness, it is a documented fact. Even though many of our psittacine patients can talk, they are not able to verbally communicate what is abnormal or diseased in their body. Decades of science, research and clinical observation have allowed veterinarians treating avian species to command a better understanding of their disease states, diagnostic testing strategies and treatment modalities. We have better knowledge of disease prevention and a much clearer picture of how proper nutrition enables the avian patient to battle certain disease states and how nutritional supplements may be inherently critical to organ system homeostasis. We have reference books, conference proceedings, formularies, and scientific papers. We have testing profiles specific for our avian patients and in many cases, specific values for the species in question. We even have research validating specific dosages for medical therapy of certain disease states. With all the tools and knowledge that we have, what we don't have is the perfect patient with the perfect disease state that responds perfectly to the treatment regimen written about in the textbooks or talked about at conferences. Perfect is nice, but it is elusive and sometimes boring. What we have is a combination of advanced educational experiences and best of all, intuition. Hence, we begin the long and winding road to diagnostic testing.

Avian medicine and surgery is an ever-evolving science. Our past experiences, our zest for answers, and an ever-growing volume of answers to our questions allows us, veterinarians treating and caring for avian patients, a better chance to resolve disease states, syndromes and maladies affecting the avian species. Every veterinarian treats patients slightly differently. We all need to start the road with a diagnostic plan, which allows a foundation to illuminate problems at hand during the course of treatment and finishes with a documented return to normalcy once treatment is complete. An astute veterinarian will cast a wide net with diagnostic testing so as to document the abnormalities the patient is facing, but not so wide as to waste time, the patient's limited blood supply, or the client's financial resources. Once the diagnostic plan is formulated, the test results will allow one to set up a treatment plan or request more testing. Once a treatment plan is designed, a time frame should be established to recheck the test results to ascertain if the patient's condition or disease has resolved. Like any medical case, there must be a beginning and an ending; otherwise, you are just traveling down the winding road with no direction and no end in sight.

Diagnostic Tools

We have many diagnostic tools available to us: complete blood counts (CBC), blood chemistry profiles, radiographs, MRI, CT scans, ultrasound, cytology and histopathology. The most important tool is the knowledge and understanding of what is essential among all tests to provide us a path, or in some cases an absolute answer, to the issues presented. Again, not every case will be perfect, but with our knowledge base, we can perfect the path to a diagnosis and eventually a treatment road that is less treacherous and more straightforward.

CBCs for avian species produce the following information: red blood cell count, packed cell volume (PCV), white blood cell count, a differential, a platelet estimate, and red and white blood cell morphology. Microscopic examination of the blood smear will allow identification of extracellular and intracellular parasites if present, intracellular bacteria in cases of sepsis and in the rare case of leukemia, neoplastic white blood cells. These are pointed, specific diagnostic pictures. We commonly use the CBC to assess what varies from normal in the hemogram. The PCV helps to assess the presence or absence of anemia, but it does not tell us the cause or origin of abnormal value. Microscopic analysis of the blood smear will allow documentation or assessment as to whether the anemia is regenerative or non-regenerative. The white blood cell count allows us to place the patient into one of the following categories: normal, inflammatory disease process or an infectious disease process. The differential enables the microscopist the ability to determine the toxicity of the heterophils and the reactivity of the lymphocytes, as well as the percent each cell line represents in overall total white blood cell count, known as the differential. The numbers that each count delivers are crucial to establish the category that the practitioner places the patient in, but the microscopic analysis of the blood smear is essential to help elucidate the fine points of the severity of the patient's status. Putting all this data together allows the clinician or practitioner to recommend specific disease testing (*Chlamydia*, polyomavirus, *Aspergillus*, *Sarcocystis*), request further testing or begin therapy if the results deliver enough information to begin the road to treatment of the clinical abnormality. The CBC allows the path some clarity. Rarely, is a diagnosis rendered from just a CBC, but the results allow for placing the patient in a category of normal, exhibiting an infectious hemogram, an inflammatory hemogram, anemic or normocythemic. Treatment strategies need this information. Diagnosing a return to normalcy requires this information as the starting point.

Blood chemistry profiles come in a variety of combinations that can be performed either in-house or sent to a reference lab. The analytes contained in a blood chemistry profile, the causes for increases, the causes for decreases and what body functions may be affected have been discussed in scientific proceedings and textbooks for years. Each and every case should be analyzed on its own merits to decide whether a standard testing panel will suffice or if an advanced profile is necessary and whether the in-house lab (if one is present) will supply the necessary results for a health or disease status diagnosis or if the sample should be sent to an avian reference lab. Like with a CBC, we need as much information as possible to help obtain a diagnosis of the health status of the patient, but not too much to waste the patient's blood supply on unnecessary testing or deplete the client's funds on results that are meaningless. Leaving room to add tests after the first battery of test results are obtained is often a wise option. As with the CBC, the results of the blood chemistry profile often tell us what is outside the normal range, what organ system may be affected or what analytes are abnormal, but they rarely diagnose a specific disease. True, elevated bile acids in a 15-year-old blue and gold macaw tells us there is abnormal liver function, but it does not name a specific liver disease. Elevated bile acids could be from fatty liver disease, hepatic fibrosis, cirrhosis, polyomavirus, Pacheco's disease, and a variety of other overwhelming liver diseases. A definitive diagnosis will require a liver biopsy or specific disease testing. Even if the chemistry profile reveals all normal results, we can only inform the client that the blood chemistry profile was normal. It does not signify a perfectly healthy patient. Again, the winding road continues, but has more signs to point you in the right direction. It is still up to your examination, your intuition, your knowledge, personal experiences, and assessment of the results so far, to lead to a diagnosis and a treatment strategy or the need for more testing. Various textbooks have formulas for testing profiles for suspected liver disease, suspected renal disease, nutritional deficiencies and excesses, and immunoglobulin abnormalities. The patient's history, dietary history, and physical exam will help clarify which road we should travel.

Diagnostic imaging can certainly pinpoint organ size, fluid or soft tissue or metal densities and structure defects, but like other diagnostic testing, the results may give us a morphologic diagnosis, such as hepatomegaly, but it cannot differentiate among diseases such as fatty liver disease, chlamydiosis, and iron storage disease. Even in cases of metal-dense ventricular foreign bodies noted on radiographs, we are unable to tell if the metal is lead, zinc, cadmium, or aluminum without combining signs, symptoms and in most cases, toxicologic testing. With the advent of more private veterinary hospitals owning and utilizing ultrasound and CT scans on avian patients, the morphologic diagnostic picture is clearer and in some cases, can offer a concrete diagnosis. As in the case of a patient with an elevated bile acid and radiographic evidence of hepatomegaly, an abdominal CT scan or ultrasound may be able to elucidate a homogeneous density in all liver lobes compatible with hepatic fibrosis. The road to a complete diagnosis becomes clearer in many cases with these advanced diagnostic-imaging tools. Yet, in some cases we still have to use our knowledge, other test results, or further testing to arrive at the destination of an absolute concrete diagnosis and a detailed treatment plan.

Cytology and histopathology of lumps, bumps, tissue lesions, organs and masses are usually definitive for a concrete diagnosis, but even these advanced diagnostic tools have their limitations. There are cases where special stains utilized on sections of liver tissue that originally diagnosed hepatic fibrosis, now reveal chlamydial organisms in the hepatocytes completing the road trip to its final destination of liver malfunction from hepatic fibrosis caused by chronic chlamydiosis.

Treatment Plan Considerations

Ok, great, we have a diagnosis. Once again we embark on the long and winding road to a treatment plan. The veterinary clinician or practitioner must consider costs, risk to humans and other pets in the household, client compliance with treatment plan, patient compliance with mode of drug delivery and the proper diagnostic plan for determining the success or cure the treatment plan was designed to deliver. Offering options to the owner is always wise, but these options must be combined with a cost per option and a percent of success for each option. If treatment for the *Chlamydia* case above in a blue and gold macaw has a 90% chance of resolution with 2 weeks of hospitalization, injectable doxycycline, oral milk thistle and oral denosyl at a cost of $570 with 5 more doxycycline injections over the next 5 weeks for an extra $110 or at-home treatment of doxycycline in the drinking water for 45 days for a total cost of $135 but a 30% chance of resolution, then the discussion must be had that a serious disease like *Chlamydia* and a valuable, long-lived pet certainly deserves the treatment course with the highest degree of resolution. Discussion of the options takes time, but there will always be a place for offering the best option, irrespective of cost, first. After all, this is why we, veterinary practitioners, go to school for 4 years or longer, buy textbooks, attend conferences and read journals, so we can provide the most up-to-date, optimal treatment course to deliver a complete resolution for the case at hand. Cost should only be part of the discussion if the owner cannot afford the optimum treatment. At this time, you can work on Option 2 or 3. With the increase knowledge of organ-support pharmaceutical agents, nutritional supplements, homeopathic/herbal/alternative medical therapy, we need to be aware of the full spectrum of treatment options available for avian patients and the vast array of conditions that require our medical expertise. Treatment plans should have an ending that accomplishes our goals. Just the fact that a patient is eating better and has gained weight is not enough to say the case is 100% resolved. Blood values, radiographs, biopsy results, or specific disease testing will be tantamount to our ability to accurately communicate that the patient has been cleared of the disease, injury or imbalance.

During the initial meeting with the client as well as subsequent office visits during the course and end of treatment, it will be imperative for our veterinary team to discuss proper nutrition, adequate caging, safe toys, training and foraging techniques. As with the case of the blue and gold macaw with hepatic fibrosis secondary to chronic chlamydiosis, it would be deleterious for the health of the liver if the patient remained on a diet of sunflower seeds, peanuts and Ritz crackers. We must be knowledgeable about the need for proper nutrition for every avian patient, healthy or otherwise. In addition, it is imperative for us to educate our clients about toxins like avocado and PTFE, as well as the safety issues regarding harmful items such as cedar shavings as a cage substrate. We must be a proactive voice for the pet's health before, during and after disease conditions are addressed. The Association of Avian Veterinarians (AAV) has brochures about diet, feather loss, signs of illness, among others. There are numerous texts that highlight proper diets, dangers of certain toys or cage adornments and behavior modification strategies. In addition, there are several valuable Internet sites that we can refer clients to that may provide them valuable information about diets, care and behavior (ie, AAV.org, VSPN, Goodbirdinc, Lafeber.com).

Role of Staff

As with all trips down a long and winding road, it is best for us to travel with a partner. Hence, it is extremely important for our staff to understand the variety of species we intend to treat, the diets they require, the diseases and maladies that befall them and be taught to answer questions about our treatments, surgery or therapy recommendations. A well-educated, well-versed staff in avian medicine and surgery will help the entire process of client education and communication proceed with a more professional experience for the client and allow the veterinarian to accomplish more during our busy and often hectic day.

Conclusions

It sounds like an insurmountable task for us to assess the patient in its cage, perform a physical exam, determine what tests are necessary to diagnose the condition, design a treatment plan, reassess the patient during treatment, test for the endpoint of the treatment regimen and provide client education materials to allow the patient to have a healthier life afterwards, but it is what we do and we do it every day. It just behooves us to try to navigate the winding road better each time we travel down its path. Always be open for ways to allow the path to be smoother and clearer. Find out what medications or supplements will help the patient improve quicker, find out ways to help the home environment be more engaging for the bird, and always make sure the client provides the right fuel for the vehicle (food for the patient).

Recommended References

1. Altman RB, Clubb SL, Dorrestein GM, Quesenberry K. *Avian Medicine and Surgery.* Philadelphia, PA: Saunders;1997.
2. Campbell TW. *Exotic Animal Hematology and Cytology.* 4th ed. Hoboken, NJ: Wiley-Blackwell;2015.
3. Fudge AM. *Laboratory Medicine: Avian and Exotic Pets.* Philadelphia, PA: WB Saunders;2000.
4. Graham JE. *Blackwell's Five-Minute Veterinary Consult: Avian.* Ames, IA: Wiley-Blackwell;2016.
5. Rosskopf WJ, Woerpel RW. *Diseases of Cage and Aviary Birds.* Baltimore, MD: Williams and Wilkins;1996.
6. Ritchie BW, Harrison GJ, Harrison LR, eds. *Avian Medicine: Principles and Applications.* Lake Worth, FL: Wingers Publishing Inc;1994.

7. Rupley AE. *Manual of Avian Practice.* Philadelphia, PA: WB Saunders;1997.

8. Speer BL. *Current Therapy in Avian Medicine and Surgery.* St. Louis, MO: Elsevier Saunders;2016.

Section 2

Aviculture

Greg Burkett, DVM, DABVP (Avian Practice)
Moderator

An Introduction to Non-psittacine Aviculture

Michele D. Goodman, VMD, MHS, CWR

Session #2008

Affiliation: From Elmwood Park Zoo, 1661 Harding Boulevard, Norristown, PA 19401, USA and Board of Directors, National Wildlife Rehabilitators Association, 2625 Clearwater Road, St. Cloud, MN 56301, USA.

Abstract: There is so much more to aviculture than psittacine birds. Aviculturists exist in nearly every community, breeding and rearing waterfowl, cranes, pheasants, flamingos and many other species. This master class will provide insight into this growing community of bird enthusiasts and will offer strategies for partnering with aviculturists to promote better avian health and welfare.

Introduction

There are many reasons why people maintain various species of birds in captivity: enjoyment, public education, captive breeding and conservation to name a few. By maintaining birds in captivity, people have developed a better understanding of avian taxonomy, feeding and courting behaviors, growth, development and reproduction. Through captive breeding, reintroduction efforts have led to the reestablishment of critically endangered species back in their native habitat. There is a wide variety of non-psittacine, non-passerine species maintained in captivity, with unique husbandry requirements and health needs.

Non-psittacine Aviculture

According to Jean Delacour, aviculture is defined as "the worldwide hobby of keeping and breeding numerous species of wild birds in captivity to maintain their numerical status in nature with a view of forestalling their extinction by supplying aviary raised stock." Since the early 1900's, numerous avicultural societies have been formed worldwide with this definition in mind. Since the early 1900's, numerous species of birds from all corners of the earth have been collected from the wild with the aim of breeding these birds in captivity. Traditionally there is a huge interest in captive psittacine and passerine birds; these birds often are kept as pets in addition to being maintained in zoos and by private aviculturists for breeding and conservation. There are numerous other groups of birds that are generally not maintained as pets but are frequently kept by private aviculturists and zoos for enjoyment, entertainment, and conservation purposes.

Getting to Know the Birds

When working with any group of birds, it is essential to understand something about their natural history as this often provides information pertaining to climate tolerance, natural diet items, husbandry requirements, behavior and reproductive strategies. Aviculturists and zoos can be excellent resources for general husbandry and breeding information. There are a number of veterinary texts that provide excellent information on caring for diverse avian species and on administering medications. Suggested veterinary texts include Ritchie, Harrison and Harrison (1994), Harrison and Lightfoot (2005), Samour (2016), Speer (2016) and Carpenter (2012). See Table 1 for suggested natural history references.

Table 1. Recommended natural history and identification resources.

Bird	Resources
All birds	Holland G. *Encyclopedia of Aviculture*. Blaine, WA: Hancock House Publishers;2007. Rutgers A, Norris KA. *Encyclopedia of Aviculture*, Vol 1. London, UK: Branford Press;1970:183-243.
Waterfowl	Baldasarre G. *Ducks, Geese, and Swans of North America*. Baltimore, MD: Johns Hopkins University Press;2014. Kear J. *Ducks, Geese and Swans*. Oxford: Oxford University Press;2005. Madge S, Burn H. *Waterfowl: An Identification Guide to the Ducks, Geese and Swans of the World*. Chicago, IL: Houghton Mifflin;1998.
Cranes	Ellis DH, Gee GF, Mirande CM. *Cranes: Their Biology, Husbandry and Conservation*. Washington, DC: US Department of the Interior;1996.
Gallinaceous birds and currasows	Johnsgard PA. *The Pheasants of the World: Biology and Natural History*, 2nd ed. Washington, DC: Smithsonian Institution Press;1999. Potapov R, Sale R. *Grouse of the World*. Chatswood, AU: New Holland;2014. Jones DN, Dekker RWRJ, Roselaar CS. *The Megapodes. Bird Families of the World*. Oxford, UK: Oxford University Press;1995.

Getting to Know your Local Non-psittacine Aviculturist

As with other groups of breeders, Aviculturists are a diverse group of individuals ranging from hobbyists to large-scale operations. The majority of aviculturists maintain collections of birds for enjoyment; many will breed birds on a small scale and aim to sell or trade birds locally. Many smaller facilities do not solicit help from veterinarians and depending on the value of their collection may not be able to afford any level of veterinary care. In addition to the small private aviculturists, there are several distinct groups of larger avicultural facilities. There are game bird breeders that raise and sell birds for the primary purpose of hunting (generally ring-necked pheasants, certain species of quail and some species of waterfowl). There are also large private aviculture facilities, some of which are nonprofit organizations, that maintain and breed large numbers of birds for other private aviculture facilities or for zoological collections. The perceived and actual veterinary needs of these facilities vary widely depending both on objectives and available resources.

Perceived veterinary needs

Generally speaking, aviculturists reach out to veterinarians with a limited number of requests: health certificates, medications for general use and the occasional emergency. Veterinarians may not have a pre-existing relationship with the aviculturist and have limited knowledge of their facility, level of training or any other information necessary for establishing a Veterinary-Client-Patient-Relationship (VCPR).

Actual veterinary needs

There is a long list of services that a veterinarian can provide to an aviculture facility. These services include establishing herd health protocols, parasite management, developing vaccination protocols, managing emergencies, providing diagnostic services (including necropsy), helping to improve husbandry practices, providing reproductive assistance and, of course, writing health certificates.

There are several things that a veterinarian needs to consider when working with aviculturists. In the majority of cases, the birds have a particular value which, depending on a particular facility includes the amount paid for the bird, shipping costs, feed costs, breeding potential and value of offspring. This animal value often prevents aviculturists from seeking veterinary care or from trying to establish a relationship with a veterinarian to meet their actual veterinary needs. Most aviculture facilities will benefit from a "herd health" approach to care and will not be able to afford individualized care for each bird in their facility. By taking this sort of approach, a relationship can be developed to benefit both aviculturist and veterinarian. In some cases, aviculturists can be trained similarly to many livestock producers to collect samples from birds for clinical pathology, to administer injections, oral medications, fluids and gavage feedings. See Table 2 for detailed recommendations on providing veterinary services to aviculturists.

Table 2. Recommendations for veterinary services provided to aviculturists.

Critical care and emergency services
Description: Bird should be evaluated by a veterinarian for diagnostics and customized treatment plan. *Suggested pricing:* Generally, prices are individualized to patient needs; basic bloodwork (PCV, TP) and imaging often are required. *Comments:* In many cases, aviculturists can benefit from training to better detect debilitated or injured birds and to provide temporary stabilizing care until transportation to a veterinarian.
Diagnostic sample collection and processing
Description: Collect swabs for pre-ship testing, obtain cultures, and collect blood and fecal samples. *Suggested pricing:* If using a primary veterinary lab account, discounted services are recommended if an aviculturist is obtaining samples; otherwise charge a flat interpretation fee depending on sample type. *Comments:* Depending on the size of the facility, it can be advantageous to both parties to set up a separate account with a specialized avian lab so that the aviculturist is responsible for collecting and submitting samples directly to the lab. Aviculturists should be trained in aseptic technique prior to collecting samples; for routine parasite screening, individual animals do not necessarily have to be examined. Submit pooled samples by enclosure or aviary. Pooled fecal samples should be submitted quarterly for at least the first year to assess parasite burden of the collection.
Fluid therapy and nutritional support
Description: Obtain patient weights, assess hydration status, and administer fluids and gavage feedings. *Suggested pricing:* Pricing is highly variable depending on veterinarian comfort level, state regulations and aviculturists. Charges are often for initial training and then for communication with aviculturist about patient progress. *Comments:* In many cases, training to administer oral fluids and to gavage nutrition can be extremely valuable. Veterinarians are rarely set up to provide appropriate husbandry for many of these groups of birds, so outpatient care often is required. When determining appropriate critical care diets there are two general types: (1) elemental or hydrolyzed formulas designed for herbivores, omnivores, carnivores or piscivores and (2) soaked and blended captive diet normally fed to specific bird.
Health certificates
Description: For the majority of intra-state transactions, a CVI or Health Certificate is required; it requires a physical exam of bird(s) being shipped. Additional testing may be necessary depending on state requirements. *Suggested pricing:* Charge based on whether a house call is required or if birds are brought to the clinic. A general recommendation is to charge per bird (ideally discounted exam rate) and then add an individual health certificate charge depending on the time needed to research requirements, interpret testing results, among other things.
Husbandry evaluation services
Description: Evaluate diet, rearing conditions, and provide aviary and pond design recommendations as well as substrate, perching and landscaping recommendations. *Suggested pricing:* Charges are based on consulting fees. *Comments:* When making recommendations it often is best to base recommendations on management concerns (eg, prevalence of pododermatitis, poor body condition, poor feather condition).

Table 2 Continued. Recommendations for veterinary services provided to aviculturists.

Infectious disease surveillance and recommendations
Description: Provide facility-tailored recommendations for vaccinations, flock deworming. *Suggested pricing:* Charges are based on vaccine costs and medication costs, then charge for a site visit if the veterinarian is administering vaccines and medications. *Comments:* Recommendations should be based on species susceptibility to infectious diseases and based on routine parasite screenings. In many cases, aviculturists can be trained to administer vaccines and any necessary deworming medication. Deworming medication often is mixed into daily feed rations (birds are generally not fed the night before so they consume medicated feed rapidly).
Pathology services
Description: Perform gross necropsy and submit tissues for histopathology. *Suggested pricing:* State poultry diagnostic labs often can provide valuable diagnostic services at an affordable rate. *Comments:* Whenever possible, tissues should be sent to a lab with experience dealing with avian species.
Provision of medications for general use
Description: Depending on state requirements, many aviculture facilities will benefit from having a limited number of basic medications on hand for emergencies. Medications should come with weight-based dosing chart for ease of use. *Suggested pricing:* Aviculturists should be required to contact a veterinarian when use of these medications is required and they can be charged a consult fee. Aviculturists then pay for medications provided. *Comments:* Common medications that an aviculturist may require include: isotonic fluids, non-steroidal anti-inflammatory medications, non-steroidal antibiotic eye drops and anthelminthic medications. Due to the risk of antibiotic resistance at many facilities, broad-spectrum antibiotics should be provided on an as-needed basis for specific patients unless managing a heard health issue. Facilities maintaining aspergillosis-susceptible species (eg, sea ducks, cranes) may benefit from having second- or third-generation triazole medication on hand. For species that can be considered food animals (eg, wild waterfowl, pheasants or quail with a hunting season, domestic waterfowl), consult the Food Animal Residue Avoidance Databank (http://www.farad.org) and avoid using medications with prohibited off-label use.
Reproductive services
Description: Generally manage birds with reproductive difficulties (eg, egg-binding) as well as make breeding management recommendations, primarily in terms of nutrition. *Suggested pricing:* Egg-binding and related conditions should be managed as emergencies and charged accordingly. Nutritional recommendations and other breeding management recommendations can be charged as consultation fees. *Comments:* Aviculturists often are well informed on techniques for getting birds to breed, particularly in terms of making nest boxes and putting birds in breeding pairs.

Methods for Marking and Wing-clipping Captive Birds

When working with captive-bred birds, there are a number of marking conventions used by aviculturists to distinguish captive-bred birds from their wild counterparts (USFWS requires that migratory species be marked by one of these methods and individual states may require marking of all captive-bred species regardless of whether they are native to a particular area).

The three methods for marking captive-bred birds are: pinioning, amputation of the hallux, and application of a close-ring band during growth. Pinioning is the amputation of the metacarpal bones at the carpometacarpal joint that leaves the alula intact. This procedure often is performed by the aviculturist on birds less than 3 days of age. Pinioning commonly is performed on geese, swans, cranes and flamingos. Alternatively, the hallux may be amputated; this also is generally performed by the aviculturist on birds less than 3 days of age. This method is used commonly in ducks and gallinaceous birds that are to remain fully flighted in a closed-top enclosure. It is important to note than in many countries, any form of amputation at any age must be performed by a licensed veterinarian under either local or general anesthesia.[1] Closed-ring banding is gaining popularity particularly among larger private aviculture facilities; here a seamless band is applied to a bird during growth (generally between 2-5 weeks of age depending on species). The bird will grow into the band such that the band cannot be physically removed once the bird is fully grown but remains freely moveable between the hock and foot joints. If the sex of the bird is known when applying a closed-ring band, then the band is applied to the right leg on males and to the left leg on females. For non-native species, application of a metal seamed leg band may be an acceptable form of identification for certain species; with these types of bands there is an increased risk of leg injuries due to band constriction.

For fully flighted birds where wing-clipping is used as a method of flight restriction, a different approach is generally used than that for psittacine birds. For waterfowl, cranes and flamingos, a unilateral wing clip is recommended whereby all 10 primary feathers are clipped up to the level of the primary covers (making sure the feathers aren't in blood). This procedure is then performed at least on an annual basis depending on molt strategies to continue to restrict flight. Wing-clipping (or any method of flight restriction) of currasows or gallinaceous birds is not recommended. Many species are perching birds; they can become quite stressed if not able to fly to roost. Many species will utilize some form of flight or dance as part of courtship displays.

Basic Husbandry for Non-psittacine Species

Nearly all the species maintained by aviculturists are housed in outdoor enclosures or aviaries. Gallinaceous birds, currasows, and small waterfowl are generally kept in completely enclosed structures with roofs constructed of either wire or heavy-duty poultry netting. Enclosures should be predator proof, which generally means that the wire or other building materials are buried 12-24 inches underground or that these materials completely cover the entire aviary floor. Many facilities utilize electrified wire (hot wire) to provide additional security to birds and to prevent wildlife or feral cats from climbing up on top of the cages and defecating or causing stress to the birds inside the aviary. Larger birds, including flamingos, cranes, geese and swans are generally flight restricted and can be kept in enclosed or open-top aviaries. Flight restriction is performed in these larger species to prevent birds from striking aviary walls. Even in open-top enclosures every effort should be made to prevent predator access, which means burying perimeter wire and lining a perimeter fence with hot wire. Birds should have access to large ponds to roost on or near; many species of raptorial birds will not hunt over water.

Many species of gallinaceous birds have an increased susceptibility to parasite infection. The provision of dirt substrate is thought to increase parasite prevalence. Gravel, sand or other materials that drain well may provide a better alternative but still require frequent cleaning. Waterfowl and flamingos are prone to pododermatitis often associated with unvaried substrates that are difficult to clean and do not drain well. General recommendations include providing a variety of substrates including lawn, river stones and a limited amount of sand.

Cranes, flamingos and all waterfowl species require water for swimming and wading. Ponds can be natural dirt ponds depending on the landscape or can be made of concrete. There are numerous designs available for artificial ponds. Key features are that they be easy to clean and disinfect and that the pond edges be gradually sloped to allow easy entry and exit. Water must be designed to stay open through the winter season. Cold tolerant species require access to swimming water in order to prevent frostbite.

There is a wide variety of captive diets specifically formulated for non-psittacine birds. Mazuri diets (Mazuri Exotic Animal Nutrition, PMI International, Coffeyville, KS, USA), which are fed routinely in zoological institutions, are formulated for gallinaceous birds, waterfowl, cranes and flamingos. While many aviculture facilities utilize commercial feed-mill products, they often require custom-milled feeds in order to maintain healthy birds over their lifetime since feeds designed for commercial poultry are based on production costs and are designed for birds with short life expectancies. See Table 3 for husbandry recommendations by bird group and for clinically relevant taxonomy.[2-6]

Veterinarians and Aviculturists: Building Relationships

Regardless of species maintained in care, aviculturists keep various species of birds for enjoyment, breeding, and conservation potential. The majority of private aviculturists do not have established relationships with veterinarians, which is often due to cost constraints and because of some veterinarians' unwillingness to work with non-psittacine avian species in a "herd health" management framework. Veterinarians can provide critical services to non-psittacine aviculturists to promote animal health and welfare and may in some cases be able to contribute to genetic preservation of species in decline. Veterinarians can begin by reaching out to aviculturists (eg, regional, state and national aviculture organizations) and offering valuable services to these individuals. In many cases, it will take time to grow this relationship and it will take time for many aviculturists to realize the value of veterinary services and be willing to pay for them.

References

1. Vinke CM, van Zeeland YRA, Schoemaker NJ, Meijboom FLB. As free as a bird on the wing: some welfare and ethical considerations on flight restraint methods in birds. In: Speer BL, ed. *Current Therapy in Avian Medicine and Surgery.* St. Louis, MO: Elsevier Saunders;2016:683-705.

2. Wyss F, Wenker C, Hoby S, et al. Factors influencing the onset and progression of pododermatitis in captive flamingos (Phoenicopteridae*). Schweiz Arch Tierheilkd.* 2013: 497-503.

3. Kelley C, Hartup BK. Risk factors associated with developmental limb abnormalities in captive whooping cranes. *Proc North Am Crane Workshop.* 2008;119-124.

4. Heard D. Calliformes and Columbiformes. In: West G, Heard D, Cauklett N, eds. *Zoo Animal and Wildlife Immobilization and Anesthesia.* 2nd ed. Ames, IA: Blackwell Publishing;2007:473-480.

5. Nicholson DS, Lochmiller RL, Steward MD, et al. Risk factors associated with capture-related death in Eastern wild turkey hens. *J Wildl Dis.* 2000;32:308-215.

6. Höfle U, Millán J, Gortázar C, et al. Self-injury and capture myopathy in net-captured juvenile red-legged partridge with necklace radiotags. *Wildl Soc Bull.* 2004;32:344-350.

Table 3. Clinical features and diets of taxonomic groups.

Bird	Primary diet and clinical features	Cold weather tolerance
Flamingos	Specialized filter-feeders. Fed a pelleted diet and supplemental aquatic invertebrates; diets provide carotenoids for coloration. Can be maintained in small colonies. Long-lived birds. Prone to pododermatitis on inappropriate substrate.[a] Head and legs should be restrained during handling.	In captivity, most species are housed indoors due to concerns with ice accumulation around pond edges.
Cranes	Fed a pelleted crane or waterfowl diet. Diverse group with huge size variation. Many species listed as vulnerable or endangered. Most species susceptible to capture myopathy and angular limb deformities during growth, particularly if not housed appropriately.[b] Head should be restrained at all times during handling. Legs should be restrained to prevent kicking and scratching but should be handled carefully as skin is delicate and prone to bruising.	Both species of crowned cranes are cold intolerant.
Currasows and turkeys	Fed a pelleted diet designed for wild gallinaceous birds or waterfowl; can be offered insects, vegetables and some fruits. Extremely strong birds challenging to capture and restrain. Produce excessive tracheal mucous under anesthesia that can lead to endotracheal tube obstruction. Currasows have elongated tracheas and are prone to regurgitation under anesthesia.[c] Both may undergo a fright molt and can become hyperthermic during handling. Turkeys prone to capture myopathy during handling.[d]	Currasows are not winter hearty; some species of turkey are not winter-hearty.
Pheasants, guineafowl, grouse, quail and partridge	Fed a pelleted diet designed for wild gallinaceous birds; can be offered insects, leafy greens, vegetables and some fruits. Many species susceptible to capture myopathy[e] and will undergo a fright molt during handling.	Wild species of guineafowl are cold intolerant. Some quail and partridge species are cold intolerant.
Waterfowl (ducks, geese and swans)	Fed a pelleted diet designed specifically for wild waterfowl; depending on species specifics birds may be offered insects, aquatic invertebrates, leafy greens, vegetables and some fruits. Prone to hyperthermia associated with handling. Produce excessive tracheal mucous under anesthesia that can lead to endotracheal tube obstruction.	Provided there is sufficient open water, many species are winter hearty with several exceptions: most whistling ducks, Cape Barren geese, Magpie geese and Hartlaub's ducks

[a]Wyss F, Wenker C, Hoby S, et al. Factors influencing the onset and progression of pododermatitis in captive flamingos (Phoenicopteridae*)*. *Schweiz Arch Tierheilkd.* 2013:497-503.

[b]Kelley C, Hartup BK. Risk factors associated with developmental limb abnormalities in captive whooping cranes. *Proc North Am Crane Workshop.* 2008;119-124.

[c]Heard D. Calliformes and Columbiformes. In: West G, Heard D, Cauklett N, eds. *Zoo Animal and Wildlife Immobilization and Anesthesia*, 2nd edition. Ames, IA: Blackwell Publishing; 2007:473-480.

[d]Nicholson DS, Lochmiller RL, Steward MD, et al. Risk factors associated with capture-related death in Eastern wild turkey hens. *J Wildl Dis.* 2000;32:308-215.

[e]Höfle U, Millán J, Gortázar C, et al. Self-injury and capture myopathy in net-captured juvenile red-legged partridge with necklace radiotags. *Wildl Soc Bull.* 2004;32:344-350.

Aviculture and Avian Veterinary Medicine: Understanding their Relationship

Sharman Hoppes DVM, DABVP (Avian Practice), and
Rick Jordan

Session #2112

Affiliation: From the Department of Small Animal Clinical Sciences, College of Veterinary Medicine, Texas A&M University, 4474 TAMU, College Station, TX 77845, USA (Hoppes), Hill Country Aviaries, Dripping Springs, TX 78620, USA (Jordan).

Abstract: The broad definition of aviculture is the practice of keeping birds (Class Aves) in captivity using controlled conditions, normally within the confines of an aviary, for hobby, business, research and conservation purposes. Avicultural goals include breeding birds for zoological collections, breeding and preservation, or for the companion pet bird trade. Aviculturists encourage and often fund conservation, provide education about avian species, provide birds for the public (zoological collections and pet trade) and often assist in research on disease processes and behavior issues in captive birds. Avian veterinarians provide medical care, diagnostics and surgery to the avian patient. The avian veterinarian is essential in educating the pet bird owner to help keep birds healthy and happy and to aid in prevention of future behavioral problems. It is important for aviculturists and avian veterinarians to work together to provide and maintain healthy and well-socialized birds to pet bird owners, to preserve species and birds in captivity, and to assist in conservation of birds in the wild. Aviculture is a struggling community at this time. Many larger aviaries have closed and many are struggling to remain open. This lecture discusses the issues affecting aviculturists today and the importance of the aviculturist and captive breeding to the avian community and to the avian veterinarian. Without aviculture in the United States, the future of avian medicine, as a specialty, will be severely limited.

Background

Aviculture is the practice of keeping birds in captivity using controlled conditions, normally within the confines of an aviary or zoo, for hobby, business, or research and conservation purposes. Aviculturists provide birds for the pet trade and zoological collections and often fund conservation projects. Avian veterinarians provide medical and surgical care to the avian patient. Both the avian veterinarian and the aviculturist are essential in educating the pet bird owner to help keep birds healthy and happy and to aid in prevention of behavioral problems. It is essential for aviculturists and avian veterinarians to work together to provide and maintain healthy and well-socialized birds to pet bird owners, to preserve species and birds in captivity, and to assist in conservation of birds in the wild.

Aviculture and avian medicine were at their peak in the 1980s and 1990s. Aviculture was thriving with a seemingly unlimited supply of imported birds and avian veterinarians were struggling with multiple infectious diseases of parrots. Aviculturists and avian veterinarians worked together to fight these diseases and to develop diagnostics and treatment protocols.[1] In 1980, the Association of Avian Veterinarians (AAV) was established. During this time, a new industry evolved, specializing in exotic birds. The industry was multifaceted, developing and providing diets, engineering safe cages and enrichment, building brooding and incubation equipment, along with avicultural and veterinary supplies.[1,2] The science behind aviculture expanded and soon laboratories began offering specialized testing for exotic bird diseases such as psittacine beak and feather disease and avian herpes virus. Between 1970 and 1992 many species of psittacines were bred in captivity for the first time. Aviculturists and veterinarians collected and published biological data and lectured at veterinary and avicultural conferences, often sharing important husbandry and medical information.

In 1992 the Wild Bird Conservation Act limited the importation of wild caught birds listed in the appendices of CITES (the Convention on International Trade in Endangered Species of Wild Fauna and Flora). Initially this was a boost to the aviculture community; they began actively breeding to keep up with consumer demand for captive bred parrots. But the public sentiment toward keeping wild birds spilled over into captive aviculture as well, and the keeping of parrots as companion animals began to dwindle over the next decade.[2]

Due to this new public attitude, aviculture has seen a decline in the number of breeding operations and in the demand for captive parrots. Sale prices for pet birds is stagnant or in some cases reduced. Membership in aviculture organizations and conference attendance is down. As older experienced aviculturists retire, there are no young enthusiastic individuals to take their place.[1] Factors that have led to this decline include the abundance of unwanted parrots that have ended up in rescue organizations and the new attitude of many people in the avian community, including veterinarians, who don't believe parrots are good pets or should be kept as pets. Animal rights groups openly oppose parrots as pets and are very vocal in the media citing inflated statistics of overpopulation, abuse and neglect. These stories are shared by some rescue and sanctuary organizations who oppose aviculture and the breeding of parrots.

For many aviculturists, conservation has been a key part of their job, using the money made from selling captive birds to help birds in the wild. Consumer demand has resulted in increased breeding of some species and reduced breeding of those that may be at risk in the wild, but don't have pet appeal. In 1974, the United States joined CITES. CITES regulates international trade in species listed in their appendices I, II, and III.[3] Most pet parrots are listed in either CITES I or II, affecting aviculturist's ability to get rarer birds. The Wild Bird Conservation Act of 1992 also made it harder to obtain and maintain rarer species of birds not previously kept in captivity in the U.S. State and local regulations that use the federal lists may actually create a "local ban" on keeping certain species. Since most laws are created to protect wild bird populations, aviculture is often caught in the middle, and aviculturists can no longer breed, maintain, or sell listed species.

Avian veterinarians are seeing a reduction in the number of birds coming into their practice.[1,3] There has been a shift in the presenting complaint of these patients. In the 1970s-1990s, nutritional and infectious diseases were a constant threat, but with the advent of diagnostic tests, successful treatment protocols, formulated diets and more educated owners, the number of infectious diseases have declined and the issues now are behavioral and geriatric diseases. Many young and newly graduated avian veterinarians have limited to no experience with avicultural medicine and often have a negative attitude towards breeders in general. These newer graduates do not understand the concept of flock medicine or the issues that face aviculturists. Many aviculturists feel the expectations of the avian veterinarian are unrealistic given they are barely able to make ends meet financially so they begin to diagnose, medicate and treat their own flocks.

Other factors affecting the number of pet birds in homes is the number of birds with behavioral issues (feather destructive behavior, biting and excessive screaming). Many birds are given up to sanctuaries or rehomed due to behavioral problems. Behavioral issues are multifactorial in parrots and may include: genetics, poor husbandry and nutrition, inadequate socialization, boredom and stress. Many of these problems can be prevented or minimized if owners are better educated on the welfare needs and normal behaviors of parrots prior to taking one home.

As aviculture and pet bird ownership declines so will the need for avian veterinarians and the industry that supports them. A side effect of the decline in aviculture and pet birds is a lessened interest in parrots in the wild or their conservation. There will be reduced support of conservation projects and therefore less interest in protecting and preserving wild populations. Avian medicine will become static and the industry will be negatively impacted. With reduced numbers of parrots, there will be less demand for diets, cages, toys, educational materials, research, disease prevention, vaccines, specialized equipment such as brooders and incubators and even avian surgical equipment.

Goals for the Future

The future of parrots both in the wild, and in the caged environment, depends on the collaboration of aviculture and veterinary medicine. Aviculturalists and avian veterinarians need to work toward understanding and supporting each other. They need to educate the community on the intelligence and social qualities of parrots and work together to provide a healthy well-adjusted and well socialized bird to prospective pet owners. Veterinarians need to recognize the importance of aviculture to our businesses and recommend good breeders to clients. Both should support the industry, recommending appropriate diets, cages, foraging toys and enrichment items. Pet bird owners need to be educated on avian behavior, how to prevent behavioral problems and how to maintain a parrot in their home. This education should begin with the breeder and be reinforced by the avian veterinarian with the common goal of keeping birds in the home and minimizing the need for rescue organizations. Rescue organizations that provide a refuge for temporarily unwanted birds and assist in the rehoming of birds should be supported.

Avian welfare issues such as appropriate caging requirements, the pros and cons of deflighting, the need for exercise to promote health, and environmental enrichment and training need to be discussed with new bird owners when they buy a chick or adopt a rescued bird, and when they come to the veterinarian for their wellness checks.[4]

The avian community needs to share with others what they know; parrots are amazing pets. They are beautiful, entertaining animals, many of which can learn our language and often sing songs and dance. The human avian bond that forms between a well-adjusted parrot and the educated owner is a joy that few other animals can provide. Some parrots exhibit the intelligence of a young child. They are social and bond closely with their human flock and they can live for many decades. The human avian bond can cross generations and lead to a lifetime of companionship and love.

References

1. Clubb SL. Laws and regulations affecting aviculture and the pet bird industry. In: Altman RB, Clubb SL, Dorrestein GM, Quesenberry KE, eds. *Avian Medicine and Surgery*. Philadelphia, PA: Saunders;1997.

2. Jordan R. Is there a future for aviculture? *Parrot Festival*. 2011;28-30.

3. Koch V, Briscoe J. Animal welfare legislation and its influence on avian welfare. In: Speer BL, ed. *Current Therapy in Avian Medicine and Surgery*. St. Louis, MO: Elsevier Saunders;2016:677-682.

4. Endenburg N. The human avian bond. In: Speer BL, ed. *Current Therapy in Avian Medicine and Surgery*. St. Louis, MO: Elsevier Saunders;2016:709-714.

Section 3

Poultry

Cheryl Greenacre, DVM,
DABVP (Avian Practice),
DABVP (Exotic Companion Mammal Practice)
Moderator

Ducks, Geese, and Swans: An Introduction to Waterfowl Medicine

Michele D. Goodman, VMD, MHS, CWR

Session #2009

Affiliation: From the Elmwood Park Zoo, 1661 Harding Boulevard, Norristown, PA 19401, USA and Board of Directors, National Wildlife Rehabilitators Association, 2625 Clearwater Road, St. Cloud, MN 56301, USA.

Abstract: Waterfowl (ducks, geese, and swans) can be frequent visitors to veterinary clinics. Waterfowl can be from backyard flocks, native birds in need of rehabilitation or exotic species maintained by aviculturists or zoos. This master class will discuss identification, nutritional requirements, housing strategies (both for in hospital and for clients at home), common reasons for presentation and tips to promote flock health.

Part I. Waterfowl Basics

Introduction and regulations

The term "waterfowl" refers to a wide range of both wild and domesticated birds. For the purpose of this paper, waterfowl will be divided into three categories: wild, captive-bred and domestic. Wild waterfowl refers to native birds living in the wild that may be presented to veterinarians for rehabilitation. These birds are protected both by the Migratory Bird Treaty Act (MBTA) of the United States Fish and Wildlife Service (USFWS) and also by state regulations imposed by state wildlife agencies. State and federal permits are required for rehabilitation, although veterinarians are often allowed to provide emergency care and stabilization to injured or orphaned wild waterfowl prior to transfer to a licensed wildlife rehabilitator. It is essential to understand that wild birds must be capable of surviving in the wild following treatment. Captive-bred waterfowl refers to birds that are maintained and bred in captivity by aviculturists, propagators or zoos; species native to the United States are regulated by USFWS while species non-native to the United States may or may not be regulated by state wildlife agencies. Domestic waterfowl refers to ducks and geese that have been domesticated over many generations for meat, eggs or show and are being kept as pets or in backyard flocks with increasing frequency. Domestic waterfowl may be regulated by state Departments of Agriculture and by the United States Department of Agriculture (USDA).

Taxonomy of Anseriformes

Knowledge of natural history is an essential component of providing veterinary care to waterfowl species. For detailed natural history information on wild waterfowl species see Baldasarre (2014) and Kear (2005).[1,2]

Ducks, geese and swans are extremely diverse in terms of their anatomy, habitats that they occupy and foods that they eat. Swans have a high requirement for swimming water, feed almost entirely on the water and are prone to pododermatitis when not housed on water. Geese are hearty bird that primarily graze on land. Different species of ducks are adapted to widely different habitats. Whistling or tree ducks have long legs and spend considerable time perching, while sea ducks and stifftails are truly aquatic and are poorly adapted to land. Many of the more aquatic species of ducks are prone to aspergillus infections, particularly when housed out of water or when treating with an antibiotic.

Age and sex determination

There is a wide variety of methods for determining sex and age of a variety of species of waterfowl.

Methods for determining sex: Many waterfowl species are sexually dimorphic; the sex of adult birds can be determined by plumage. This does not apply across all species: swans, true geese, whistling ducks and others are sexually monomorphic. Body weight may be used to estimate the sex of a bird although this can often be misleading depending on time of year. Vocalizations may also be used to determine the sex of adult waterfowl although this requires some experience. A reliable method for determining the sex of waterfowl is by "vent sexing" as unlike most other birds, waterfowl have an intromittent phallus. The phallus is kept inverted within the phallic sac located within the cloaca until copulation, at which time a lymphatic erectile mechanism leads to eversion into the female reproductive tract. Vent sexing requires both training and practice; the technique is easier to master on smaller species of ducks examined during the breeding season. This procedure is often performed in young waterfowl between 3-7 days of age or on adult waterfowl of breeding age. Vent sexing full-size but sexually immature birds can be challenging (particularly swans); in these cases, blood sexing can be performed.

Methods for determining age: Plumage can be used to determine the age of waterfowl. Birds can be aged as either "hatch year" or "after-hatch year." To make this distinction, tail feathers can be evaluated to look for the presence of notches at the tips of the feathers. The presence of a notch indicates that a downy plume was attached to the tip of this feather, which only occurs in young birds (ie, hatch year birds). The absence of a notch indicates that the bird has already been through one full molt and is then characterized as an after-hatch year bird. Certain sea ducks and swans have elaborate molting patterns as they age; this can allow accurate aging up until they reach breeding age. The age at which downy young acquire contour and later flight feathers varies by species and nutritional status. As a general guideline, ducklings will begin to get some contour or covert feathering between 2-4 weeks of age and will develop primary flight feathers between 4-10 weeks of age.

Molting in waterfowl

Waterfowl are unique in their molting patterns as compared to other groups of birds. All species of waterfowl undergo a complete or eclipse molt during which they lose all of their primary feathers and are flightless for a period of time (ducks and geese generally 2-3 weeks, swans generally 3-4 weeks). It is important to realize that this is a normal stage in the annual cycle and does not necessarily indicate injury. Eclipse molting generally occurs at some point during the breeding season; swans may not molt every year while the majority of ducks and geese will undertake an eclipse molt annually. Replacing all primary feathers simultaneously is metabolically expensive; concurrent illness or injury can easily put a bird into a catabolic state. As with other avian species, care must be taken when handling waterfowl during molt so as not to bend or break developing blood feathers. In addition to the eclipse molt, many species will undergo a body molt which for most species allows them to develop their breeding plumage but can also serve to replace worn feathers.

Aviculture conventions (marking and wing-clipping)

When working with captive-bred waterfowl, there are a number of marking conventions used by aviculturists to distinguish captive-bred birds from their wild counterparts (USFWS requires that species protected by the MBTA be marked by one of these methods and individual states may require marking of all captive-bred species regardless of whether they are native to a particular area). The three methods for marking captive-bred birds are: pinioning, amputation of the hallux (digit I) or application of a close-ring band during growth. Pinioning is the amputation of the proximal metacarpal bones just distal to the carpometacarpal joint that leaves the alula intact. This procedure is often performed by the aviculturist to birds less than 3 days of age. Pinioning is commonly performed on geese and swans, but can be performed on any bird that will be housed in an open-top enclosure.

Alternately, the hallux may be amputated; this is also generally performed by the aviculturist to birds less than 3 days of age. This method is commonly used for ducks or other birds that are to remain fully flighted in a closed-top enclosure. It is important to note than in many countries any form of amputation at any age must be performed by a licensed veterinarian under either local or general anesthesia.[3] Closed-ring banding is currently gaining popularity particularly among larger private aviculture facilities; here a seamless band is applied to a bird during growth (generally between 2-5 weeks of age depending on species). The bird will grow into the band such that the band cannot be physically removed once the bird is fully grown but remains freely moveable between the hock and foot joints. There are general guidelines for marking when the sex of the bird is known (vent sexing is generally performed by aviculturists prior to marking). Male birds are marked on the right and female birds are marked on the left, using one of the three methods mentioned above. This can be an easy way to determine the sex of a marked bird provided that the aviculturist adheres to this convention.

Wing clipping waterfowl differs from psittacine birds, where a bilateral wing clip is typically performed. For waterfowl, a unilateral wing clip is recommended for flight restriction when needed; all ten primary feathers should be clipped up to the level of the primary coverts (make sure feathers aren't in blood). This procedure will have to be repeated annually or after each eclipse molt.

Part II. Managing Waterfowl in a Clinical Setting

Exam room rules

Most waterfowl will present for veterinary care in some sort of container; the exception, of course, is those individuals with harness trained domestic ducks who may just walk into the clinic. As with most avian patients, waterfowl should be brought directly into an exam room to minimize stress. Waterfowl are prone to hyperthermia during handling and examination. Domestic waterfowl can be provided water during the exam; wild waterfowl are generally not provided water as they are often flighted. With all waterfowl, the veterinarian should have all examination equipment organized to minimize handling time.

Obtaining a history

Waterfowl often require that a slightly different type of history be obtained, compared to other avian patients. Below is a list of information that should be included in a waterfowl history.

1. *Patient information:* Species (breed, if domestic), age, sex.

2. *Nutrition information:* Ask not just what, but how, items are fed; wild waterfowl may require being fed on the water. Staple diet, supplemental food items, and nutritional supplements.

3. *Housing information:* Enclosure information (inside or outside, dimensions, materials and substrate). Water source (bowl, small litter pan, pond including pond dimensions and depth, above- or in-ground). Shelter or nest box information along with nesting material type. Supplemental housing information (heat lamps, heat mats, misting system, etc.).

4. *Cleaning Information:* How often are the dishes cleaned and disinfected and with what, how often is the enclosure cleaned (substrate cleaned or changed)?

5. *Other birds or animals:* Other species or breeds housed with patient (ideally including sexes and ages), other birds in the household (poultry, psittacines, passerines, etc.).

6. *Reproductive information:* Normal mating behavior, last egg laid and was it a normal egg.

7. *Medication information:* Any medications given including type, dose, volume, frequency; include seasonal medications.

8. *Presenting complaint:* Type, duration, other birds affected, etc.

Restraint and safety precautions

Waterfowl are fairly easy to handle and restrain and generally don't pose the same safety risks to handlers that raptors, parrots and other groups of birds may pose. Waterfowl can use their bills, wings, claws and in rare cases spurs to try to defend themselves; restraining the head (for swans and larger geese) and wings (all species) will generally prevent the bird from injuring itself or handlers.

As with other groups of birds, avoid compressing the keel as this can lead to suffocation. Depending on the size of the bird, either the hands or a towel can be wrapped around smaller waterfowl to prevent flapping; the bird can alternately be placed underneath the restrainers arm to hold the wings against the body. When placing a bird underneath the arm, the tail should be facing forward; this reduces the risk the bird trying to bite the handlers face. For swan restraint, it is often helpful to perform the exam on the floor and to have the restrainer straddle the bird such that the knees and thighs serve to prevent the wings from flapping; make sure that the restrainer doesn't bear any weight on the bird. The method of restraint can vary by species or by examiner preference and can be useful not only for conducting a physical examination but also for administering treatments or collecting diagnostic samples.[4]

Physical examination

Waterfowl should only be examined after a medical history has been taken to avoid excessive handling. Whenever possible, a visual physical exam should be performed prior to restraint to evaluate respiratory rate, posture, alertness and to look for anatomic asymmetries that may not be apparent once the bird is in hand. Waterfowl, particularly wild birds, will conceal injury or illness for as long as possible and may only give clues about what their underlying problem is when they think no one is watching.

With the bird in hand a physical exam should be conducted thoroughly but quickly, particularly in high stress species or in unstable patients. Birds that are open mouth breathing should not be examined until stable and should ideally be maintained in an oxygen cage. Waterfowl are prone to overheating due to their dense layer of down and this can lead to open mouth breathing as a method of cooling; a temperature can be taken by placing only the tip of a thermometer into the cloaca. Birds with a temperature of > 41° C (106° F) should not be examined; alcohol can be applied with a cotton ball or gauze square under the wings and on the legs and feet. Keep in mind that hyperthermic birds in some form of respiratory distress may have an underlying respiratory issue; oxygen supplementation is recommended until stable.

Considerations for performing a waterfowl physical exam: Hydration can be challenging to assess in species that have little to no exposed skin. The inside of the mouth can be examined for the presence of stringy mucous and an eyelid skin tent can be performed (the skin should return to normal in <1 second in euhydrated animals).

Pay particular attention to the umbilicus in neonates; this can be a site for myiasis in debilitated birds and often is neglected. Evidence of inflammation or infection may be present, particularly if strict sanitation isn't followed by the breeder or aviculturist for incubation and hatching.

Waterfowl have an easily visualized uropygeal gland; there are often small down feathers that emerge from the tip of the gland. Special attention to the keel and feet should be afforded to any captive bird or to any wild bird that has the potential to have a leg injury. Keel lesions are common in waterfowl, often due to inappropriate husbandry following a leg or nervous system injury. Pododermatitis can be common in waterfowl and is often associated with inappropriate husbandry.

Euthanasia considerations

When working with wild waterfowl, it is important to understand federal regulations that require the euthanasia of any migratory bird that "cannot feed itself, perch upright, or ambulate without inflicting additional injuries to itself where medical and/or rehabilitation care will not reverse such conditions. You must euthanize any bird that is completely blind, or any bird that has sustained injures that would require amputation of a leg, a foot, or a wing at the elbow or above (humero-ulnar joint), rather than performing such a surgery."[5] While there is a stipulation that allows a veterinarian to petition USFWS for an exemption, the author adheres to these federal regulations as written so as not to subject an animal to a poor quality of life or to an increased susceptibility to captivity associated lesions.

When working with captive-bred or domestic waterfowl, these federal regulations do not apply. While there are occasional exceptions, waterfowl requiring the amputation affecting more than 25% of the surface area of the foot (in general, more than one digit with associated webbing) are at an increased risk of developing pododermatitis and of accumulating ectoparasites as toe nails are removed. Amputations of the foot are not recommended; there are anecdotal reports of prosthetic devices being used; in the authors' experience these are rarely effective. As technology with 3D printing improves, prosthetic devices (namely feet and possibly bills) may be considered for pet waterfowl, but should not be considered for wild or captive-bred waterfowl as the bird may not survive if the prosthetic falls off. Amputation of the carpal joint is commonly performed in neonatal waterfowl and may be surgically performed in adult captive-bred or domestic waterfowl when required for flight restriction. Prior to performing an amputation of the carpal joint in a wild duck, goose, or swan, placement opportunities as well as patient suitability for placement should be carefully considered.[6] Amputations at or proximal to the humero-ulnar joint can put the bird at an increased risk of drowning depending on the pond configuration and the climate (cold climates have more ice around pond edges which can making exiting a struggle). A pet domestic duck that lives primarily indoors may be able to adapt to a complete wing amputation. If an amputation is necessary and deemed appropriate, it should be performed at the joint whenever possible.

There are a number of other conditions where euthanasia is recommended. Euthanasia is recommended for any waterfowl with severe myiasis such that larvae have penetrated either the skull or coelom or have sufficiently damaged soft tissues associated with the appendicular skeleton requiring amputations as discussed above. Euthanasia is also recommended in patients with compound fractures where bone fragments are severely contaminated or necrotic and would require an amputation as discussed above. Euthanasia is recommended for patients that exhibit a loss of deep pain indicating a spinal injury. There is anecdotal information of custom-made wheelchair carts; these should not be utilized with wild or captive-bred waterfowl due to welfare concerns but in rare cases may be considered for domestic pet waterfowl when alternatives have been exhausted.

Diagnostics and anesthesia

The approach for diagnostics and treatment in waterfowl is similar to other avian species; information on drug dosing can be found elsewhere.[7,8] As with other birds, blood can be collected from the jugular, cutaneous ulnar or medial metatarsal veins. When larger volumes are needed, the jugular vein is preferred but often requires a blind stick (plucking feathers is discouraged in waterfowl as this will disrupt waterproofing). Waterfowl have an easily accessible medial metatarsal vein that is ideal for collecting smaller blood samples and for intravenous catheterization. The cutaneous ulnar vein is prone to hematoma formation and is generally avoided. Intravenous catheterization can be achieved using any of the aforementioned vessels; the author prefers the medial metatarsal vein.

Radiographs or other imaging modalities should only be performed on stable patients. The author prefers to sedate with butorphanol (1-2 mg/kg IM) and midazolam (1-2 mg/kg IM) or anesthetize waterfowl for radiographs as this reduces the risk of injury and the potential for exertional myopathy following manual restraint. Gas anesthesia can be used in addition to sedative medications; the author avoids using only gas anesthesia in waterfowl as they can become apneic at higher concentrations of gas.

Waterfowl anesthesia is fairly straightforward. For procedures longer than 15 minutes, waterfowl should be intubated with an uncuffed or Cole endotracheal tube, taking care not to traumatize the crista ventralis, or septum, which is present in certain species and extends upwards from the ventral aspect of the glottis. There are a wide variety of anesthetic protocols that have been successfully used in waterfowl. The author premedicates with butorphanol (2 mg/kg IM) and midazolam (2 mg/kg IM) for most species and procedures. Waterfowl are prone to forming thick tracheal mucous plug during anesthesia; anticholinergic medications may act to thicken mucous, increasing the likelihood of endotracheal tube obstruction.[9] Spare endotracheal tubes should be available during any procedure; if increased noise is heard during respiration, the endotracheal tube should be replaced. Atropine or glycopyyrolate can be used to correct bradycardias as in other species. Many species of waterfowl are prone to apnea during anesthesia and require intermittent positive pressure ventilation. Depending on size, most waterfowl require 6-12 breaths per minute and should receive approximately 2 breaths per minute even if breathing spontaneously.[10] There are many anesthesia monitors available; minimum parameters to monitor are heart rate, respiratory rate and body temperature. Waterfowl are well insulated and can become hyperthermic at any point during a procedure.

Initial care, temporary housing and nutrition

Initial care: Initial care should include rehydration, medication administration and patient stabilization. This can also include wound management and temporary fracture immobilization. An important aspect of initial care for waterfowl is to provide appropriate temporary housing while under veterinary care.

While the vast majority of approaches to initial care in waterfowl are similar to other avian species, there are a few differences that will be highlighted. Fluid therapy can be achieved with any isotonic rehydration solution; depending on the level of dehydration and overall patient status, the author prefers to administer intravenous boluses with a butterfly catheter twice daily at 15-20 ml/kg q12h. Waterfowl generally have less subcutaneous space than other avian species which can make administration challenging. If administering subcutaneous or oral fluids, a dose of 100 ml/kg/day often is appropriate. Species that spend the majority of their time in the water (primarily the sea ducks), often require higher doses of fluids if not able to be given access to swimming water.

Common medications administered during initial care include pain medications (generally an opioid in conjunction with non-steroidal anti-inflammatory drugs administered after the patient is hydrated), antimicrobials (may consist of an antibiotic and an anti-fungal medication) and vasodilators (isoxsuprine).[7,8] Please keep in mind that many species and breeds of waterfowl are considered food animals (minor producing species); regulations pertaining to off-label drug use particularly in species that will be released must be adhered to. Consult the Food Animal Drug Residue Avoidance Databank (http://www.farad.org) for current information on drugs used in food animals.

Management of fractures in waterfowl is similar to other avian species and is covered elsewhere.[10] The challenge of fracture management in waterfowl is preventing captivity associated lesions (pododermatitis and keel lesions). Ideally, the bird will be given access to swimming water during convalescence, which is often contraindicated with orthopedic hardware. Adaptive splinting with waterproof, porous materials, such as porous thermoplastic splinting material can be extremely helpful in managing fractures while maintaining the bird in a more normal environment.

Wound management in waterfowl is also similar to that in other avian species. Primary closure should only be attempted in clean, fresh wounds. If primary closure is an option, pluck the bare minimum of feathers just to prevent them from being tracked into the suture line. Plucking additional feathers, particularly in wild species, will disrupt waterproofing such that the bird will not be able to swim following wound closure. Oil-based topical products should be avoided in all species due to the risk of feather contamination. The author has good success with hydrogels, medicinal honeys and hydrocolloid products depending on the type, stage and location of a wound; these products can be useful at promoting debridement and granulation tissue formation and do not alter feather condition when changed appropriately (generally every 24-48 hours). Absorbent materials should be avoided if the bird will have access to swimming water. In some cases, it may actually be better for a bird to leave a wound undressed if that means that the bird can swim and eat normally during recovery.

Temporary housing: Most veterinary clinics are not well equipped for long-term housing of waterfowl and other aquatic birds; waterfowl are generally managed on an out-patient basis. There are a number of ways to safely house waterfowl in a veterinary clinic for short periods of time. Designated housing setups should be available for waterfowl that may present in various conditions and of different ages.

Neonatal waterfowl should be housed in an opaque plastic storage container with screening inset into the lid. In nearly all cases, neonates are dehydrated and should receive oral or subcutaneous fluids. Neonates must be weighed twice daily until maintaining weight. Neonatal waterfowl, despite being precocial, may require supplemental gavage feeding (generally administered through shortened urethral catheters). The gavage diet will vary by species. Feeding 0.1 ml/10 g bird 4-6 times daily often is sufficient until they start self-feeding. Neonates will often naturally feed on the water. Providing several short daily swims with aquatic invertebrates and waterfowl starter can both help to establish waterproofing and self-feeding behavior. It is a myth that neonates are not waterproof until they are fully feathered. Access to clean water for supervised swimming daily will enable birds to become fully waterproof within a few days after hatching.

For most adult waterfowl, a metal or similar-type hospital cage can be provided for short-term housing. Species that are ambulatory should have adequate padding provided either with bath towels or with unbacked vinyl coiled matting; padding and traction will help to protect their feet. Species that are poorly adapted to land or individuals with leg or neurologic injuries should be housed on a net-bottom pen to avoid the development of keel or food lesions. A net-bottom pen is constructed from PVC pipe connected by elbows to form the square or rectangular shape that fits into the hospital cage. Knotless aquaculture or poultry netting is then stretched tightly over the PVC frame and secured with either twine or screws.[11] Waterfowl require spill-proof bowls (eg, heavy ceramic) for both feed and water. Most species will eat better when food is offered in water dishes.

Nutrition: Waterfowl nutrition can be complicated, especially given the diversity of species that may be encountered. A balanced, pelleted diet should be offered ad libitum. Natural food items also should be offered to meet the needs of the particular species or breed in care. With the increasing interest in pet poultry, there is a myriad of commercial diets available that claim to be appropriate for waterfowl. Many of these products are marketed as "all-flock" diets that are intended to be fed to multiple groups of birds. Waterfowl have unique nutritional requirements compared to other groups of poultry-type birds, including a relatively low requirement for protein and a relatively high requirement for B vitamins. Commercial diets designed for other groups of birds will rarely meet the needs of waterfowl. Feeding an inappropriate commercial diet can lead to developmental issues in neonates as well as reproductive problems, poor feather quality and condition, and numerous other problems in adult birds.

Currently, the only product designed in the United States to meet the nutritional requirements of wild waterfowl is the Mazuri line of products (Mazuri Exotic Animal Nutrition, PMI International, Coffeyville, KS, USA), which includes a waterfowl starter, maintenance, breeder, and sea duck diet. All of these products float on water (aside from a sinking waterfowl maintenance diet that sinks). The starter diet is recommended primarily for neonatal waterfowl, but also can be fed to debilitated adult waterfowl due to the small pellet size and increased protein level compared to the maintenance diets. All of the Mazuri waterfowl diets can be soaked and blended into a gavage formula when necessary. In addition to the Mazuri line of products, there are a number of specific duck grower or duck pellets that are regionally available depending on local feed mill availability. Avoid using medicated feeds. These pellets are rarely extruded and therefore do not float and form a mushy paste when mixed with water. They generally contain appropriate levels of B vitamins and have low protein concentrations compared to many of the all-flock or gamebird products. These products can be appropriate for certain species of waterfowl, including geese, some swans and dabbling ducks. It can be advantageous to offer a debilitated bird a variety of commercial diets to increase the likelihood of self feeding.

In addition to commercial diets, waterfowl often require the addition of natural food items depending on species. Below is a table summarizing some commercially available food items that can be incorporated into wild waterfowl diets to stimulate self feeding.

Table 1. Natural food items for waterfowl diets.

Food item	Recommended species	Comments or special instructions
Greens (red leaf, green leaf, and romaine lettuce)	All species except for true sea ducks	Can be fed as whole heads, longitudinally cut in half or chopped.
Duckweed	Swans, all species of ducks except for true sea ducks	Harvest fresh from natural ponds; does contain live aquatic invertebrates, which may harbor intermediate life stages of parasites, but does not seem to be a clinical problem.
White millet	All species except for true sea ducks	May present digestive challenges if fed to critically debilitated birds.
Whole corn	All species except for true sea ducks	May present digestive challenges if fed to critically debilitated birds; avoid feeding cracked corn.
Mealworms (live)	Primarily offered to ducks of all species	Gut load prior to feeding; can stimulate natural feeding behavior when sprinkled on top of a food dish or on water.
Krill (frozen)	Diving ducks, mergansers, stifftails, sea ducks	Provide in drinking or swimming water. Can chop if offering to neonates. Should not be left in water for more than 15 minutes to reduce the risk of feather contamination.
Fish (live)	Mergansers	Stimulates self-feeding behavior when offered in swimming water; fish size should vary depending on age of ducklings.
Aquatic invertebrates (either freeze-dried or frozen)	All species of ducks	Generally fed to neonates to stimulate self-feeding behavior. Offer in swimming water and ensure that the water is changed several times daily to reduce the risk of feather contamination.

Recommendations for preventive maintenance and common presentations

Most captive-bred and domestic waterfowl are maintained in flocks and require routine healthcare. Given that cost is often an issue for clients that keep large numbers of birds, herd health management techniques often have to be utilized. Depending on the size and type of a given collections, recommendations may have to be customized. Veterinary care needs for captive-bred and domestic waterfowl often include parasitology, infectious disease prevention, reproductive assistance, trauma, developmental disorders, toxin ingestion and management of captivity-associated lesions.

Parasites: Waterfowl living outdoors are susceptible to a wide variety of parasite infections depending on access to natural water sources and food items that may harbor intermediate host life stages and soil. Clients should be encouraged to drop off fresh pooled fecal samples from each separate aviary of birds quarterly to assess parasite burden and devise a management plan. The majority of parasites can be detected by routine fecal examination (should include direct smear, flotation and sedimentation). Not all species of cestodes possess genital pores to release eggs; the absence of cestode eggs in a fecal sample does not rule out the possibility of cestode infection. Parasites with complex life cycles are observed rarely in a captive setting; however, parasites with direct life cycles are encountered frequently.

Syngamus trachea: This presentation is frequently observed in neonatal goslings and can cause mortality if not promptly treated. Hemoparasites also are encountered in captive waterfowl with certain species being more susceptible than others; *Plasmodium* spp. can lead to debilitation and even death and often requires treatment, particularly in young birds. Seasonal blood sampling of the collection to look for hemoparasites is recommended whenever possible; specific treatments and dosages are discussed elsewhere.[7]

Infectious disease prevention: There is a growing list of infectious diseases that affect both wild and captive waterfowl; detailed information is available elsewhere.[12-14] One key pathogen affecting captive waterfowl flocks is mycobacteriosis. This disease often is challenging to diagnose antemortem; clinical signs are often associated with subcutaneous granuloma formation and wasting despite a good appetite.[15] Many waterfowl species are thought to be resistant to clinical disease despite infection; these infected birds can contribute to a large environmental pathogen burden.[15] Veterinarians should educate clients about the disease and discourage obtaining new birds from collections with known problems. If *Mycobacterium* already exists at a particular facility, attempts should be made to minimize stress in captive birds and to promptly remove droppings in enclosures (often involves changing to a substrate that allows for drainage to remove access to soil). Current treatment options are cost prohibitive and ineffective.

Highly pathogenic avian influenza (HPAI) has been intermittently documented in wild waterfowl in the United States since 2014. This group of viruses presents a huge threat to both commercial food production systems and to individuals who maintain birds in captivity. It is important to note that not all strains of HPAI cause clinical disease in waterfowl, which can complicate recognition. Veterinarians should direct clients to resources available through the United States Department of Agriculture Animal and Plant Health Inspection service website (https://www.aphis.usda.gov) and should themselves be informed about clinical signs associated with infection. Clients should be encouraged to report sick birds to their state veterinarian and should understand basic biosecurity to reduce the risk to their own collection. Individuals with valuable private collections should avoid areas where domestic waterfowl and poultry are sold or exhibited and should reduce contact between wild birds and captive birds as much as possible (this can be especially important for aviculturists who are also hunters).

Reproductive assistance: As with other captive birds, breeding and egg-laying can be problematic. Injuries can occur during breeding season and will be discussed in the next section on trauma. The term "egg-binding" represents a complex set of abnormalities that can result in difficulty for a bird to lay a normal egg. In a very broad sense, breeding waterfowl require the ability to increase their own calcium intake according to their needs. In some cases, egg-laying may not be desired by the client, but in most cases individuals maintaining waterfowl do so either for eggs or to produce young. Appropriate egg-laying can be accomplished by feeding a high quality pelleted diet and by offering oyster-shell grit in a separate pan or spread onto the substrate. Breeding waterfowl should also be provided with appropriate nesting material; this often means some sort of nesting box, platform or other structure depending on species. Untreated pine shavings should be provided within the nesting box or structure to encourage egg laying. Hay and straw are generally not recommended due to the presence of fungal spores.

Waterfowl presenting with signs consistent with egg-binding (tail-bobbing, straining and sometimes open mouth breathing) require immediate stabilization. The presence of a formed egg within the coelom can interfere with the passage of urates and feces; this leads to an accumulation of wastes and affected birds will be in some state of shock.[16] Birds should receive intravenous fluid therapy at 10-15 ml/kg/h, anti-inflammatory medications and calcium gluconate (diluted and given with intravenous fluids). If an egg is palpable within the coelom, sterile lubricating jelly can be carefully infused into the cloaca to aid in passage. Diagnostic imaging can be performed on birds if stable. Advanced surgical intervention may be required, particularly in cases of egg-yolk peritonitis. Surgery of the avian reproductive tract is discussed elsewhere.[17]

Trauma: There are a number of ways that captive birds may injure themselves in captivity. If spooked, birds can fly into the sides of their enclosure leading to soft-tissue or orthopedic injuries or head trauma. Birds can fight within enclosures, either during breeding season or at other times of the year and can sustain soft tissue or orthopedic trauma. Forced copulation is a strategy employed by many species of waterfowl during the breeding season and can lead to soft tissue injury over the cranium, neck and dorsum and in extreme cases (particularly with domestic breeds) can lead to spinal injury. Management of trauma cases should all follow the same general pattern: patient stabilization should always come first with fluid therapy and pain management; specific injuries can then be managed on a case-by-case basis. For mating-related scalping injuries, a cervical flap can be created to close the exposed skull.[18] If mating-related injuries are commonly seen in a particular flock, the client should increase the number of females in the collection (for large domestic breeds, 3 female ducks often need to be provided for every male) or to separate aggressive male birds into "bachelor" pens.

Developmental disorders and considerations for neonates: Developmental disorders in waterfowl are well-characterized and specific treatments are discussed in detail elsewhere.[10] The two primary conditions seen in captive-bred and domestic waterfowl are "angel-wing" and perosis. It is important to note that unlike in commercial poultry, there is no evidence for a genetic contribution to these conditions. Husbandry seems to be the critical factor in the development of these conditions, primarily nutrition and enclosure space. A pelleted product designed for captive wild waterfowl such as Mazuri is recommended for all species and breeds of waterfowl, including domestic waterfowl. This starter pellet should be offered for 4-8 weeks depending on species and growth rate and should be accompanied by natural food items. Downy waterfowl should also be given plenty of room to exercise during their development; overcrowded conditions restrict exercise activity which can negatively affect bone, tendon and joint development.[10] Shortly after hatching, waterfowl also should be given supervised access to swimming water.

Specific recommendations for the treatment of developmental disorders:

Angel wing. As soon as it is evident that the carpal joint is deviating laterally or drooping, and as soon as the wing is large enough to hold a bandage, the wing should be taped with paper tape (generally doesn't leave a residue on feathers) in a figure-of-8 fashion. This tape wrap should be left in place for 5 days (replace if needed in rapidly growing species) and then removed for 2 days of rest before re-bandaging. If after two rounds of wrapping the wing still deviates laterally, the condition is generally permanent. Splinting will not be effective in birds that already have fully developed primary feathers (ie, shafts no longer in blood). In severe cases of "droop" wing (where the growing primary feathers drag on the ground), primary feathers in blood may have to be plucked under sedation or general anesthesia. There is a risk of damaging feather follicles with this procedure but the author has not experienced any negative outcomes. Removal of the growing primary feathers generally gives the affected bird additional time to develop strength in the ligaments and joints of the wing before the feathers regrow.

Perosis. In this condition, the gastrocnemius tendon deviates either medially or laterally out of the condylar groove over the hock joint. Euthanasia is recommended with bilateral presentation. While methods for surgical correction have been described,[12] the author has had good results with corrective splinting. For detailed information on corrective splinting see Goodman, 2017.[10]

Toxin ingestion. There are a number of toxins that waterfowl can ingest depending on the conditions in which they are maintained in captivity or what they encounter in the wild. Information on common toxicities in waterfowl have been covered elsewhere.[19] Wild waterfowl presenting for rehabilitation can be affected by various toxins including heavy metals, algal blooms and botulism. Captive-bred and domestic waterfowl should in theory have less exposure to toxins but heavy metal toxicity and botulism can occur. While lead toxicity occurs commonly with wild waterfowl, zinc toxicity is more likely to occur in captive bred waterfowl due to the fact that most aviary construction involves zinc staples that birds can ingest. In captive birds, botulism cases often involves poor husbandry, poor water quality and over-crowding.

For cases of heavy metal toxicity, both radiographs and bloodwork should be performed to look for the metallic objects (note that the absence of radiographic evidence of metallic objects in the gizzard does not rule out the possibility for toxicity). Affected birds require gastric lavage if metallic objects are found in the gizzard and also require chelation therapy with either calcium EDTA or with meso-2,3-dimercaptosuccinic acid (DMSA).[7,8] Birds also require intensive supportive care including fluid therapy and gavage feeding.

Birds presenting with botulism toxicity require intensive supportive care and fluid therapy. Activated charcoal should be provided on intake and should be followed with aggressive oral fluid therapy. Intravenous fluid therapy is often recommended for more severe cases, particularly in birds with paralysis of the third eyelid. The eyes should be kept lubricated and the bird should be placed in a towel donut to maintain the head in a normal position and to prevent aspiration following oral fluid administration. The use of a net-bottom pen can be helpful for these cases to keep the birds from sitting in their own fecal material but a towel donut is still necessary. Food and water should not be provided until birds can maintain head carriage and can ambulate appropriately.

Management of captivity-associated lesions:

The two most common captivity-associated lesions are keel lesions and pododermatitis. Both lesions often are related to inappropriate husbandry, particularly following a leg injury or neurologic injury that prevents normal ambulation. Prevention is essential because these conditions are far more difficult to treat than they are to prevent. The provision of a net-bottom pen can go a long way towards preventing these lesions; similarly, the provision of swimming water particularly for diving ducks can also help to prevent lesions. Keel lesions require surgical intervention; a detailed surgical approach can be found in Duerr, 2016.[20]

Pododermatitis is more difficult to manage. While surgery often is performed in other avian species, waterfowl present unique challenges such that surgery should be performed only as a last resort. Waterfowl are messier than many other groups of birds and live primarily on the ground and as a result, any surgery of the feet carries a high risk of infection. In addition, bandaging following surgery on the pads of the feet is not waterproof. Post surgery, birds cannot be given access to swimming water, which means that they are forced to spend all of their time on land, resulting in exacerbation of pressure lesions. Alternatives to surgery include treatment with appropriate antibiotics if the lesions are infected, changing the substrate to provide padding, traction and drainage and increasing access to swimming water if at all possible. Client education is essential when managing cases of pododermatitis related to poor husbandry. A diet change may be required and general living conditions likely require improvement. While it can take several months for lesions to resolve with husbandry changes alone (ie, no additional medications or therapies), lesions can resolve completely with sufficient husbandry modifications.

Conclusions

Waterfowl represent a unique group of birds that can be a challenge to manage in a veterinary setting. An understanding of natural history, nutrition, and husbandry requirements can provide a solid foundation for optimizing veterinary care.

References

1. Baldasarre G. *Ducks, Geese, and Swans of North America.* Baltimore, MD: Johns Hopkins University Press;2014.

2. Kear J. *Ducks, Geese and Swans.* Oxford: Oxford University Press;2005.

3. Vinke CM, van Zeeland YRA, Schoemaker NJ, Meijboom FLB. As free as a bird on the wing: some welfare and ethical considerations on flight restraint methods In: Speer BL, ed. *Current Therapy in Avian Medicine and Surgery.* St. Louis, MO: Elsevier Saunders;2016:683-705.

4. Backues KA. Anseriformes. In: Miller E, Fowler ME, eds. *Fowler's Zoo and Wild Animal Medicine.* Vol 8. St. Louis, MO: Elsevier Saunders;2015:116-126.

5. US Fish and Wildlife Service. Regulation 50 CFR Part 21.31. Available at https://www.ecfr.gov/cgi-bin/text-idx?SID=3b4e5a2ad53084333969fc59323689f6&mc=true&node=se50.9.21_131&rgn=div8. Accessed March 13, 2017.

6. Lacy K. Selection *Process for Non-Releasable Birds: The First Step in Bird Welfare. International Association of Avian Trainers.* 2017;1-8.

7. Miller EA, Goodman MD, Cox S. *NWRA Wildlife Formulary*, 4th edition. St Cloud, MN: National Wildlife Rehabilitators Association;2017.

8. Hawkins MG, Barron HW, Speer BL, et al. Birds. In: Carpenter JW, ed. *Exotic Animal Formulary.* 4th ed. St. Louis, MO: Elsevier Saunders;2013:183-437.

9. Grosset C, Sanchez-Migallon Guzman D, Waymire A, et al. 2013. Extraoral surgical correction of a lingual entrapment in a Chinese goose (*Anser cygnoides*). *J Avian Med Surg*. 2013;27:301-308.

10. Goodman MD, Schott R, Duerr RS. Waterfowl (Ducks, Geese, and Swans). In: Duerr RS, Purdin, GJ eds. *Topics in Wildlife Medicine: Orthopedics.* St. Cloud, MN: National Wildlife Rehabilitators Association;2017:99-119.

11. Holcomb J. For the new rehabber: net-bottom caging for waterfowl. *Wildl J.* 1988;11(1):3-4.

12. Olsen GH. Bacterial and parasitic diseases of Anseriformes. *Vet Clin North Am Exot Anim Pract.* 2009;2: 475-190.

13. Hess JC, Paré JA. Viruses of waterfowl. *Semin Avian Exot Pet Med.* 2004;13:176-183.

14. Friend M, Franson JC. *Field Manual of Wildlife Diseases: General Field Procedures and Diseases of Birds.* Madison, WI: USGS;1999.

15. Riggs G. Avian mycobacterial disease. In: Miller RE, Fowler ME, eds. *Zoo and Wild Animal Medicine.* Vol 7. St. Louis, MO: Elsevier Saunders;2011:266-274.

16. Romagnano A. Avian obstetrics. *Semin Avian Exot Pet Med.* 1996;5:180-188.

17. Mison M, Mehler S, Echols MS, et al. Selected coelomic surgical procedures. In: Speer BL, ed. *Current Therapy in Avian Medicine and Surgery.* St. Louis, MO: Elsevier Saunders;2016:645-656.

18. Stroud PK, Amalsadvala T, Swaim SF. The use of skin flaps and grafts for wound management in raptors. *J Avian Med Surg.* 2003;17:78-85.

19. Degernes LA. Toxicities in waterfowl. *Semin Avian Exot Pet Med.* 1995;4:15-22.

20. Duerr RS. Surgical repair of keel lesions and lacerations in aquatic birds. ExoticsCon. 2016;83-88.

Backyard Poultry Cases: Let's Work Them Up Together!

Cheryl B. Greenacre, DVM, DABVP (Avian Practice), DABVP (Exotic Companion Mammal Practice)

Session #2074

Affiliation: From the College of Veterinary Medicine, Department of Small Animal Clinical Sciences, 2407 River Drive, C-247, University of Tennessee, Knoxville, TN 37919, USA.

Introduction

Backyard poultry are becoming popular and presenting to veterinary practices for individualized care more frequently. Described below are several potential case presentations you are likely to encounter in private practice, and following are outlines of some of the more common diseases of backyard poultry organized by system.

Case 1

Six days prior to presentation 17 adult (2 years old) barred rock hens and 3 adult (2 years old) barred rock roosters were purchased from a man who had said they were raised on his farm. Two days prior to presentation the chickens exhibited varying degrees of rales, mild coughing, periorbital swelling and edema of the head. A mild serous nasal discharge also was present and some of the hens had laid eggs with abnormal shells. At presentation approximately half the birds had died and the owner brought three for euthanasia and testing. While under isoflurane anesthesia, but prior to euthanasia with pentobarbital IV, a choanal swab was taken and blood was collected from the right jugular vein.

Case 2

Three 2.5-month-old female black Orpington chicks presented dead with a history of 2-3 days of lethargy and lameness in one leg as well as dragging of the foot. One also had a 2- to 3-day history of torticollis. On gross post mortem they were considered mildly thin (2/5), weighing 0.67 kg. Two chicks had an enlarged ischiadic (sciatic) nerve. One had a firm, white, smooth, round pedunculated mass (2 × 0.5 × 0.5 cm) left of the thoracic inlet. The liver was moderately enlarged and extended beyond the breast plate. The spleen was markedly enlarged (2 × 1.3 × 1 cm), triangular, and red-brown. The kidneys were bilaterally markedly enlarged (4 × 1 × 2 cm), white to tan, and friable.

Case 3

A 3-year-old speckled Sussex hen presented with sudden onset of lameness. She had not laid eggs in about 2 weeks. The only physical examination abnormalities included that she is mildly thin and had a doughy coelomic cavity.

Case 4

A 2-year-old buff Brahma chicken presented with chronic lameness due to an abscess on the plantar aspect of the foot.

Case 5

A bronze turkey hen presented with an open fresh wound on her entire dorsum after being attacked by toms. She was quiet, tachypnic, and weak.

Respiratory Disease

Mycoplasmosis (most common)

There are 3 different species of *Mycoplasma* that can infect chickens. *Mycoplasma gallisepticum* causes respiratory disease in chickens, but an infectious sinusitis in turkeys. *Mycoplasma meleagridis* causes an air sacculitis and skeletal deformities in turkeys, as well as reduced egg hachtability, sinusitis, airsacculitis, poor growth, abnormal feathering, and skeletal abnormalities (eg, osteomyelitis, osteodystrophy). *Mycoplasma synoviae* causes air sacculitis and synovitis or lameness in chickens. In both turkeys and chickens it causes arthritis, synovitis, breast blisters, respiratory signs, egg drop (eggshell abnormalities), tenosynovitis, salpingitis, and airsacculitis, chronic respiratory disease, prostration, egg drop, poor egg quality, sinusitis, keratoconjunctivitis, airsacculitis, tenosynovitis, and salpingitis. *Mycoplasma gallisepticum* (MG) is seen in backyard flocks and is of concern because it can easily spread to nearby commercial flocks and cause economic devastation for that commercial flock. Most commercial flocks are MG free. To participate in the National Poultry Improvement Plan (NPIP) a flock needs to be MG free. Transmission takes place through fomites and vertically through eggs. Clinical signs of MG in chickens includes upper respiratory disease with swelling of the infraorbital diverticulum of the infraorbital sinus with caseated pus. The best prevention is to depopulate and repopulate with clean stock. Treatment can be attempted with antibiotics (spectinomycin, lincomycin, erythromycin, or tylosin), but birds remain carriers for life.

Infectious bronchitis virus

Infectious bronchitis virus, caused by a coronavirus, affects only chickens and has a worldwide distribution. Younger, immunosuppressed chickens show worse clinical signs than older immunocompetent chickens. Clinical signs include upper respiratory signs including sneezing, gasping, rales, and nasal discharge, also facial swelling, purulent ocular and nasal discharge, swollen wattles, sneezing, dyspnea, loss in condition, drop in egg production of 10-40%, and inappetence. Young are affected worse than adults showing gasping and labored breathing. In an affected flock of chicks, all birds will typically develop clinical signs within 36-48 hours and the clinical disease will typically last approximately 4 days (longer if secondary infections develop). Older chickens show a 5-10% drop in egg production for about 10-14 days. The reproductive tract can also be affected by this virus resulting in irregular and roughened eggshells with watery albumin and decreased egg production. Also newer strains may affect the kidneys. Tests include virus neutralization, hemagglutination inhibition or ELISA. The best method to control is to disinfect, repopulate and use live vaccine. This disease is highly contagious that easily spreads via airborne particulate matter and via fomites. There is no treatment, but antibiotics can be given to prevent secondary bacterial infection, especially with infectious coryza. The virus is easily destroyed by disinfectants, sunlight and heat. Increasing the environmental brooder temperature by 5 °F helps chicks to recover. There is a vaccine available, but it is not used in backyard flocks since there are numerous serotypes.

Infectious coryza

Infectious coryza affects mainly chickens (it can affect pheasants and guinea fowl as well). The causative agent is *Haemophilus paragallinarium*, a gram-negative rod. The incubation period is 1-3 days and the course of the disease is approximately 4-12 weeks. Clinical signs include upper respiratory signs and swelling of the face with foul smelling (hallmark) and sticky nasal and ocular discharge, dyspnea and rales. Clinical signs also include decreased egg production. Birds that recover are lifetime carriers. Transmission is through direct or indirect contact. Mortality is about 20-50%. Treatment consists of giving antibiotics such as sulfa drugs, erythromycin or tetracycline, but since affected birds are lifetime carriers, the only way to control the infection is through depopulation and leaving the premises vacant for at least 30, preferably 60 days after cleaning and disinfecting before repopulating. Avoid mixing flocks or mixing different ages and sources of birds. There is a vaccine available, but it is seldom used in backyard flocks.

Infectious laryngotracheitis

Infectious laryngotracheitis (ILT) affects only chickens, and pheasants and peafowl). The causative agent is a Herpesvirus. Chickens older than 14 weeks are more affected than younger chickens, so the disease is usually seen in mature chickens. There is a mild form in the US that is associated with decreased egg production, conjunctivitis, nasal discharge, swollen infraorbital sinus and in more severe cases moist rales. Shaking of the head and flinging necrohemorrhagic material from the trachea is a hallmark of this disease including an inspiratory dyspnea and death. Milder strains in the United States may not present with hemorrhagic tracheitis. At gross necropsy a mucoid to necrohemorrhagic tracheitis is present. Diagnosis is confirmed via virus isolation, ELISA, PCR, or indirect fluorescent antibody test. Prevention is through the use of a live vaccine. The disease can be spread by fomites. Properly dispose of dead birds to prevent spread (incinerate). Antibiotics can be used to control secondary infections, but it is better to depopulate and then vaccinate new birds.

Avian influenza (need to know about this one due to potential for zoonosis)

Avian influenza affects many species. The causative agent is an *Orthomyxovirus*. It is also known as fowl plague. It is easily transmissible by airborne droplets. The clinical signs are variable since there are low pathogenic and high pathogenic forms, but signs include anorexia, decreased egg production and respiratory disease with the mild form to respiratory distress, facial swelling, diarrhea and neurologic signs with the highly pathogenic form. Usually the mild form of the disease is associated with high morbidity and low mortality. Coughing, sneezing, and sudden death are the typical signs of this disease, but there are no pathognomonic signs, and other signs can include depression, swollen face, cyanosis of comb/wattles, diarrhea, paralysis, sudden loss of appetite, and decreased vocalization. This a highly contagious disease associated with high mortality of both domestic and wild birds. There is no treatment. Testing usually involves virus isolation. This is a reportable disease. A killed vaccine is available, but not used in backyard flocks. Depopulation is usually the recommended action during an outbreak and the disease should be reported to the state veterinarian.

Newcastle's disease (need to know about this one)

Newcastle's disease affects many species of birds. The causative agent is a *Paramyxovirus*. There are 4 different forms that vary in severity. The least pathogenic form is the lentogenic form which causes a mild upper respiratory disease and usually only affects the young. The mesogenic form also causes a mild upper respiratory disease, decreased egg laying, and has a low mortality. The neurologic/velogenic form causes a sudden onset of upper respiratory disease followed by neurologic signs with approximately a 50-90% mortality. The worst form, which is a foreign animal disease for the United States, is the exotic Newcastle's disease (END), also known as the viscerotropic velogenic form that is associated with neurologic signs and high mortality. In the United States there is a vaccine for the first 3 forms listed here; however, we do not vaccinate against END—we test and eradicate since it is a foreign animal disease. There is a serology test available at California's San Bernardino County Laboratory for PMV 1, 2 and 3, a paramyxovirus. There are many different kinds of Newcastle disease, mostly mild in poultry, and birds are vaccinated at one day of age; it causes mild conjunctivitis in people. Testing is performed at US quarantine stations. Exotic Newcastle's disease (ie, VVND or viscerotropic-velogenic Newcastle's disease) is a foreign animal disease and millions of dollars are spent to eradicate all exposed birds when there is a US outbreak; the last outbreak was in California 2002-2003 and involved 3.5 million birds.

Musculoskeletal Disease

Marek's disease (most common)

Marek's disease affects only chickens, or turkeys that are in close proximity to affected chickens. Causative agent is a herpesvirus. Clinical signs are generally seen in birds that are 12-20 weeks of age, but can be seen as young as 3 to 4 weeks of age. This highly contagious disease is associated with a high morbidity and low mortality. The incubation period is typically 4-12 weeks, clinical signs are seen in 10- to 12-week old chickens. Transmission is mainly via virus shed in skin and feathers. The virus can persist in the environment indefinitely (feathers and dander in poultry houses and yards). Clinical signs of the paralytic form include an asymmetric paralysis since the virus affects the "sciatic" (technically the ischiadic) nerve and presents as a bird that cannot stand because one leg is pointed forward while the other leg is pointed backward. A clinical sign of the ocular form is "gray eye" where the color of the iris gradually changes from brown to gray and the pupil becomes irregularly shaped due to infiltration of neoplastic cells. There is no treatment. Prevention is by administering a polyvalent vaccine in the egg or after hatching at 1 day of age. Understand that no vaccine is100% protective and that vaccine administration may not be 100% at the hatchery since it is typically a spray on vaccine as the chicks go by on the conveyor belt. Marek's disease must be differentiated from lymphoid leucosis based on age of clinical signs, and gross and histopathologic lesions (ischiadic nerve enlargement).

Lymphoid leukosis

Technically LL is a complex of viral diseases manifesting as neoplasia and presents as such. Affects only chickens. Clinical signs usually occur in older chickens at approximately 24-40 weeks of age. The clinical signs are non-specific and include anorexia, weakness, pale comb and distended abdomen. At gross necropsy grey to white tumors are observed in the liver and other organs. The clinical signs are sometimes difficult to differentiate from those of Marek's disease, but lymphoid leukosis does not occur before 14 weeks of age, whereas Marek's disease usually occurs at 10-12 weeks of age. The best prevention is to test and cull positive breeder birds.

Pododermatitis

Ulcerative or non-ulcerative pododermatitis tends to occur if the chicken is overweight, on a roughened surface, or if one leg and foot bears more of the body weight than the other, or a combination of all these factors. There are varying grades of ulcerative pododermatitis from mild with hyperemia of the skin, to severe with osteomyelitis of underlying bone. A radiograph is the best method to determine if there is underlying osteomyelitis, a condition that requires long term antibiotic therapy and probably debridement of necrotic tissue. Most cases of ulcerative pododermatitis are somewhere in between mild and severe and consist of a thickened area of skin on the plantar surface of the foot usually over the metatarsal pad, but can also be seen on the phalangeal pads. There are differing opinions as to how aggressive to be with debriding the tissue or not – usually if there is necrosis, then debridement is necessary. Soaking the foot will greatly soften the tissue and ease surgery. Surgery should be performed under general anesthesia with administration of a pain reliever such as butorphanol since this is inherently a painful procedure. An aspirate or tissue sample is often needed to culture the area and prescribe the appropriate antibiotics. The substrate should be made as soft as possible and kept clean. Underlying lameness should be corrected. Pain relief should be addressed.

Egg-related Peritonitis

Commonly at necropsy hens can have egg-related peritonitis. Generally it is mild and not as severe as we are used to seeing in parrots. Diagnosis is based on typical clinical signs of lethargy, radiographs, or endoscopy. Occasionally antibiotics and/or surgery is necessary.

Intestinal and Hepatic Disease

Coccidiosis

There are many different species of coccidia in domestic fowl: 9 in chickens, 7 in turkeys, and at least 4 in quail. They are host specific and not zoonotic. A flock may develop resistance to one species only to be infected with another species. The typical clinical signs of coccidiosis are diarrhea, unthriftiness, and variable levels of mortality. Cecal coccidiosis is worse in that it typically causes bloody droppings and is associated with higher mortality, whereas intestinal coccidiosis is typically more chronic in nature and is associated with a lower mortality. The clinical signs of coccidiosis are more severe in young (4-16 weeks of age) chickens by having bloody diarrhea, pale combs, lethargy, tendency to huddle, partial anorexia, weight loss, dehydration, and death. As chickens get older they become more resistant and show little to no clinical signs, but can act as carriers to later expose young chicks. Transmission is through direct or indirect contact with droppings from infected birds (fomites, free-flying birds, insects and rodents). The oocysts shed in feces are not immediately infective, they have to first go through a maturation phase (sporulation) which can take as little as 1-3 days in warm, damp litter. The disease is most common in the springtime. Diagnosis is based on a fecal float. At necropsy, if *Eimeria tenella*, the cecal coccidian, is the cause, then hemorrhage of the ceca is seen. The other species (*E. acervulina and E. necatrix*) affect the intestine and are less severe. The key is prevention. Wet litter, poor sanitation, poor nutrition, and concurrent immunosuppressive diseases are the most common triggers of a coccidiosis outbreak. Treatment is with a coccidiostat such as amprolium or sulfamethazine. The recommendation to prevent the disease is to feed medicated feed between the ages of 0 and 16 weeks. There is a coccidia vaccine available for use in 1-3 days old chicks, but it is only useful in certain poultry operations, since it uses live organisms and re-ingestion at 4-25 days is necessary as a booster.

Dermatologic Disease

External Parasites

External parasites such as mites and lice are common in poultry. Prevention, checking your flock periodically for external parasites and treating early helps prevent a larger flock outbreak. Fall and winter are the most common times to see a louse problem. The lice are big enough to see with the naked eye and they move fast in comparison to mites. Nits (louse eggs) are laid in clumps at the base of feathers along the ventrum. Treatment of lice and mites consists of either pyrethrin powder or ivermectin. Follow label directions for withdrawal times in food producing poultry. If selling the eggs commercially, sometimes the only allowed treatment is diatomaceous earth. Poultry lice and mites are host specific and are not zoonotic.

The northern fowl mite, also known as the feather mite (*Ornithonyssus sylvarium*) is the most common external parasite in poultry, especially in cool weather climates. This mite spends its entire life cycle on the chicken. Clinical signs include soiled feathers around the vent, tail and rear legs.

The chicken mite, (ie, the red roost mite, *Dermanyssus gallinae*) is associated with more clinical signs than the northern fowl mite and is more difficult to diagnose because this mite feeds on poultry at night and then remains secluded during the day within the poultry house. This mite can live off the bird for 2-3 weeks. Clinical signs include mild weight loss and decreased egg production.

The body louse (*Menacanthus stramineus)* is very common. Clinical signs include hyperemia and irritation of the skin of affected birds with small scabs. They are chewing ice, chewing on skin fragments and feather debris on the surface of the skin. They spend their entire life cycle on the chicken.

References

1. Saif YM. *Diseases of Poultry.* 12th ed. Ames, IA: Blackwell Publishing;2008.

2. Carpenter JW. *Exotic Animal Formulary.* 4th ed. St. Louis, MO: Elsevier Saunders;2013.

3. Damerow G. *The Chicken Health Handbook.* North Adams, MA: Storey Communications, Inc;1994.

4. Damerow G. *A Guide to Raising Chickens.* North Adams, MA: Storey Books;1995.

5. Ussery H. *The Small-Scale Poultry Flock.* White River Junction, VT: Chelsea Green Publishing;2011.

6. Cattoli G, Susta L, Terregino C, Brown C. Newcastle disease: a review of field recognition and current methods of laboratory detection. *J Vet Diagn Invest.* 2011;23(4):637-56.

7. Greenacre CB, Morishita TY. *Backyard Poultry Medicine and Surgery: A Guide for the Veterinary Practitioner.* Ames, IA: John Wiley & Sons, Inc;2015.

Immunization of Poultry

Bruce A. Singbeil, DVM,
George Boggan, VMD,
Rocio Crespo, MS, DVSc, DACPV,
Mark Jackwood, PhD

Session #2066

Affiliation: From Bellevue Crossroads Vet 15600 NE 8th ST Ste 0-6 Bellevue, WA 98008, USA (Singbeil), Ceva 8735 Rosehill Road, Ste. 300 Lenexa, KS 66215, USA (Boggan), Avian Health and Food Safety Laboratory, Washington State University, 2607 West Pioneer, Puyallup, WA 98371, USA (Crespo), Department of Population Health, Poultry Diagnostic and Research Center, College of Veterinary Medicine, 953 College Station Road, University of Georgia, Athens, GA 30602, USA (Jackwood).

Abstract: This presentation is given to provide insight into the considerations of vaccination of poultry. Given the broad diversity of species and diseases only *Gallus* species are represented. A list of diseases currently vaccinated for is provided, but considered not absolute at publication. Examples of various viral genetics for demonstration are provided between enveloped, non- enveloped and immunosuppressive diseases. Appreciation of genetics of both viral and species for immunization are presented.

Introduction

The immunization of poultry is a complex and ever-evolving subject. This is because the viral genetics and breed genetics are always in a state of flux. Serotypes and pathotypes differ among the same viruses. Geographic differences are common. Current molecular genetics are a great aid for understanding viral mutations/differentiations but don't always give answers for the right immunization approach. This often requires laboratory testing for cross protection studies with different viral strains.

An appreciation for immunology, avian genetics and molecular virology for the protection of poultry from disease is needed to be successful to perform immunization; vaccination is just a step in the process. The strain of pullet for egg production varies widely in the acceptance for vaccination. The same is true for meat type poultry. There are many pharmaceutical companies that produce a multitude of products for providing immunization. All are licensed for poultry but incorrect strain or timing of administration will lead to illness and mortality. I, personally, have seen more illness from improper vaccination than the diseases themselves. To demonstrate immunization in poultry, we will cover 4 viruses as examples of immunizations and the influences of breeder vaccination programs, timing, routes and administration. Examples of administration of common vaccinations will be demonstrated in the presentation.

Immunization in poultry is performed using attenuated live, chemically mutated, gene deleted, recombinant or adjuvant mediated killed pathogens. Methods of administration include in-ovo, gel droplet, water, eye drop, spray, wing web and needle injection.

Poultry Vaccines Available in the United States

Viral

Vaccines are available for Marek's, infectious bursal, Newcastle disease, infectious bronchitis, infectious laryngotracheitis, fowl pox/pigeon pox, avian encephalomyelitis and chicken infectious anemia.

Bacterial

Vaccines are available for *E. coli*, *Salmonella*, *M. gallisepticum*, cholera (*Pasteurella*), and coryza (*Haemophilus* spp.).

Protozoan

Vaccines are available for coccidiosis.

Marek's Disease

Marek's vaccination is the most ubiquitous in distribution and need for vaccination. This enveloped herpes virus causes lymphoproliferative disease in many forms. These are dermal lesions seen in broilers, paralysis (most common), neurologic, ocular, degenerative inflammatory and immunosuppression, often are presented in multiple forms in the same bird. Immunization needs to be performed in the hatchery via in-ovo or injection, as once exposed to field virus the vaccine is ineffective. While the infection is ubiquitous the disease is not. This speaks to the genetics of both the virus and the breed. Some avian breeds have genetic resistance; others do not. These genetic lines are still susceptible to vvMarek's strains (very virulent). Transmission of the virus is not vertical (hen to chick). It is transmitted via inhalation of fluff at hatch, fomites and people working in the environment. There is a genetic and age resistance to this virus. Passive immunity, (maternal antibody from hen to chick), does not occur as "sterile immunity," in that it lasts a very short while but progeny is still susceptible. Many backyard breeders believe if the breeders are vaccinated chicks are protected. Maternal antibodies diminish the cytolytic infection and may decrease the efficacy of low titer cell associated vaccines and cell free HVT if used but they do not prevent infection. Current vaccines are of serotype 1 such as SB1, RB1B and HVT. Serotype 2 virus from the Netherlands has been used to control problem areas in the United States, using CVI988 strains, but can induce losses to A strain avian leukosis virus B cell tumors. Immunization in poultry is complex.

Infectious Bursal Disease Virus

Infectious bursal disease virus (IBDV) is a non-enveloped Birnavirus double stranded RNA virus which allows for transformation of trimeric surface binding proteins. Vertical transmission of this virus has not been demonstrated. There are many serologic and pathogenic types of this virus. Clinical signs may be seen in non-apparent changes in flock development, mild non- pathologic immunosuppression, and major mortality due to vvIBDV (very virulent). Immunosuppression is the hallmark of this disease due to the depletion of B cells in the bursa. IBDV is used in hatchery and field vaccination programs in both broilers and laying flocks as well as the breeder programs. Its counterpart, chicken infectious anemia (CIA), is a positive sense DNA non-enveloped virus in the *Gyrovirinae* genus. This affects the thymus T cells and other lymphoid tissue. It is transmitted horizontally and vertically. Both viruses cause major immunosuppression and due to their non-enveloped structure are very difficult to control by sanitation. The question to ask in any poultry disease is, is there an underlying cause for the current infection? Both diseases are best addressed as part of the breeder vaccination program which allows for passive transfer of maternal antibodies. CIA is only addressed this way.

Infectious Bronchitis Virus

Infectious bronchitis virus (IBV) is the most common serologic positive disease that I observe, whether it is causing symptoms or not. IBV is a Coronavirus, enveloped, + sense RNA which lends itself to translational and transcriptional mutations even amongst the vaccine viruses. Incubation periods as short as 18 hours are possible. Most commonly, this causes respiratory problem, eliminating the protective tracheal lining which leads to secondary infections and pneumonia. In addition to respiratory disease it is notorious for causing reproductive declines and egg shell problems. Renal disease from exposure causes increased mortality and production losses. Corona viruses in most species are intestinal problems. In poultry, successful isolation is from cecal tonsil. Early onset may be found in tracheal swabs but the cecal tonsil is consistent. Knowing the strain for your geographic area to immunize for this disease is paramount. There are currently 3 strains in the US that are not controlled by current vaccines. Immunization is best using breeder vaccinations for passive immunity, vaccine administration in hatchery and field exposure.

Conclusions

Utilizing all available resources in your area to best protect against diseases that may be controlled with vaccination is necessary. Universities in many states have diagnostic and poultry research programs. All manufacturers of biologics used in poultry have technical service veterinarians willing to share knowledge of use of their products. Many of them work with commercial producers in your area and have experience of current disease control programs. State departments of agriculture work closely with commercial producers and diagnostic labs for disease surveillance and control and have public meetings for discussions about this.

References

1. Swayne D. *Diseases of Poultry.* 13th ed. Hoboken, NJ: Wiley-Blackwell;2013.

Reproductive Health Considerations for Back Yard Flocks

Bruce Singbeil, DVM, and
Rocio Crespo, MS, DVSc, DACPV

Session #2067

Affiliation: From Bellevue Crossroads Vet 15600 NE 8th ST Ste 0-6 Bellevue, WA 98008, USA (Singbeil), Avian Health and Food Safety Laboratory, Washington State University, 2607 West Pioneer, Puyallup, WA 98371, USA (Crespo).

Abstract: This presentation examines reproductive health in back yard flocks which have many influences. These may begin as early as day 18 in ovo with Marek's vaccination. Genetic differences, husbandry and disease exposure play a role in our ability to maximize healthy reproduction. Many of these flocks have multiple breeds and ages adding complexity to our ability maintain reproductive health. Reproductive health also changes with age. A review of factors that influence good reproductive health given these considerations is presented.

Introduction

Raising back yard poultry has been an ever increasing activity as the trend for urbanization continues. For many, this is their first experience in attempting to produce eggs for personal consumption. The ability to facilitate and maximize reproductive health given the multi-factorial environment of this effort requires our review. The growth and care of birds to bring them into production requires attention to genetics and husbandry and is somewhat of an art as much as it is a science. Considerations for the prevention of infectious diseases that may cause problems with reproduction are presented.

Chick Quality

Reproductive health begins with chick quality. The establishment of the National Poultry Improvement Plan (NPIP) was to control the transmission of diseases at the breeder flock level. These agents include *Salmonella gallinarum*, *Salmonella pullorum*, *Mycoplasma gallisepticum*, and *Mycoplasma synoviae*. Since these diseases are vertically transmitted, chicks hatched from positive flocks are compromised and thrive poorly, which has a negative impact on egg production. Vaccination either in-ovo or at hatch for Marek's disease prevents losses due to immunosuppression or neurologic mortality. Ideally, chicks would be purchased from flocks where these activities are routine. Breeder immunization programs may have a positive effect on chick health particularly for those diseases that maternal antibodies aid in protection against. Unfortunately breeder vaccination programs, if they exist, are difficult obtain for most breeds in back yard flocks.

Flock Management

Flock management during the grow-out period has profound effects on reproductive health. Factors include: type of facility, the time of year the chicks are placed, if the chicks are to be raised in one facility and moved to another (integration), timing of feed changes, and the genetic strain(s) of the chicks. I currently see many back yard growers raising 3 or 4 different breeds. These may include standard breeds, heritage breeds and sex-link breeds. Grown in a facility that uses natural light and off season chick placement in October the sex-link breeds will come into production in 5 months. This is 1 month ahead of the other breeds. Since the feed needs to be changed 1 month before production from developer to layer. The layer diet should be integrated at month 4. In 15 years of raising 18 different breeds together I have not seen a negative effect on the development and production of the non sex-link slower-maturing birds by changing feed at that time. Waiting until the birds are in production to change to layer feed leads to less than optimal performance and may result in skeletal deformities. Integration of birds into existing flocks in production needs to be done carefully. The pecking order is hard wired. If possible to expose the pullets to the hens without contact for a few days by fencing or crating eases this process. Release the pullets at night after the exposure period so they begin the new day mingling. It may take a week to get settled so integrate flocks 2-3 weeks before production to minimize the stress of this process.

Lighting

Lighting management has the greatest influence on bringing flocks into production and maintaining healthy reproduction. In my experience this is the least understood aspect in back yard flocks. In commercial production, the use of artificial illumination is mandatory. This true in an open-sided or closed house. There are ways to use natural light in back yard flocks but an understanding of the chicken light requirements and of their physiologic orientation for egg production helps to maximize reproductive health in doing so. The lighting requirements to stimulate egg production is a minimal 11- to 12-hour day and 1 foot candle of light. A rough estimate of 1 foot candle is that if you hold a newspaper at arms length and you are able to read it, this is enough light to stimulate production. Light increases in controlled environments should be 2 weeks prior to the start of lay and increased from 12 hours to 16 hours by weekly increases of 15 minutes. Excessive intensity leads to vent picking and cannibalism. Different wavelengths of light affect egg production differently (eg, red light increases age of sexual maturity and increases egg production while yellow light increases egg size for example). The white light bulbs provide enough of light in those spectrums that birds see those longer wavelengths. Once stimulated, egg production reaches its peak in 3-4 weeks. Maintenance of production is then influenced by light, breed and environment. The majority of back yard chick placement is in the Spring. Since it takes an average of six months to come into production, daylight is rapidly declining. These flocks will continue to produce eggs through the winter without being lighted. Spring arrives with rapidly increasing daylight at a time where the birds have reached their annual lay cycle and would benefit from a molt but light stimulus overrides this. As a result, at least in our practice, we see a higher incidence of reproductive disease in April through July in birds that are 1.5 years old and again in 2.5 years old birds. These present as adenocarcinoma with or without ascites, salpingitis, egg yolk peritonitis and egg bound cases. There are two solutions to prevent or reduce these problems. One is to brood flocks off season. Start chicks in or around October and let the natural daylight bring them into production in the spring. They will naturally molt in December to rest for the next year. For first-year flocks the other option is to begin light supplementation when daylight drops below 12 hours to maintain 14-hour days. Keep these birds in lay until mid to late January and shut the lights off abruptly. They will molt and have a rest until daylight increases and then allow them to follow the natural daylight cycle. One benefit to this program is you have eggs for the holiday season whereas with off season flocks they are out of production.

Reproductive Health

Reproductive health issues would not be complete without some discussion of shell quality and broodiness. The normal interval of about 23-26 hours between ovulations is required to maintain optimum shell quality. When disrupted you will see an increase of eggs produced with sandpaper texture, white bands, calcium splashing and chalky white deposits. Other factors may cause deterioration in eggshell quality and their influence may or may not be due to inadequate supply of carbonate ions. Shell quality is correlated with shell thickness. Factors to consider are: age, the hen's ability to produce enough calcium carbonate in latter part of lay cycle, environmental heat, eggs laid in the morning (these typically have poorer quality than those laid between 0600 and 0800 hours, are misshapen and have body checks), stress on flock, certain diseases (eg, infectious bronchitis, Newcastle disease) and certain drugs. I encourage the use of oyster shell supplementation for any egg production.

Broodiness described as when a hen wants to maintain her position on a nest to incubate the eggs. This is a concern if you do not want to hatch chicks or if you have limited nests available and other hens need to use them. Physical removal of the hen and discouraging the activity may work over time. Another approach practiced is to place the bird in a cage where it is unable to get to the nest. At some point you may release the bird and it stops this behavior. In commercial operations birds are trained with trap nests to lay only when nests are open.

Diseases that affect reproductive health may be infectious or metabolic. Diseases may be primary causes (eg, infectious bronchitis) or secondary due to an infection of an immunosuppressive virus-like infectious bursal disease. Husbandry problems may not be thought of as diseases but if the birds don't receive adequate water or appropriate feed, then reproduction is impacted. As previously stated, breeder health and disease status impacts health, which ultimately affects reproductive health. The NPIP program was implemented to control four diseases previously described but doesn't address others such as the flock status for lymphoid leukosis. Immunosuppressive chicken anemia virus is only controlled in progeny by breeder flock vaccination. Immunosuppressive infectious bursal disease virus protection in progeny is benefited in producing flocks with high titers of maternal antibodies. Marek's disease is an immunosuppressive neurologic disease that has some limited protection with good breeder flock status but unpredictable. In-ovo or day-of-age vaccination against this virus is considered beneficial for all poultry.

Other flock diseases such as fowl pox, laryngotracheitis and avian encephalomyelitis are known to adversely affect egg production. Vaccination with the newer recombinant vaccines for these may be of benefit in at-risk flocks or show birds.

Infectious bronchitis virus (IBV) and Newcastle disease virus (NDV) may cause primary reproductive disease. IBV is, by far, the most common seropositive disease I see in back yard flocks in my region. That said, the incidence of disease associated with this is low. NDV, like IBV, has many strains. Vaccination for these at the back yard flock level is not recommended as the vaccines themselves may cause morbidity and mortality. Please consult your state diagnostic lab, Department of Agriculture, technical service veterinarians and possibly your states commercial industry when considering vaccinating for these diseases.

We are seeing some interesting developments in hatchery vaccination for *E. coli*, which is commonly found in egg yolk peritonitis. Dietary-induced obesity and fatty liver syndrome negatively affect reproduction. Ectoparasites do as well. There is no substitute for good biosecurity to control disease in poultry.

Conclusions

In conclusion, there are a multitude of factors to consider to maintain reproductive health in back yard flocks. Some of these we have more control over than others. All require some consideration relative to each unique environment the birds are placed in. It is possible to raise multiple breeds and ages together in good reproductive health.

References

1. Swayne D. *Diseases of Poultry.* 13th ed. Hoboken, NJ: Wiley-Blackwell;2013.

2. North MO, Bell DD. *Commercial Chicken Production Manual*. 4th ed. New York, NY: Springer;1990.

Section 4

Sunday
Master Classes

Parrot Cognition and Behavior: 40 Years and Counting

Jan Hooimeijer, DVM, CPBC, and
Irene Pepperberg, PhD, SB, MA

Session #2022

Affiliation: From the Consultancy Practice for Birds, Galgenkampsweg 4 7942 HD Meppel, The Netherlands (Hooimeijer) and Department of Psychology, Harvard University, Cambridge, MA 02138, USA (Pepperberg).

Abstract: It is important to acknowledge and to deal with the intelligence and cognitive abilities of parrots to prevent behavior problems. The first part of the presentation (Pepperberg) provides evidence of parrot cognitive processes (probabilistic reasoning, visual working memory) that are as good as or better than those of 6- to 7-year-old human children. The second part of the presentation (Hooimeijer) examines how to use this intelligence to avoid and correct behavior problems. Ignoring parrots' abilities to respond to negative effects of human body language and human attitude can contribute to behavior problems. A 5-step protocol is presented as a tool to show respect for parrots as intelligent prey animals, create and reward desired behavior, reduce the stress of handling, and create mutual trust. Definitions of normal behavior, unwanted behavior, desired behavior, enforced behavior, and displacement behavior are described as tools to understand parrot behavior. Using displacement behavior is described as a method to deal with parrots demonstrating behavior problems. Wing clipping can make it possible to offer parrots more freedom as companion birds and to become part of the family flock inside as well as benefit outside from the social interaction, exercise, sun and fresh air. The disadvantages of having a parrot on the shoulder are discussed.

Part I. Cognitive Processes

The first study replicates work performed with children by Piaget. We examined whether a nonhuman might succeed in an experiment based on probabilistic reasoning, specifically, the ability to make inferences about a sample based on information about a population. A grey parrot (*Psittacus erithacus*) previously trained to use English labels referentially to identify objects, observed a human researcher deposit two different types of items in a 3:1 ratio (ie, three corks and one piece of paper) into an opaque bucket. One item was then randomly withdrawn while hidden from the parrot's view. When asked to identify the still-hidden object, the parrot's vocal responses tracked this 3:1 ratio over a large number of trials. Some levels of probabilistic reasoning therefore are not limited to humans, nonhuman primates, or even mammals.

The second study involves abilities to store and manipulate information in visual working memory (VWM), which stands at the cornerstone of intelligent behavior. Demonstration of a 3- to 4-item storage limit in humans and other species implies a homologous storage mechanism. In contrast, evolutionary bases underlying VWM manipulation remain unexplored. Here, we used a task akin to the "shell game" to examine whether the ability to manipulate information is uniquely human. We compared human performance on a computerized version of this task with that of a grey parrot on a live version adapted for animal testing, administered by a human experimenter. In both tasks, a varying set size (2-4) of colored objects was presented briefly, after which objects were covered by occluders. Occluders either remained stationary (storage: 0 swaps) or pairs of occluders swapped positions a number of times (manipulation: 1-4 swaps). A target color was subsequently presented and participants had to indicate the location of the cued color. Storage trials require only that color-location bindings be stored in VWM; whereas, manipulation trials additionally require that these bindings be updated as objects swap positions. Performance accuracy on storage trials exhibited by humans and the parrot were both consistent with a 3- to 4-item storage limit; the parrot also was perfect on 5 items (humans were not tested). Manipulation-related performance was near ceiling for both groups on set size 2 trials; however, humans experienced a systematic decrease in accuracy as swaps increased for set sizes 3 and 4. Interestingly, the parrot generally remained near ceiling for all set size 3 dynamic trials, and matched human decrement in ability for set size 4 at 3 and 4 swaps. These results suggest that VWM manipulation is not unique to humans, inviting further investigations into whether VWM manipulation is supported by common versus separate mechanisms across species, and whether differences exist for computer-based versus live trials.

Part II. Behavioral Interventions

Manuscript adapted from: Hooimeijer J. Welfare and behavior in parrots. *Proc Annu Conf Assoc Avian Vet.* 2013;289-297.

Introduction

In general, there is still a lack of awareness that parrots are nondomesticated prey animals being kept in captivity under unnatural circumstances as highly intelligent animals with remarkable cognitive abilities. Without understanding normal behavior and underestimating the high intelligence and cognitive abilities of parrots a wide variety of behavioral problems are apt to occur depending upon the individual parrot, species, and circumstances surrounding the bird.

Behavior

Behavior in nature is based on innate behavior and founded on learning, experience, intelligence and cognitive abilities. Behavior is driven by the urge to survive as individual and as species.

Dealing with parrots in captivity we need to understand that the behavior of parrots is determined by different factors including hereditary factors. To understand behavior we have to understand and learn about the principles of ethology. Looking through the eyes of parrots we have to consider that our body language and attitude determines the body language and behavior of parrots.

Parrots are built and behave as prey animals

Parrots are prey animals in nature. The fear of being killed as a prey animal determines a major part of normal behavior in nature. Fear of getting killed as prey animals also can be expected as normal behavior within captivity.

A typical anatomic feature of parrots as prey animals is the positioning of their eyes. The eyes of parrots are positioned at the side of their head enabling parrots to observe the whole environment.

The eyes of predators are positioned so as to enable the animals to watch straight ahead using binocular vision. The eyes of humans are positioned as in predators such as dogs, cats, owls and birds of prey, which also have binocular vision. Humans that are talking and smiling to the parrot, showing their teeth, can be perceived by a parrot as being intimidated by a predator. Dealing with parrots without understanding the consequences of the specific characteristics of parrots as prey animals is a likely reason for parrots to develop insecure/defensive behavior.

Parrots are intelligent birds with amazing cognitive abilities

From studies demonstrating the intelligence and cognitive abilities of parrots, we can infer that they understand human intentions, posture and attitude. Parrots learn by observing other birds, other animals and humans as part of their environment. Parrots evaluate information, make inferences, and behave accordingly. Dealing with parrots in a way resembles dealing with children who are eager to show and demonstrate what they have learned. The Model/Rival technique, as described by Pepperberg, acknowledges this intelligence and observational learning skills of parrots. Allowing myself (JH) some anthropomorphism: "From the viewpoint of parrots, it must be a frustrating experience not to be appreciated for their intelligence and talents but just to be considered as beautiful, funny and cute."

Consultancy Practice for Birds: Definitions of Behavior

Normal behavior

Normal behavior is species-typical individual behavior that is not meant to manipulate the caretaker. Examples are eating, drinking, grooming, playing, sitting in the cage, perching on the hand. It is the experience of the author (JH) that the outcome of rewarding normal behavior is that the bird will develop additional behavior patterns that are then used to manipulate the caretaker to get attention. These additional patterns can be considered undesired behavior.

Undesired behavior

Undesired behavior is behavior that is intended to manipulate the behavior of the caretaker. This may include screaming, talking, biting, feather picking. A human response of fear, frustration, anger, concern and laughter might be considered a reward by the bird for such behavior and thus as encouragement to act in this manner. Consequently, demonstrating that a bird's behavior is problematic may actually increase its occurrence.

Desired behavior

The definition, used within the Clinic for Birds, for desired behavior is behavior that is created and manipulated by the caretaker. It is behavior that the caretaker would like to see repeated in the future. Desired behavior is not behavior that is the result of a command or a request. Desired behavior also is not behavior the bird is showing to get a treat. Desired behavior involves (but is not limited to) actions such as the bird demonstrating an interest in novel toys or actions. Rewarding desired behavior, including rewarding intelligence, is a way to show respect for the bird and can aid in creating self-esteem, mutual respect, and might prevent undesired behavior.[1]

Enforced behavior

Enforced behavior is the result of a command, any kind of force, or behavior motivated by a treat or food. Enforced behavior also is behavior out of fear of punishment.

Displacement behavior

Tinbergen, one of the founders of ethology, has described displacement behavior in gulls during territorial behavior, when a gull generally either defends its territory or cedes to another individual (eg, fight or flight).[2,3] Displacement behavior can be considered as the ultimate alternative behavior when a fight or flight response are not the best choices or even impossible. Instead of a fight or flight response, the gulls started to preen themselves, started to pull at grass as if they were collecting materials to improve the nest, and looked away from their competitors. Note that a "freeze" response also might occur.[4,5] In a conflict situation in nature as well as in captivity, parrots also have the choice between a fight or flight response and displacement behavior. Displacement behavior seems to be completely "out of order" but has the effect that the other parrot or the human caretaker becomes distracted and a new situation is created. Typical examples of displacement behavior are nail biting, scratching, grooming, flapping the wings, shaking the tail, shaking the head, and looking up with the head tilted sideways. Grooming and preening have been associated with normal behavior that birds perform when they are comfortable but can also be a form of displacement behavior that is displayed in a conflict situation.[6] When humans respond to displacement behavior in a way that is rewarding for the bird, this behavior, which is normal in a conflict situation, becomes undesired behavior.

When dealing with parrots, we can expect to see displacement behavior in different conflict situations. The most common conflict situation is when parrots as prey animals are approached by humans in an intimidating manner; that is, face forward with the typical characteristics of a predator with their eyes in the front of their face, just like dogs, cats, owls and birds of prey.[1] We also observe displacement behavior of parrots that are inside or up on their cage showing territorial behavior (eg, defending their space in an aggressive manner). Under those circumstances it is important to know how an avian species analyzes and combines its perceptual experiences to understand what is happening in the surrounding world.[7] There are reasons to hypothesize that parrots utilize extensive information processing (eg, at least to some degree, they do not simply react mindlessly to environmental stimuli) but they actually process the stimuli and choose to react in a certain ways.[8]

When displacement behavior is not recognized or not understood and caretakers are responding to displacement behavior in a way that is considered as rewarding, displacement behavior can easily result in abnormal repetitive behavior (also known as maladaptive behavior) without the primary trigger that was causing the conflict situation to begin with.[9] It can be one of the explanations when we are dealing with complex behavioral problems like feather destructive behavior. It can be assumed that the type of displacement behavior has a genetic component and that the type of displacement may therefore differ between species. It can help explain why some species are more prone to feather destructive behavior.[10] It can be postulated that the way displacement behavior is expressed also will be influenced by environmental factors and the observational learning ability of the parrots. Because of that, it has been argued that hand feeding baby parrots is contraindicated because birds need to learn by observing normal behavior of their parents, including normal displacement behavior in conflict situations.[11]

How to deal with displacement behavior: It is important to recognize and acknowledge displacement behavior and to consider that the parrot is showing this behavior because it perceives itself to be in a conflict situation. Any direct response to the bird at that moment can potentially be rewarding to the animal and can create maladaptive behavior. By improperly reacting to displacement behavior when in the role of caretaker or veterinarian, we are interrupting normal behavior; however, displacement behavior can be an important tool for humans to use when dealing with parrots.

5-Step behavioral protocol to deal with and manipulate a parrot

The most important aspect of dealing with parrots is the attitude of the person or persons caring for the parrot. A basic protocol for veterinarians and caregivers of parrots can be summarized in 5 steps. This protocol also can be used with any parrot showing any behavior problem.

Step 1: Show the parrot that you have no negative intentions by acknowledging the fact that parrots are prey animals. For example, avoid staring directly at the bird. Allow the bird to observe you and your actions from a short distance, as though it was in the audience of a play. Relax your body, show that you are not (overtly) attending to the bird, and playfully interact with objects in your environment. For example, toss and catch a parrot toy, shake the toy, fiddle with your pen, tear up pieces of paper, rub a towel on the side of your face. In that way, be a positive role model. Engage in these actions without any interaction with the parrot; don't look directly at the bird. Note via peripheral vision whether the bird is attending and try different playful actions until it does. Parrots are more likely to feel comfortable with people who are comfortable with parrots and with themselves. Your actions (eg, positive, confident, or nervous, insecure) around the parrot will provide the bird with the information it needs in order to decide how to react. People often have to act and play a role around the parrot. What we create by doing so is desired behavior.

Step 2: This step can be considered rewarding the behavior that has been created and manipulated by your attitude in performing step 1. You have created an environment that encourages the bird to exhibit desired behavior—interest in your actions—and you reward this interest to by telling the parrot that it's the most beautiful creature on this planet and how precious the bird is to you.

Step 3: Establish yourself as the best teacher the parrot can wish for. You do this by showing respect for the bird's intelligence by describing, in detail, what there is to see around it (eg, birds in the trees, toys on the table, their colors, shapes, sizes, materials). You also can describe the color of the tail or beak. Parrots truly enjoy observing and learning. Pepperberg has shown that birds will request labels for novel items in their environment.[12] Reward the bird for its intelligence in a nonintimidating way (eg, be aware of the direction of your gaze).

Step 4: After rewarding the intelligence, allow the bird to touch your pen, toy, paper or towel and praise the bird enthusiastically. Only let them touch items when you tell it is okay. When they reach out for anything, just take it away without a word and present it later after telling that now it's okay to touch or even bite into the item. Then they are in general very gentle touching with their tongue. Ask the bird to participate in the learning process by offering him new, small, unthreatening objects. Tell the bird that it may touch and feel the objects and allow it to investigate. Praise each positive move. You also can present your hand and tell the bird that it may step up, and praise the bird for doing so. This is the opposite of giving a command or making the bird step up to get a treat.

Step 5: Because, in the previous steps, you have shown the parrot that you are safe to be around, you can now extend the experience to new situations. When the parrot shows fear, do not reassure the bird. Reassurance can easily exacerbate the bird's sense of fear and insecurity as it can function as a reward for the behavior. It is the experience of the author (JH) that after steps 1-4, a situation of mutual respect and mutual trust is established. When the parrot fears a place or object like the towel, simply start over again with step 1 of the protocol. The bird will eventually realize that there is no reason for fear. Then tell the bird how smart and wonderful it is. This will help to stimulate the self-esteem that all parrots need in order to display normal parrot behavior. The outcome is to create a situation in which a parrot can accept novel situations like an examination, taking blood, grooming, wing clipping, gently towelling and others.[1,13]

Dealing with behavior problems of parrots

Dealing with behavior problems involves looking for a diagnosis.[11] It is important to consider the circumstances and to understand the reason for the behavior and its consequences. In general, unwanted behavior of parrots in captivity can be considered behavior that started as normal behavior under unnatural circumstances. When dealing with unwanted behavior, we must consider that unwanted behavior will become more and more a problem when the consequence of engaging in that behavior results in a reward for the parrot. Any response, whether positive, negative, emotional or even subconscious, to the (unwanted) behavior, including ignoring a behavior, could be a reward for the (undesired) behavior.

Preventing undesired behavior

Preventing undesired behavior requires understanding the reason for the undesired behavior: looking at the circumstances, the consequences, the natural behavior, and body language of parrots as a social, intelligent prey animal. Preventing undesired behavior involves optimizing health and welfare by making sure that birds benefit from certified organic pelleted food, sunlight, fresh air and exercise. Preventing undesired behavior is accomplished by creating an enriched environment in which the parrot is allowed to express its intelligence and skills, providing toys, and providing food in a way that it stimulates parrots to express normal foraging behavior. Preventing undesired behavior is enabled with social interactions with other birds, other animals and humans. Taking a bird outside for a walk, a bike ride, a picnic or a family visit (with appropriate safety measures) will prevent undesired behavior. Preventing undesired behavior also includes understanding that the cage of a bird can be considered by the bird as its nesting site. In nature, birds spend their time during the day and during the night at the same location when they are breeding as part of their reproductive cycle. Being at what it sees as its nesting site, the bird can feel threatened by interlopers and try to defend its nest site by displaying territorial behavior. To prevent undesired behavior, an owner can create a more natural day-night routine in which the bird gets 10-12 hours of sleep in a special small sleeping cage that is located in another room of the house, away from the location the bird spends the daylight hours. The underlying reason for this is that birds in nature do not sleep at night where they spend their daytime hours, foraging and drinking. Preventing undesired behavior is about understanding that birds prefer to sit in a high position when they feel insecure. For that reason, allowing the bird to sit on the shoulder is not encouraged. This point will be discussed in further detail during the talk.

Redirecting undesired behavior, displacement behavior

We have to consider that every response to undesired behavior can be perceived as a reward. Instead of responding to the behavior of the parrot and instead of ignoring the behavior, the advice is to act in a way that signals to the bird that there is no perceived problem with a demonstration of displacement behavior. Part of the displacement is "looking away" (ie, using only peripheral vision) and distracting the bird with your interest in objects with which you wish the bird eventually to engage. The result of demonstrating displacement behavior is that insecurity and aggression are replaced by interest in these objects. Over the years, the author (JH) has observed this behavior both as a birdwatcher in nature and as an avian veterinarian dealing with parrots in captivity. In a situation of undesired behavior, step 1 of the 5-step behavior protocol can be considered as showing displacement behavior, resulting in desired behavior that can be rewarded as step 2. Using this protocol in this order prevents undesired behavior from being rewarded. Having the bird watch us when we are not looking at the bird, acknowledging and rewarding their intelligence and rewarding the behavior by allowing the bird to touch, feel and even bite a specific object can all be done within 30 seconds. Using this protocol, undesired behavior is redirected into desired behavior and the bird is rewarded for that.

Wing Clipping

In the wild, flying is essential for the survival of parrots, with the typical exception of the kakapo (*Strigops habroptilus*), Flying enables parrots to find food, water, and a safe place to sleep. Flying is a way to escape from predators; flying in flocks also offers protection from predators. For parrots in captivity that are kept as companion birds, flying is no longer essential for their survival. Within the Consultation Practice for Birds, clipping can be recommended for safety reasons but is mainly recommended because it makes it possible to provide the parrot more freedom and social interactions both inside the house and during the daily outside activities with the caretaker; the bird has a wider range to become more a part of the family (flock). In the experience of the author (JH), wing clipping done with a parrot-friendly attitude helps prevent behavior problems and prevents parrots from ending up in parrot rescue facilities.

Hand-rearing Baby Parrots

Basically, the necessity for hand rearing baby parrots in the past was the result of management failures concerning housing, nutrition and care complemented with a lack of knowledge about the natural breeding behavior of parrots. Over the years hand-rearing baby parrots became a commercial activity because females produced more eggs.

The Dutch Parrot Foundation, supported by the Consultancy Practice for Birds has been successfully campaigning in the Netherlands against the hand-rearing of baby parrots and the trade in unweaned baby parrots.

The outcome has been that since July 1 2014, it is not allowed to separate a baby parrot from the parents until the baby parrot is fully weaned (i.e. able to eat completely on its own). The new legislation puts parrots on the existing list with dogs, cats, rabbits, chimpanzees and other species for which the timing of separation of youngsters from the parents has been regulated for many years.

The new regulation is based on the data that have become available during the past 15 years showing the negative consequences of separating baby parrots from their parents. A range of behavioral problems and welfare issues have been recognized as consequences of separating youngsters from the parents, particularly before or during the important first imprinting phase Some of the behavioral problems can occur after months or even after years, especially when the bird becomes reproductively active. Insecure behavior, phobic behavior, biting, feather picking and self mutilating can become serious consequences of separating a baby parrot from the parents.[14,15] Some researchers have shown that enrichment at an early age is important in maintaining good interactions later in life.[16]

Conclusions

We have to learn how to interact with parrots by looking at the world through their eyes.

Parrots belong to the most charismatic, intelligent and social animals on the planet that can have a very long life expectancy when provided with the appropriate living environment. Consider and acknowledge parrots as prey animals. By adjusting our body language we show respect for parrots instead of showing an intimidating posture, creating insecure and defensive behavior. By showing respect for their social skills and intelligence, we demonstrate respect for parrots. The natural learning process involves observing the behavior and skills of their parents, other members of the flock, and humans in captivity. Showing that we have a problem with a parrot's behavior can create insecurity and behavior problems. Instead, using the 5-step behavior protocol can help create a relationship based on mutual trust and respect.

It is vital to reward desired behavior, but in the appropriate manner. It is vital to reward the intelligence and acknowledge the cognitive abilities in parrots, but, again, in the appropriate manner. It is vital not to reward normal behavior. It is vital not to reward undesired behavior. It is important to recognize displacement behavior and to use displacement behavior when we are in a conflict situation.

Taking care of the health, welfare, well-being, and conservation of parrots in the world is our primary responsibility as avian veterinarians. Parrots' behavior patterns show whether you are doing a good job, just as children show whether their teachers are doing a good job. In that way parrots have become my best teachers because they correct me when I'm wrong. I'm grateful for the parrots in my life because they force me to be a better avian veterinarian.

If Not Us, Who? If Not Now, When?

Acknowledgments: Authors would like to thank Harrison's Bird Foods, The Netherlands for sponsoring the parrot events in the Netherlands. Finally, we thank the staff of the Consultancy Practice for Birds for their support and understanding. Without them, nothing could have been accomplished over the years.

References

1. Hooimeijer J. Dealing with behavior problems of parrots. *Proc Annu Conf Deutsche Veterinärmedizinische Gesellschaft.* 2008.

2. Tinbergen N. *Spieden en Speuren in de Vrije Natuur.* Amsterdam, The Netherlands: Uitgeverij Ploegsma;1959.

3. Tinbergen N. *Curious Naturalists.* London,UK: Country Life;1958.

4. Hagenaars MA, Oitzl M, Roelofs K. Updating freeze: aligning animal and human research. *Neurosci Biobehav Rev.* 2014;47:165-176.

5. Bracha HS. Freeze, flight, fight, fright, faint: adaptationist perspectives on the acute stress response spectrum. *CNS Spectrums* 2004;9:679-685.

6. Spruijt BM, van Hooff JARAM, Gispen WH. Ethology and neurobiology of grooming behavior. *Physiol Rev.* 1992;72:825-852.

7. Zucca P. Mind of the avian patient: cognition and welfare. *Proc Annu Conf Euro Assoc Avian Vet.* 2007;357-365.

8. Pepperberg IM. The African grey: how cognitive processing might affect allospecific vocal learning. In: Balda RP, Pepperberg IM, Kamil AC, eds. *Animal Cognition in Nature.* London, UK: Academic Press;1998:381-410.

9. Garner JP. Perseveration and stereotype-systems-level insights from clinical psychology. In: Rushen J, Mason G, eds. *Stereotypic Animal Behavior; Fundamentals and Applications to Welfare.* Wallingford England, UK: CABI;2006:121-152.

10. Seibert LM. Feather-picking disorder in pet birds. In: Luescher AU, ed. *Manual of Parrot Behavior.* Oxford, England: Blackwell;2006:255-265.

11. Orosz S. Diagnostic workup of suspected behavioral problems. In: Luescher AU, ed. *Manual of Parrot Behavior.* Oxford, England: Blackwell;2006:233-245.

12. Pepperberg IM. *The Alex Studies.* Cambridge, MA: Harvard University Press;1999.

13. Hooimeijer JA. Practical behavioural protocol for dealing with parrots. *Proc Annu Conf Assoc Avian Vet.* 2003;177-181.

14. Engebretson M. The welfare and suitability of parrots as companion animals: a review. *Anim Welf.* 2006;15:263-276.

15. Schmidt R. The influence of the breeding method on the behaviour of adult African grey parrots. Inaugural Dissertation zur Erlangung der Doktorwurde der Vetsuisse-Falkultat der Universitat Bern; 2004.

16. Fox RA, Millam JR. The effect of early environment on neophobia in orange-winged Amazon parrots (*Amazona amazonica*). *Appl Anim Behav Sci*, 2004;89:117-129.

For additional references, contact info@vogeladvies.nl.

Section 5

Plenary Session

Michelle Hawkins, VMD,
DABVP (Avian Practice)
AAV President
Moderator

Differential Effects of the Environment on Feather-Damaging Behavior in Shy and Bold Parrots

Malou van der Sluis, BSc, and
Yvonne van Zeeland, DVM, MVR, PhD, DipECZM (Avian), DipECZM (Small Mammal), CPBC

Session #2017

Affiliation: From the Division of Zoological Medicine, Department of Clinical Sciences of Companion Animals, Faculty of Veterinary Medicine, Utrecht University, Yalelaan 108, 3584 CM Utrecht, the Netherlands.

Feather-damaging behavior (FDB) is a common sequela to a suboptimal living environment in captive parrots.[1,2] To optimize living circumstances, risk factors for FDB need to be identified. Several studies evaluated risk factors for FDB in different species[3-6]; however, risk factors can differ within a species (eg, according to personality or coping style).[7,8] In this study, the potential link between environmental risk factors, coping style, and onset of FDB was investigated. For this purpose, data on parrots' coping style (bold versus shy, as indicated by their responses to novel stimuli),[8] living conditions, and presence of FDB were obtained from 459 owners of grey parrots (*Psittacus erithacus*). Binary logistic regressions subsequently were performed to study correlations among environmental factors, coping style, and FDB, with P values < 0.05 considered statistically significant. Results showed that several environmental factors affected FDB differentially for bold and shy birds, including the size of housing, time spent on daily interaction with the parrot, changes in foraging enrichment, and the refilling and replacement rate of foraging enrichments and chewing materials. A factor that influenced presence of FDB, regardless of coping style, included the provision of auditory stimuli during absence of the owners. These results indicate that environmental factors may influence FDB differentially in bold and shy birds, emphasizing the importance of taking the bird's personality characteristics into account when evaluating FDB. Further prospective studies are needed to study the relationships among personality, environment, and FDB.

References

1. Garner JP, Meehan CL, Famula TR, et al. Genetic, environmental, and neighbor effects on the severity of stereotypies and feather picking in orange-winged Amazon parrots (*Amazona amazonica*): an epidemiological study. *Appl Anim Behav Sci*. 2006;96:153-168.

2. Van Zeeland YRA, Spruit BM, Rodenburg TB, et al. Feather damaging behaviour in parrots: a review with consideration of comparative aspects. *Appl Anim Behav Sci*. 2009;121:75-95.

3. Costa P, Macchi E, Tomassone L, et al. Feather picking in pet parrots: sensitive species, risk factor and ethological evidence. *Ital J Anim Sci*. 2016;15:1-8.

4. Gaskins LA, Hungerford L. Nonmedical factors associated with feather picking in pet psittacine birds. *J Avian Med Surg*. 2014;28:109-117.

5. Jayson SL, Williams DL, Wood JL. Prevalence and risk factors of feather plucking in African grey parrots (*Psittacus erithacus erithacus* and *Psittacus erithacus timneh*) and cockatoos (*Cacatua* spp.). *J Exotic Pet Med*. 2014;23:250-257.

6. Kinkaid HMY, Mills DS, Nichols SG, et al. Feather-damaging behaviour in companion parrots: an initial analysis of potential demographic risk factors. *Avian Biol Res*. 2013;6:289-296.

7. Cussen VA, Mench JA. The relationship between personality dimensions and resiliency to environmental stress in orange-winged amazon parrots (*Amazona amazonica*), as indicated by the development of abnormal behaviors. *PLOS One*. 2015;10:e0126170.

8. Van Zeeland YRA, van der Aa MM, Vinke CM, et al. Behavioural testing to determine differences between coping styles in grey parrots (*Psittacus erithacus erithacus*) with and without feather damaging behaviour. *Appl Anim Behav Sci*. 2013;148:218-231.

Validation of a Point-of-Care Coagulometer in Two Raptor Species

Susan J. Baley,
Julia B. Ponder, DVM, MPH,
Leslie C. Sharkey, DVM, PhD, DACVP (Clinical Pathology)

Session #2039

Affiliation: From the College of Veterinary Medicine (Baley), Raptor Center (Ponder), and Department of Veterinary Clinical Sciences (Sharkey), University of Minnesota, 1365 Gortner Ave, St. Paul, MN 55108, USA.

Methods

Prothrombin (PT) clotting time of frozen plasma from great horned owls (*Bubo virginianus,* GHOW) and red-tailed hawks(*Buteo jamaicensis,* RTHA) were measured on a point-of-care (POC) mechanical clot detection coagulometer at the University of Minnesota (UMN) and were compared with PT times at the Cornell Animal Health Diagnostic Center. A mixture of healthy and unhealthy raptors was used in the study.

Results

Precision of the UMN assay was considered fair to good. Bland-Altman (BA) analysis of GHOW PT revealed a mean negative bias of 73.9 seconds for the Cornell method compared with the UMN values; however, the effect was proportional, being greater at longer times. The limits of agreement were 47.4-195.3 seconds. In contrast, BA analysis of RTHA PT showed a mean positive bias of 28.1 seconds for the Cornell method versus UMN that was also proportional and greater at high values; however, the proportional change was lower than for GHOW, perhaps due in part to more random scatter in the data. The limits of agreement for this assay were 65.3-9.2 seconds. The methods were not considered comparable for either species.

Conclusions

Although the true coagulation status was not known for the birds studied, data indicate a need for further optimization of the UMN method if the Cornell method is considered closer to a reference standard, particularly at longer clotting times. Options for the optimization include modification of the stoichiometry of reaction components and the concentrations of thromboplastin in the reaction time.

Acknowledgments: This study (D16ZO-828) was made possible through support from Morris Animal Foundation, the global leader in supporting science that advances animal health.

Sinus Trephination and Sinoscopy in a Swainson's Toucan with Rhinosinusitis (*Ramphastos ambiguus swainsonii*)

Brian L. Speer, DVM, DABVP (Avian Practice), DipECZM (Avian),
David Sanchez-Migallon Guzman, LV, MS, DipECZM (Avian), DipECZM (Small Mammal), Dipl ACZM,
Amberly Sokoloff, VMD,
Rachel Baden, DVM,
Michelle Hawkins, VMD, DABVP (Avian Practice)

Session #2037

Affiliation: From the Medical Center for Birds, 3805 Main Street, Oakley, CA 94561, USA (Speer), Department of Medicine and Epidemiology (Sanchez-Migallon Guzman, Hawkins), William R. Pritchard Veterinary Medical Teaching Hospital (Sokoloff, Baden), School of Veterinary Medicine, University of California, Davis, One Shields Avenue, Davis, CA 95616 USA, Airport Animal Hospital, 6209 N. 9th Ave, Pensacola, FL 32504, USA (Sokoloff).

A 6-year-old female Swainson's toucan was presented for examination with a 1- to 2-year history of nasal discharge and abnormal growth of the ramphotheca. Nasal cytology and culture were consistent with mixed aerobic and anaerobic bacterial rhinitis. Radiographic and CT imaging demonstrated a deviated septum of the nasal diverticulum, multiple soft tissue densities in the caudal aspects of the maxillary and mandibular infraorbital diverticulae of the infraorbital sinus. Rhinoscopy and biopsy confirmed bacterial rhinitis. Treatment included multiple nasal lavage with saline, gentamicin and amphotericin B, systemic ciprofloxacin and tobramycin nebulization for several weeks. The clinical signs improved markedly, however there was recurrent clear nasal discharge noted in the fall of each of the three years following completion of treatment. Repeat radiographic imaging 4 years later showed resolution of most of the soft tissue opacities within the maxillary and mandibular diverticulae previously noted, but persistence of three areas of soft tissue dense material within the maxillary sinus diverticulum. A sinus trephination procedure was performed, using two portals of access through the maxillary bone, and rigid endoscopic assistance, for exploration and sample collection. Histopathology of the samples was consistent with keratin granulomas. Amikacin-impregnated polymer gel was applied into the maxillary sinus diverticulum, and healing of the bone, dermis and keratin at the surgical site was uneventful. Aerobic bacterial cultures from the granulomas were negative, and further antibacterial treatments were discontinued. Two years following the procedure, the bird is fully healed, and there have been no further significant events of nasal discharge noted.

References

1. King AS, McLelland J. Skeletomuscular system. In: *Birds: Their Structure and Function.* 2nd ed. London, UK: Baillière Tindall;1984:43-51.

2. King AS, McLelland J. External anatomy. In: *Birds: Their Structure and Function.* 2nd ed. London, UK: Baillière Tindall;1984:9-10.

3. Tattersall GJ, Andrade DV, Abe AS. Heat exchange from the toucan bill reveals a controllable vascular thermal radiator. *Science.* 2009;325(5939):468-470.

4. Baumel JJ, Witmer LM. Osteologia. In: Baumel JJ, Kings AS, Breazile JE, et al, eds. *Handbook of Avian Anatomy: Nomina Anatomica Avium.* 2nd ed. Cambridge, MA: Nuttall Ornithological Club;1993:80-81.

5. Baumel JJ, Witmer LM. Aparatus resiratorius. In: Baumel JJ, Kings AS, Breazile JE, et al, eds. *Handbook of Avian Anatomy: Nomina Anatomica Avium.* 2nd ed. Cambridge, MA: Nuttall Ornithological Club;1993:264-265.

6. King AS, McLelland J. Respiratory system. In: *Birds: Their Structure and Function.* 2nd ed. London, UK: BaillièreTindall;1984:110-118.

7. Speer B, Echols MS. Surgical procedures of the psittacine skull. *Proc Annu Conf Assoc Avian Vet.* 2013;99-109.

8. Gartrell BD, Alley MR, Kelly T. Bacterial sinusitis as a cause of beak deformity in an Antipodes Island parakeet (*Cyanoramphus unicolor*). *NZ Vet J.* 2003;51:196-198.

9. Raidal SR, Butler R. Chronic rhinosinusitis and rhamphothecal destruction in a Major Mitchell's cockatoo (*Cacatua leadbeateri*) due to *Cryptococcus neoformans* var *gattii*. *J Avian Med Surg.* 2001;15(2):121-125.

10. Speer BL. Beak deformities: form, function, and treatment methods. *Proc Annu Conf Assoc Avian Vet.* 2014;213-219.

Effects of Housing Variables and Season on Bone Density in Parrots

M. Scott Echols, DVM, DABVP (Avian Practice),
Boston Clark Terry,
Samer Merchant,
Doug Folland, DVM, DABVP (Avian Practice),
Crystal Wilcox, LVT,
Rebecca Bishop,
Alex Echenberg,
Fred Kromm,
Kate Lyngle-Cowand,
Michiko Wallace, CVT,
Ed Hsu, PhD

Session #2111

Affiliation: From Echols Veterinary Services, Salt Lake City, UT 84054, USA and Medical Center for Birds, Oakley, CA 94561, USA (Echols), Department of Bioengineering, University of UT, Salt Lake City, UT 84112, USA (Terry, Merchant, Hsu), Parrish Creek Veterinary Clinic, Centerville, UT 84014, USA (Folland, Wilcox), Tracy Aviary, Salt Lake City, UT 84105, USA (Bishop, Echenberg, Kromm, Lyngle-Cowand, Wallace).

Manuscript presented previously at: iCARE 2017.

Bone density has been recognized as an indicator of activity, nutritional status and various diseases. Currently, computed tomography and dual-energy x-ray absorptiometry are the only noninvasive means for measuring bone density. These systems are typically only validated for adult humans and they are expensive and inaccessible. Recently a novel method of standardizing tissue density scores from digital radiographs was introduced.[1] This method was applied to characterize bone density in sun conures (SC, *Aratinga solstitialis*, n = 11) and Quaker parrots (QP, *Myiopsitta monachus*, n = 10), housed indoors and outdoors, during fall and spring and living within Salt Lake, Davis, and Weber counties of Utah, USA.[2] After calibration, intensities were quantified in 14 matching regions of interest (ROIs). Correlation coefficient (R) was calculated for density against weight, age and sex for outdoor SC and QP in fall and spring. There was a modest correlation between bone density and weight ($R = 0.493$ average) for the outdoor QP in spring. While there were a few strong correlations ($R > 0.7$ or < -0.7), among all the regions examined, most correlations were weak. For outdoor SC, there was no significant difference in ROI density regions for fall and spring. For outdoor QP, 4 of 14 ROIs were significantly higher in the fall. For indoor SC and QP, fall ROI readings were generally higher; however, due to low patient numbers (3), the values were not statistically significant. These findings represent the first systematic and quantitative investigation of bone density in captive birds and correlations with housing variables.

References

1. Echols MS, et al. Validation of, and improvement upon, a novel radiograph density determination process-a multicenter study. *Proc iCARE.* 2017;369-370.

2. Echols MS, et al. Bone density determination using μCT and a novel digital radiograph system in parrots. *Proc ExoticsCon.* 2016;67.

Comparison of Conscious Whole-Body Dorsoventral Radiographs and Standard Anesthetized Radiographs in African Grey Parrots (*Psittacus erithacus*)

Graham Zoller, DV, IPSAV (Zoological Medicine),
Lucile Chassang, DV, IPSAV (Zoological Medicine),
Olivia van Steyvoort, DV,
Minh Huynh, DV, MRCVS, DipECZM (Avian)

Session #2077

Affiliation: From the Exotic Pet Department, Centre Hospitalier Vétérinaire Frégis, Arcueil, 94110, France.

Introduction

Screening of coelomic disorders generally relies on whole-body radiographs with ventrodorsal (VD) and lateral projections. Proper positioning may require sedation and can be deleterious in sick birds. The aim of this crossover study was to evaluate the diagnostic value of unrestrained dorsoventral (DV) projection in naturally standing African grey parrots (*Psittacus erithacus*) compared to standard views obtained under anesthesia (n = 15).

Materials and Methods

DV views were obtained by placing the conscious bird in a polycarbonate box. VD views were obtained using a standard positioning under anesthesia. Subjective assessment and objective measurements of various anatomic regions were performed by three veterinarians.

Results

Major rotational positioning was present in 4/15 conscious birds and in 0/15 anesthetized birds. Unrestrained DV view allowed the assessment of 43% of the air sac surface area visible in the anesthetized VD view. The proventriculus and ventriculus were considered significantly more visible in DV view. The heart, liver, and cardiohepatic waist width appeared significantly larger in the naturally standing DV view compared to the VD view obtained under anesthesia.

Discussion

The cardiac silhouette was equally well delineated using unrestrained DV view or VD view obtained under anesthesia. The former allowed a better assessment of the proventriculus and ventriculus but seemed to be significantly limited when evaluating the respiratory system and the liver when compared to the latter.

Comparative Evaluation of Ocular Reference Values in Growing Broiler Chicks Using Ultrasonography and Optical Coherence Tomography

Rüdiger Korbel, Prof Dr med vet, DipECZM (Avian),
Annika Schneeganss, med vet,
Bianca Reisenbichler, med vet,
Kristina Moser, Dr med vet

Session #2032

Affiliation: From the Clinic for Birds, Small Mammals, Reptiles and Ornamental Fish, University of Munich/Germany, Sonnenstrasse 18, D-85764 Oberschleissheim, Germany.

Inadequate light conditions are known to cause the so-called light-induced glaucoma (LIAG) complex[1,2] in commercial poultry, which is an animal welfare issue, considering the background of the present "20 lux discussion" and the EU Directive 2007/43/EG. We were aiming for a tool to measure pathologic ocular conditions (LIAG) induced by various light intensity conditions based on measuring ocular reference sizes in growing broilers kept under 20 lux (according German legislation) and alternate light programs. To determine the influence of light intensity and duration of the daily light period, a control group (n = 100) fattened using a conventional light program (20 lux) was compared with a test group (n = 100) using a program with light intensities gradually reduced to 5 lux. The eyes of chickens (n = 10) were investigated on 1, 2, 3, 5, 7, 10, 15, 20, 30, and 40 days of age using ultrasonography and optical coherence tomography (OCT) with techniques described before.[3-5] Preliminary results based on partial statistical evaluation of the datasets (about 28,800 single measurements) revealed changes in ocular structures in chickens kept under a discontinuous light program of less than 20 lux. These included no effect on the total corneal thickness measured using OCT ($P = 0.3605$, $P > 0.05$). Using ultrasound, a significant increase in the depth of the anterior eye chamber ($P = 0.0000$, $P < 0.05$), increase of the axial lens ($P = 0.0000$, $P < 0.05$), and axial bulbus length ($P = 0.0019$, $P < 0.05$) was observed (Figs 1 and 2). Intraocular pressure was not affected significantly ($P = 0.0642$, $P > 0{,}05$). A potential impact on animal health and animal welfare should be clarified by further investigations.

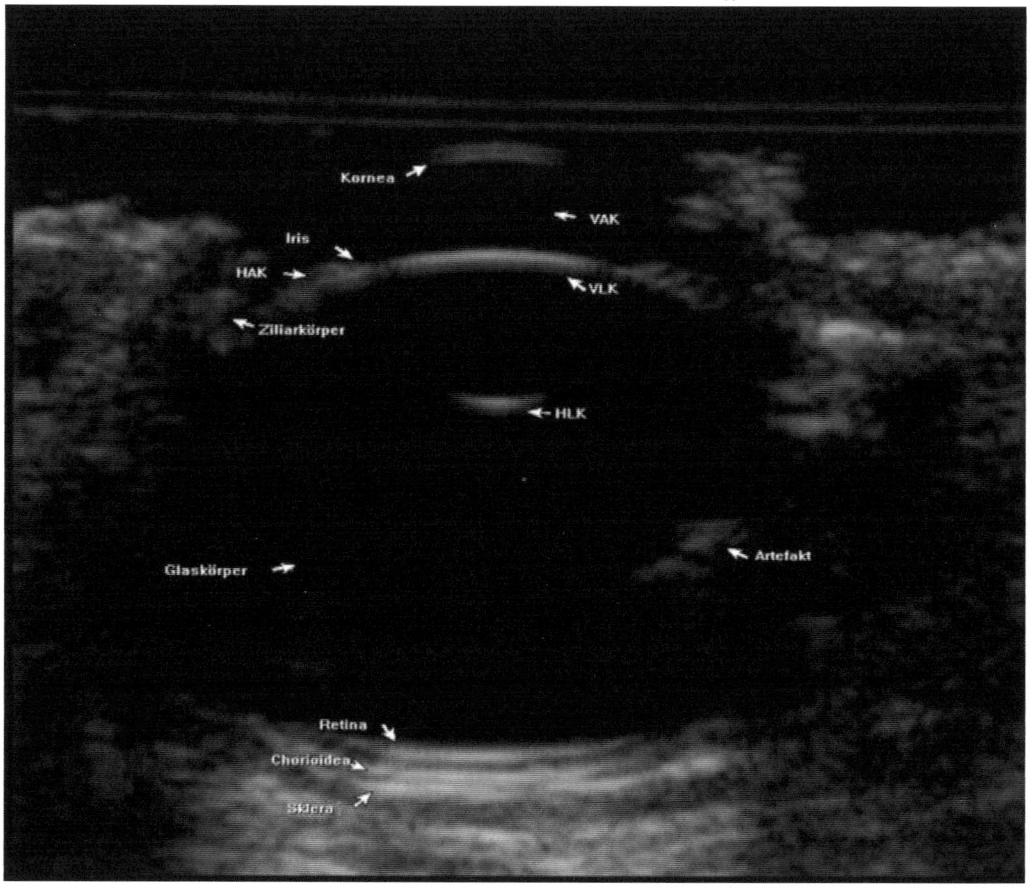

Figure 1. Ultrasonographic image demonstrating physiologic conditions in a broiler chick (day 40, oculus sinister [OS], control group) with the cornea, anterior eye chamber, iris, anterior lens capsule, posterior lens capsule, pecten oculi, vitreous (right side) and retina, choroid and sclera (bottom).

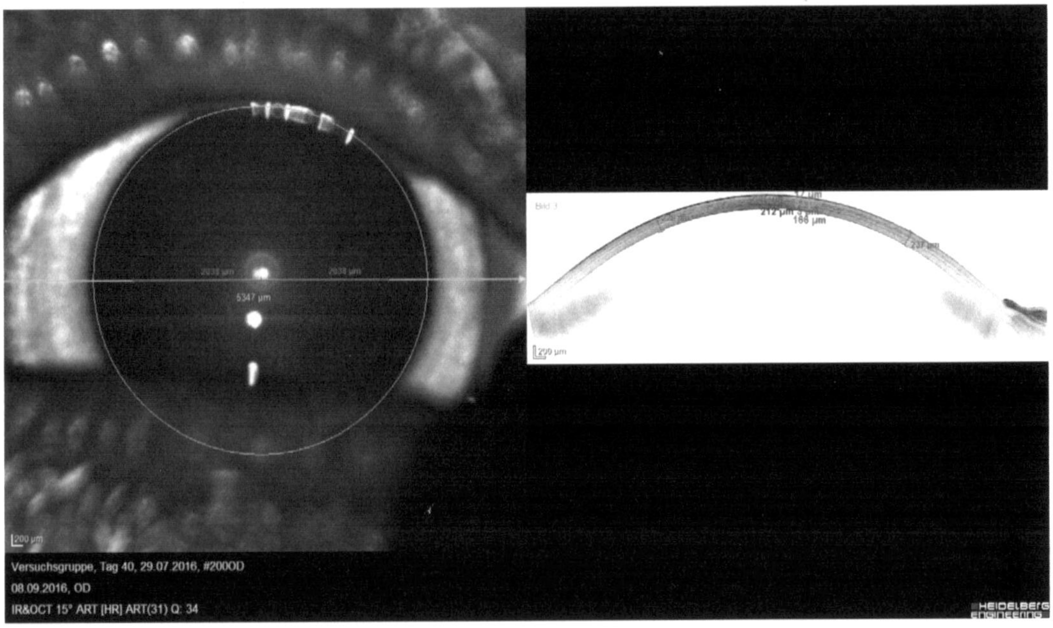

Figure 2. Optical coherence tomography (OCT) image demonstrating physiologic conditions in a broiler chick (day 40, oculus dexter [OD], control group) with frontal view (left) and optical cut through the various corneal layers (right).

References

1. Lauber JK. Light-induced avian glaucoma as an animal model for human primary glaucoma. *J Ocular Pharmacol Therap.* 1987;3:77-100.

2. Lauber JK, Kinnear A. Eye enlargement in birds induced by dim light. *Can J Ophthalmol.* 1979;14: 265-269.

3. Korbel R, Hufen H, Strobel B, Schweikl M. Light-induced influences in the turkey eye. Investigations on reference values in healthy eyes using digital scanning, ophthalmoscopy and ultrasonography. 14. Int Tag Fachgr Tierschutz DVG, Nürtingen. 2009;142-143.

4. Sigg RE. Optische Kohärenztomographie am vorderen Augensegment bei Greifvögeln. Dissertation. LMU München;2016.

5. Velasco-Gallego ML. Imaging of physiological retinal structures in various raptor species using optical coherence tomography (OCT). Dissertation. LMU München;2016.

Effects of Midazolam and Midazolam-Butorphanol on Gastrointestinal Transit Time and Motility in Cockatiels (*Nymphicus hollandicus*)

Anna Martel-Arquette, DVM,
Christoph Mans, Dr med vet, Dipl ACZM,
Grayson Doss, DVM,
Jackie M. Williams, DVM, MS, Dipl ACVR

Session #2200

House Officer Manuscript Competition Winner

Affiliation: From University of Wisconsin-Madison, School of Veterinary Medicine, 2015 Linden Drive, Madison, WI, 53706, Northstar VETS, 315 Robbinsville-Allentown Road, Robinsville, NJ 08691, USA.

Positive contrast gastrointestinal (GI) studies are frequently performed in avian medicine in order to identify GI obstruction, luminal distension, and intracoelomic mass effects; however, repeated manual restraint and radiographic positioning may result in a stress-response and associated morbidity in birds, which can be attenuated by the administration of sedative drugs. In mammals, many sedative drugs have been shown to affect GI transit times and motility. In this randomized, blinded, controlled prospective study, the effects of midazolam (6 mg/kg IM) and midazolam-butorphanol (3 mg/kg each IM) on GI transit times were evaluated in 12 healthy cockatiels. Iohexol (20 ml/kg) was administered by crop gavage 15 minutes after induction of sedation and fluoroscopic images were obtained at different time points. Both sedation protocols significantly affected GI transit times and motility. The midazolam-butorphanol protocol had more pronounced effects. Overall median (range) GI transit times were 60 (30-120), 90 (30-120) and 120 (120-180) minutes for the control, midazolam, and midazolam-butorphanol groups, respectively. Ventricular contractions were profoundly suppressed at 30 and 60 minutes after iohexol administration with both sedation protocols. Visualization of the GI tract following iohexol administration was graded significantly higher in the control group, and poorest in the midazolam-butorphanol group. The results of this study show that commonly used sedative drugs have a significant effect on GI transit time and motility in birds. Therefore, GI transit times obtained in sedated birds should not be compared to available reference transit times obtained from unsedated animals.

Effect of Restraint and Transportation on Hematologic Values and Plasma Corticosterone in Hispaniolan Amazon Parrots (*Amazona ventralis*)

Anna McRee, DVM,
Thomas N. Tully, Jr, DVM, MS, DABVP (Avian Practice), DipECZM (Avian),
Javier G. Nevarez, DVM, PhD, Dipl ACZM, DipECZM (Herpetology),
Hugues Beaufrere, DMV, PhD, Dipl ACZM, DABVP (Avian Practice), DipECZM (Avian),
Melanie Ammersbach, DVM, Stephen D. Gaunt, DVM, MS, PhD, DACVP,
Rory G. Fuller, BS,
Michael Romero, MS, PhD

Session #2013

Affiliation: From the White House Animal Hospital, Nashville, TN 37188, USA (McRee), Louisiana State University School of Veterinary Medicine, Baton Rouge, LA 70803, USA (Tully, Jr, Nevarez, Gaunt), Ontario Veterinary College, University of Guelph, Guelph, Canada N1G 2W1 (Beaufrere, Ammersbach), Department of Biology, Tufts University, Medford, MA 02155, USA (Fuller, Romero).

Increased glucocorticoids cause a characteristic "stress leukogram" in mammalian taxa.[1] It is assumed that birds exhibit a similar response, but only few studies have evaluated effects of stress on serial and simultaneously measured endogenous corticosterone and hematologic values. In a cross-over study with a 4-week wash-out in between, 18 Hispaniolan Amazon parrots (*Amazona ventralis*) were randomly subjected to the stress of transport, restraint, and common clinical procedures with serial blood samples collected at 20-minute intervals (80 total) for hematology and corticosterone levels (treatment condition) and sampling at identical intervals without further procedures (control). Longitudinal data analysis was performed with linear mixed modeling. For all hematologic analytes, the baseline value had a significant positive effect on subsequent values (all $P < 0.001$). The WBC, heterophil counts, eosinophil counts, and H:L ratio increased over time following treatment whereas it remained stable in the control group ($P = 0.016$, $P < 0.001$, $P < 0.001$, $P = 0.02$ respectively for the time*treatment effect). Lymphocyte absolute counts decreased over time, although not significantly; the decrease was significant for the relative lymphocyte count following treatment group. Monocytes and basophils were not significantly altered. Treatment resulted in an approximately 60% higher mean corticosterone level overall than the control group ($P = 0.008$). The mean corticosterone level also increased over time in both groups to a 3-4 fold increase ($P < 0.001$) by 20 minutes and then plateaued. These results demonstrate that some significant hematologic changes may arise following restraint and transportation of birds and should be accounted for in hematologic interpretation of cell counts.

Acknowledgments: The authors would like to thank the Lafeber Company and Dr. Ted Lafeber for supporting the Hispaniolan Amazon parrot research colony at the LSU School of Veterinary Medicine.

References

1. Davis AK, Maney DL, Maerz JC. The use of leukocyte profiles to measure stress in vertebrates: a review for ecologists. *Funct Ecol*. 2008;22:760-772.

Section 6

Research

James W. Carpenter, MS, DVM, Dipl ACZM
Moderator

New Investigations of Diagnostics for Aspergillosis in Birds

Guillaume Desoubeaux, PharmD, PhD, and
Carolyn Cray, PhD

Session #2063

Affiliation: From the Division of Comparative Pathology, Department of Pathology & Laboratory Medicine, University of Miami Miller School of Medicine, Miami, FL 33136, USA (Desoubeaux and Cray) and CHU de Tours, Service de Parasitologie-Mycologie-Mèdecine tropicale, Tours 37044 France, Universitè Francois-Rabelais, CEPR-INSERM U1100/ Équipe 3, Facultè de Mèdecine, Tours 37032 France (Desoubeaux).

Aspergillosis is a fungal airborne disease due to the ubiquitous *Aspergillus* sp. As in immune suppressed humans, it is an important health issue in birds and antemortem diagnostic testing options remain limited. Recent studies included the following: examination of a commercial Western blot assay for the detection of antibody,[1] mass spectrometry (MS) analysis for proteomic changes,[2] and assessment of 3-hydroxybutyrate levels as a possible adjunct test.[3] Experiments were conducted using four groups of samples obtained from African penguins (*Spheniscus demersus*) that were clinically normal or those with presumptive or confirmed aspergillosis, in convalescence post aspergillosis, or diagnosed with non-aspergillosis diseases or infections. Although the Western blot was validated in other non-human species, it could not be validated for use in penguins. Additional work is needed in this area. MS data were obtained using an iTRAQ® protocol. Results showed that 173 proteins were significantly overrepresented in *Aspergillus*-diseased penguins versus non-*Aspergillus* inflammatory controls, including 17 more than 4.0-fold increased. Most were involved in metabolic processes and cell regulation. Conversely, 77 proteins were found statistically decreased in *Aspergillus*-diseased penguins, and 39 were lowered more than 4.0-fold in comparison with non-*Aspergillus* inflammatory controls. All protein changes require further analysis for possible use in diagnostic testing. Importantly, a significant increase in 3-hydroxybutyrate levels was observed during aspergillosis versus control inflammation (1.3 versus 0.7 mmol/L, $P < 0.0001$). Overall, these results provide new insights on pathophysiologic processes during aspergillosis and can be considered as preliminary pathways towards discovery of new biomarkers of *Aspergillus* infection.

Acknowledgments: The authors thank the Association of Zoos and Aquariums African Penguin Species Survival Plan program and all the institutions that participated in this study.

References

1. Oliva A, Flori P, Hennequin C, et al. Evaluation of the *Aspergillus* Western blot IgG kit for the diagnosis of chronic aspergillosis. *J Clin Microbiol*. 2015;53(1):245-254.

2. Kumar Y, Liang C, Bo Z, et al. Serum proteome and cytokine analysis in a longitudinal cohort of adults with primary dengue infection reveals predictive markers of DHF. *PLoS Negl Trop Dis*. 2012;6(11):e1887.

3. Pappalardo L, Hoijemberg A, Pelczer P, Bailey T. NMR-metabolomics study on falcons affected by aspergillosis. *Curr Metabol*. 2014;2(2):155-161.

Pharmacokinetics of Compounded Transdermal Meloxicam in Red-tailed Hawks (*Buteo jamaicensis*)

Kaleigh M. Bush,
Marike Visser, DVM,
Dawn M. Boothe, DVM, DACVCP, DACVIM

Session #2052

Affiliation: From the Department of Anatomy, Physiology, and Pharmacology, Auburn University College of Veterinary Medicine, 1500 Wire Road, Auburn University, AL 36849, USA.

Meloxicam is an acceptable and safe non-steroidal anti-inflammatory (NSAID) commonly used as an analgesic in avian species. Multiple compounding pharmacies offer meloxicam as a compounded transdermal product, aiming to ease the administration of this drug. While this transdermal product is marketed as an alternative method of drug administration, no transdermal pharmacokinetics or efficacy data has been reported in any avian species. The purpose of this pilot study was to determine a dose for a complete transdermal meloxicam pharmacokinetic study in red-tailed hawks (*Buteo jamaicensis*, RTHA). Compounded transdermal meloxicam in pluronic lecithin organogel (PLO) was administered on the featherless tract along the right jugular vein in RTHA at 1.5 mg/kg (n = 2), 3 mg/kg (n = 2), and 5 mg/kg (n = 1). The application site was monitored for any clinical signs of contact dermatitis for 2 weeks post-application. Venipuncture (approximately 1 ml at each time point) was performed at 0 (prior to drug administration), 15, 30, 45, 60, 90, 120, 180, 240, 360, 480, 720, and 840 minutes, alternating time points between the 2 subjects in each study. No detectable plasma levels were present at 1.5 or 3 mg/kg at any time point, and inconsistent plasma concentrations were present at 5 mg/kg (78.05 µg/ml at 45 min and 47.74 µg/ml at 480 min). Furthermore, the levels detected at a dose of 5 mg/kg were far below the previously reported IV and PO meloxicam pharmacokinetics in RTHA.[1] No signs of contact dermatitis were noted during the 2-week monitoring period, but the down feathers around the tract were matted and appeared greasy. As such, transdermal meloxicam does not appear to be an effective method of achieving therapeutic meloxicam plasma drug concentrations due to poor or unpredictable absorption across the skin.

References

1. Lacasse C, Gamble KC, Boothe DM. Pharmacokinetics of a single dose of intravenous and oral meloxicam in red-tailed hawks (*Buteo jamaicensis*) and great horned owls (*Bubo virginianus*). *J Avian Med Surg*. 2013;27(3):204-210.

Establishing and Validating Stress Behaviors in Cockatiels (*Nymphicus hollandicus*)

Katherine K. Turpen, DVM,
Kenneth R. Welle, DVM, DABVP (Avian Practice),
Matthew C. Allender, DVM, Dipl ACZM,
Jennifer Trail, CVT,
Seema Patel, BS

Session #2012

Affiliation: From the University of Illinois Veterinary Teaching Hospital, Veterinary Clinical Medicine Department, Urbana, IL 61802, USA.

This study evaluated the behavior of a common pet psittacine species, cockatiels (*Nymphicus hollandicus*), during their normal daily routine and after an acute stressful event. The goal was to develop a way to noninvasively evaluate stress in this species. Non-stressed behavior budgets were established by performing quantitative ethograms using 10-minute focal animal sampling methods with point samples recorded every 5 seconds. The ethograms were then repeated after a minimum of 10 minutes of handling and restraint for physical examination and venipuncture. The stress behaviors were validated by using comparison of plasma corticosterone levels, with non-stressed samples obtained within 3 minutes of the onset of capture and restraint and stressed samples obtained after a minimum of 10 minutes of handling and restraint to allow corticosterone to rise. Plasma corticosterone was observed to increase after handling and restraint, indicating the presence of a stress response. Submissive body posturing and inactivity overall increased while locomotion, feeding, interaction with the environment, maintenance behaviors, and displays of aggression overall decreased in the stressed birds. Further studies will look at developing a behavioral stress scoring system for this species using these results.

Comparative Evaluation of Ophthalmoscopic Techniques for Birds Using Different Imaging Techniques

Rüdiger Korbel, Prof Dr med vet, DipECZM (Avian)

Session #2033

Affiliation: From the Clinic for Birds, Small Mammals, Reptiles and Ornamental Fish, Center for Clinical Veterinary Medicine, University of Munich, Germany.

Ophthalmoscopy and fundus photography are essential tools in human ophthalmology. Combined, they allow for diagnosis and photo documentation of intraocular pathologies and for sharing of collected and stored information with colleagues and patients. Binocular indirect ophthalmoscopy can be performed in birds using a headband ophthalmoscope; however, headband ophthalmoscopes must be used in combination with indirect ophthalmoscopy lenses with a refractive power of 30 diopters (raptors), 78 diopters (lager psittacines) or 90 diopters (eg, smaller psittacines, finches). Monocular direct ophthalmoscopy can be performed in birds with a pupillary diameter of greater than 7 mm using a direct ophthalmoscope such as the PanOptic Ophthalmoscope (Welch Allyn, Skaneateles Falls, NY, USA). Acquisition of high-quality fundus images requires a combination of appropriate optics and illumination, usually with a condensing lens, and the use of coaxial light sources; however, traditional ophthalmoscopes and fundus cameras are expensive and have other drawbacks including poor compatibility with other systems and lack of portability. This is in contrast to recently introduced smartphone attachments (eg, iExaminer, Welch Allyn, Skaneateles Falls, NY, USA), which are relatively inexpensive, portable, easy to operate, and compatible across various platforms and systems. We tested various traditional and smartphone-based ophthalmoscopic systems in various raptor, pigeon, and psittacine species (Figs 1 and 2). Our results showed that only a limited number of ophthalmoscopic techniques are suitable in avian ophthalmology for photo documentation, disease screening, and telemedicine. Comparing traditional and smartphone-based systems lead to the conclusion that the advantages of improved convenience and comparatively lower cost of smartphone-based systems do not outweigh the disadvantages of lower image quality and limited use in birds with very small pupillary diameters.

References

1. Reese S, Korbel R, Liebich HG. Eye. In: Koenig HE, Korbel R, Liebich HG, eds. *Avian Anatomy: Textbook and Colour Atlas.* 2nd ed. Translated and revised by C. Klupiec. Sheffield, UK: 5m Publishing Ltd;2016:159-187.

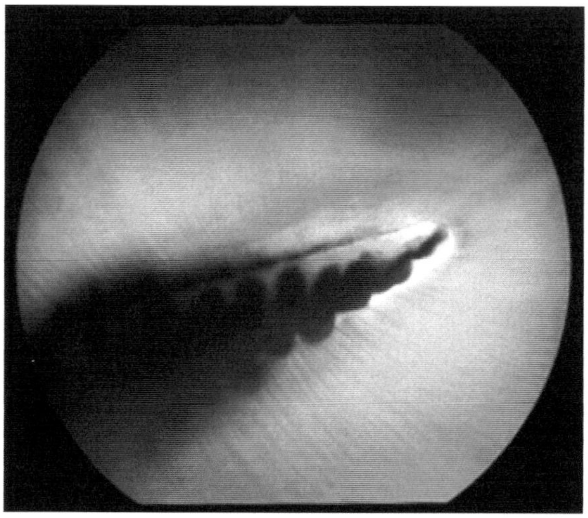

Figure 1. Fundus oculi (common buzzard [Buteo buteo], Oculus sinister [OS]) as seen with the Scanning Digital Ophthalmoscope (SDO®, Leitz Inc., Switzerland), demonstrating the diurnal bird´s fundus with typically anangiotic appearance and absent tapetum, the pectin oculi (pleated type), optic nerve head (white) and central fovea (1 o'clock position). This technique adapted to exotic animal ophthalmology allows taking still and video images through pupillary diameters as small as 4 mm.

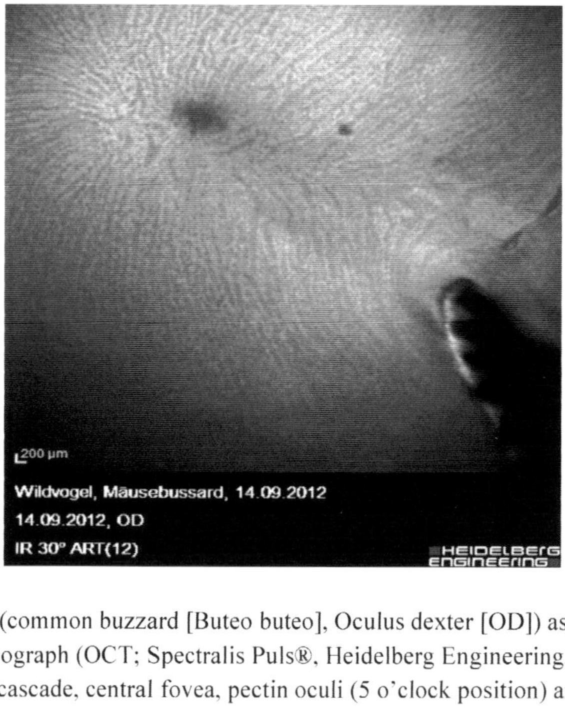

Figure 2. Fundus oculi (common buzzard [Buteo buteo], Oculus dexter [OD]) as seen with the Optical Coherence Tomograph (OCT; Spectralis Puls®, Heidelberg Engineering, Germany) with the upper and lower vessel cascade, central fovea, pectin oculi (5 o'clock position) and retinal scar formation (central and 3 o´clock position).

A Comparison of the Pharmacokinetics of Intravenous Fentanyl in Isoflurane-anesthetized Red-tailed Hawks (*Buteo jamaicensis*) and Hispaniolan Amazon Parrots (*Amazona ventralis*)

Michelle G. Hawkins, VMD, DABVP (Avian Practice),
Peter J. Pascoe, BVSc, DACVA,
Bruno H. Pypendop, PhD,
Juan C. Pavez, DVM, DACVA,
Heather K. Knych, DVM, PhD,
David Sanchez-Migallon Guzman, LV, MS

Session #2106

Affilitation: From the Departments of Surgical and Radiological Sciences (Pascoe, Pypendop) and Medicine and Epidemiology (Hawkins, Sanchez-Migallon Guzman), School of Veterinary Medicine, and the KL Maddy Equine Analytical Chemistry Laboratory (Knych), California Animal Health and Food Safety Laboratory, University of California, Davis, CA 95616, USA, Catedrático Auxiliar Programa de Tecnología Veterinaria Departamento de Biología, Universidad de Puerto Rico en Arecibo, Puerto Rico (Pavez).

The goals of this study were to compare the disposition of fentanyl citrate, following a single intravenous (IV) injection in isoflurane-anesthetized red-tailed hawks (*Buteo jamaicensis*) (RTH) and Hispaniolan Amazon parrots (*Amazona ventralis*) (HAP). Six adult RTH and six adult HAP were utilized for this study. Anesthesia was induced and maintained with isoflurane and intermittent positive pressure ventilation. The MAC_{iso} was determined for each bird using the bracketing method and a supramaximal electrical stimulus. Fentanyl (20 µg/kg) was administered IV and arterial (RTH) or jugular venous (HAP) blood samples were obtained immediately prior to fentanyl administration, and at 1, 2, 4, 8, 15, 30, 60, 120, 180, 240 and 480 min (RTH) and at 1, 5, 10, 15, 30, 60, 120 and 180 minutes (HAP) post-administration. A three-compartment model best described the disposition of fentanyl in the RTH whereas the data for the HAP fit a two-compartment model. The V_c and Vd_{ss} were 236 (166-329) ml kg^{-1} and 948 (702-1677) ml kg^{-1} for the RTH and 5108 (1963-6353) and 13079 (6907-15380) for the HAP. Clearance and the terminal half-life were 10.2 (8.9-14.2) ml/min/kg and 76 (58-120) minutes for the RTH and 199 (107-276) ml/min/kg and 51 (45-65) minutes for the HAP, respectively. These pharmacokinetic parameters provide information necessary for determination of suitable fentanyl loading and infusion doses in isoflurane-anesthetized RTH and HAP, and suggests that the significant differences in pharmacokinetics strongly discourages extrapolation between these two species.

Acknowledgments: Funding was provided by the Center for Companion Animal Health (CCAH) at the School of Veterinary Medicine, UC Davis. The authors wish to thank the California Raptor Center for facilitating the research subjects and Drs. Chalon Tiedeken and Tracy Drazenovich for technical assistance.

Pharmacokinetics of A Single Dose of Oral and Intramuscular Meloxicam in African Penguins (*Spheniscus demersus*)

Jennilee Morrison, DVM,
Cheryl Greenacre, DVM, DABVP (Avian Practice), DABVP (Exotic Companion Mammal Practice),
Robert George, DVM,
Sherry Cox, PhD

Session #2047

Affiliation: From the Departments of Small Animal Clinical Sciences (Morrison, Greenacre, George) and Biomedical Sciences and Diagnostic Sciences (Cox), College of Veterinary Medicine, University of Tennessee, Knoxville, TN 37996, USA.

Meloxicam is a non-steroidal anti-inflammatory drug (NSAID) that has been used both orally and intramuscularly in numerous avian species, but not studied, to date, in African penguins (*Spheniscus demersus*). This study describes the pharmacokinetic parameters of meloxicam after oral and intramuscular administration to African penguins (*Spheniscus demersus*). Several pilot studies were conducted initially where meloxicam (1 mg/kg, 0.5 mg/kg, 0.25 mg/kg) was given intramuscularly to 4 birds, and orally (1 mg/kg) to 2 birds. Based on pilot study results, 1 group of 8 penguins was given meloxicam 1 mg/kg PO and another group of 8 penguins was given 0.5 mg/kg IM. Plasma was harvested, stored at -80 °C until analysis, and analyzed for meloxicam concentrations using reversed phase high performance liquid chromatography method. Meloxicam $AUC_{0-\infty}$, T_{max}, C_{max}, and t½ after intramuscular administration were 238.79 h·µg/ml, 0.93 h, 9.11 ug/ml, and 32 h while oral administration produced a $AUC_{0-\infty}$, T_{max}, C_{max}, and t½ of 520.69 h·µg/ml, 13.3 h, 11.81 µg/ml, and 34 h respectively. Based on plasma meloxicam concentrations found to be therapeutic in bird species (3.45 µg/ml) and humans (0.5 µg/ml to 1.5 µg/ml), the recommended dosage and frequency for African penguins is 1 mg/kg PO q48h and 0.5 mg/kg IM q24h.

A Pilot Study: Comparison of In-box Radiographs, Standard Radiographs and Computed Tomodensitometry in Avian Patients Presenting in Respiratory Distress

Minh Huynh, DV, MRCVS, DipECZM (Avian),
Florent Modesto, DV, IPSAV (Zoological Medicine),
Graham Zoller, DV, IPSAV (Zoological Medicine),
Lucile Chassang, DV, IPSAV (Zoological Medicine)

Session #2078

Affiliation: From the Exotic Pet Department, Centre Hospitalier Vétérinaire Frégis (Huynh, Zoller, Chassang), 94110, Arcueil, France and Clinique Vétérinaire Brasseur (Modesto), Rue Deschamps 6, 7170 Manage, Belgium.

Introduction

Respiratory distress is a common presentation in avian medicine and handling for properly positioned radiographic images may increase risk of death. This pilot study compared the diagnostic quality of conscious dorsoventral "in-box" radiographs images to standard anesthetized radiographs and computed tomodensitometry, to evaluate their potential value in identifying respiratory lesions in unstable patients.

Materials and Methods

Two radiographic evaluations were compared with tomodensitometry in 18 medium-sized parrots. Eleven presented for respiratory distress and 7 for non respiratory presentations. A dorsoventral view was obtained by placing each bird conscious in a closed plastic translucid polycarbonate box (so called "in-box" radiograph) under the radiographic beam. Standard ventrodorsal, lateral views, and computed tomodensitometry (CT-scan) were then obtained under general anesthesia. Respiratory lesions were identified and recorded by each imaging modality. Sensitivity, specificity, positive predictive value and negative predictive value were calculated. Modified Fleiss kappa statistics were used to measure interobserver reliability across all observers.

Results

Pulmonary lesions were most common (14), followed by thoracic air sac (10), and abdominal air sac (5). Sensitivity of standard view and in-box view was higher (90% and 95.9%) for coelomic lesion and caudal thoracic air sac (80% and 75.7%) than abdominal air sac (42.5% and 50%) and lungs (40% and 10%). Interobserver reliability was not statistically different when in-box and ventrodorsal views are compared.

Discussion

This pilot study suggests that in-box conscious radiographs may have value when caudal thoracic airsac lesions are present but are likely to miss the majority of other respiratory lesions.

Stability of Compounded Transdermal Meloxicam Gel

Kaleigh M. Bush,
Marike Visser, DVM,
Dawn M. Boothe, DVM, DACVCP, DACVIM

Session #2053

Affiliation: From Department of Anatomy, Physiology, and Pharmacology; Auburn University College of Veterinary Medicine, 1500 Wire Road, Auburn University, AL 36849.

Transdermal application of medication via compounded pluronic lecithin organogels (PLO) has been marketed as an effective method of systemic drug delivery despite lack of scientific evidence of quality, efficacy, and safety of these unapproved products. Very few compounded transdermal products have reported therapeutic plasma drug concentrations using this delivery method. Transdermal meloxicam PLO compounded by the institutional pharmacy at 3 mg/ml, 12 mg/ml and 20 mg/ml was tested for accuracy. Three aliquots at each concentration were dissolved in methanol and meloxicam concentration measured via high-pressure liquid chromatography. Preliminary results indicate that there was significant deviation ($P < 0.05$) from the reported concentration at all three concentrations. At 3 mg/ml the average dose was 2 ± 0.12 mg/m, at 12 mg/ml the average dose was 7.27 ± 0.69 mg/ml and at 20 mg/ml the average dose was 15.40 ± 4.12 mg/ml. The variation in accuracy between the compounded concentrations highlights the challenges in the use a non-approved administration formulation. While novel administration techniques may offer perceived ease-of-administration benefits, the clinician must make every effort to determine whether the described concentration is actually achieved by submitting appropriate plasma or serum to a clinical pharmacology laboratory before and after therapy has begun.

Deslorelin 4.7-mg Implants Reduce Reproduction in Quaker Parrots (*Myiopsitta monachus*)

Andrea Winkel-Blair, DVM, MPH,
Justine Berman,
Jeann Leal de Araujo,
J. Jill Heatley, DVM, MS, DABVP (Avian Practice), DABVP (Reptile and Amphibian Practice), Dipl ACZM

Session #2059

Affiliation: From Schubot Exotic Bird Health Center, Veterinary Pathobiology, College of Veterinary Medicine and Biomedical Sciences, Texas A&M University, College Station, TX 77843, USA.

Reproductive system disease is a common cause of morbidity and mortality of captive parrots.1,3 Suprelorin is a depot-implant form of deslorelin acetate (DA), a GnRH-agonist effective for reduction of avian reproduction.2,4 This study of Quaker parrots (*Myiopsitta monachus)* with known reproductive history allowed for preliminary assessment of the safety and efficacy of DA in a psittacine model. Birds were kept in pairs at the Avian Health Complex and maintained in 6' × 6' × 4' hanging wire cages at 74-78 °F with a 12-hour light cycle. Sixteen pairs of adult Quaker parrots that had produced viable offspring were assigned to treatment (8 pairs) or sham (8 pairs) groups. After sedation (butorphanol and midazolam intranasal), birds were implanted with a 4.7-mg DA (Suprelorin® F, Virbac Animal Health, City, State, USA, IM pectoral) or had intramuscular 14-gauge needle placement and removal into the pectoral muscles. Observations of nesting behavior, egg fertility and offspring production were recorded for 7 months. Nest boxes were removed at 7 months and replaced at 1 year post implantation, to record behavior and reproduction for 90 days. Intramuscular implantation with DA appeared safe and effectively limited reproduction for 7 months. There were 57 eggs produced by sham pairs and 1 egg produced by implanted pairs ($P < 0.05$). Reproductive quiescence in DA implanted birds was short term based on successful reproduction within 1 year post-implantation in 7/8 (87.5%) implanted pairs. Based on egg necropsy, fertility of eggs was similar in implanted (71%) and sham groups (73%) 1 year post implant ($P = 0.3787$).

References

1. Ottinger MA, Baket M. Endocrinology of the avian reproductive system. *J Avian Med Surg.* 1995;9(4): 242-250.

2. Petritz OA, Guzman DS, Paul-Murphy J, et al. Evaluation of the efficacy and safety of single administration of 4.7-mg deslorelin acetate implants on egg production and plasma sex hormones in Japanese quail (*Coturnix coturnix japonica*). *Am J Vet Res.* 2013;74(2):316-323.

3. Rosen LB. Avian reproductive disorders. *J Exot Pet Med.* 2012;21(2):124-131.

4. Mans C, Pilny A. Use of GnRH-agonists for medical management of reproductive disorders in birds. *Vet Clin North Am Exot Anim Pract.* 2014;17:23-33.

Section 7

Clinical Session

J. Bruce Nixon, DVM
Moderator

Foreign Body Perforation of the Proventriculus of an Umbrella Cockatoo

Sharman Hoppes, DVM, DABVP (Avian Practice)

Session #2087

Affiliation: From the Department of Small Animal Clinical Sciences, College of Veterinary Medicine, Texas A&M University, 4474 TAMU, College Station, TX 77845, USA.

A linear perforating ventricular foreign body was diagnosed in a 19-year-old male umbrella cockatoo (*Cacatua alba*). The bird had a 7-day history of hyporexia, but otherwise seemed fine to the owner. The referring veterinarian's whole body radiographs and abdominal ultrasound revealed a dilated proventriculus, decreased serosal detail (possible effusion) and a small liver. Complete blood count and biochemistry profile revealed elevated AST and CPK activities and a PCV of 33%. The bird initially improved with supportive care (fluids, nutritional support and antibiotics) at the referring veterinarian's hospital but declined rapidly after discharge. Two days post discharge he became completely anorexic and presented to Texas A&M University's College of Veterinary Medicine at 480 grams with body condition of 1/5 with mild respiratory signs (dried exudate in nares and sneezing). A CAT scan revealed 2 linear foreign bodies (42 and 44 mm in length) in the ventriculus, with one perforating into the coelomic cavity. Exploratory surgery revealed a penetrating wound at the isthmus and a puncture wound in the caudal left lung. There was free fluid in the coelomic cavity. A ventriculotomy was performed and 3 small cedar sticks were removed. Immediate postoperative care included: Normasol R 4 ml slow IO delivery q6h (Hospira, Inc, Lake Forest, IL 60045, USA), enrofloxacin 15 mg/kg IO q12h (Baytril Bayer healthcare LLC, Shawnee Mission, KS, USA) metronidazole 15 mg/kg IO q24h (TEVA Pharmaceuticals Inc, North Wales, PA, USA), ceftazidime 70 mg/kg IO q8h (Pfizer Inc, Austin, TX, USA), butorphanol 1 mg/kg IO q4h (Torbutrol, Intervet Inc, Millsboro, DE, USA), and meloxicam 1 mg/kg PO q24h (Boehringer Ingelheim Vetmedica, Inc, St. Joseph, MO, USA). The bird was tube fed or hand fed 10 to 15 ml of Emeraid IC Omnivore diet (Emeraid,LLC, Cornell, IL, USA) q6-8h. The cockatoo was sent home on enrofloxacin 15 mg/kg PO q12h. Trimethroprim sulfamethoxazole 30 mg/kg PO q12h (Burroughs Wellcome Grenville, NC, USA), itraconazole 10 mg/kg PO q12h (Janssen Pharmaceutica Inc, Titusville, NJ, USA), tramadol 10 mg/kg PO q8h (Janssen Pharmaceuticals, Inc, South Raritan, NJ, USA) and meloxicam 1 mg/kg PO q24h. A 1-month postoperative CAT scan was performed to recheck the coelom and left lung lesion. A 25-mm linear foreign body (presumed to be another cedar fragment) was identified within the ventriculus. The coelomitis and pulmonary lesion were still present but improved. He had developed more severe respiratory issues (raspy breathing when stressed and mild wheezing). A recheck CBC revealed a 39,000 cells/µl WBC so the antibiotics were changed to amoxicillin/clavulanate 125 mg/kg PO q8h, (Clavamox. Zoetis Inc, Kalamazoo, MI, USA), and ciprofloxacin 15 mg/kg PO q12h (Bayer Healthcare LLC, Shawnee Mission, KS, USA) and the antifungal was changed to voriconazole 10 mg/kg PO q12h (Pfizer Inc). Enzymes to help dissolve the remaining foreign body were initiated. His weight 1 month post surgery was 555 grams, his appetite had improved, but the owner was still handfeeding daily. The bird is now 3 years post surgery, weighs 610 grams and has had no further gastrointestinal problems.

References

1. Metz S, Speer BL. Diagnosis and surgical management of a ventricular foreign body in an umbrella cockatoo (*Cacatua alba*). *Proc Annu Conf Assoc Avian Vet.* 2010;37-40.

2. Oglesbee B, Steinghart L. Gastrointestinal string foreign bodies in a juvenile umbrella cockatoo. *Compend Contin Educ Pract Vet.* 2001;23(11):946-950.

3. Honnas CM, Blue-McLendon A, Zamos DT, et al. Proventriculotomy in ostriches: 18 cases (1990-1992). *J Am Vet Med Assoc.* 1993;202(12):1989-1992.

4. Adamcak A, Hess LR, Quesenberry KE. Intestinal string foreign body in an adult umbrella cockatoo (*Cacatua alba*). *J Avian Med Surg.* 2000;14(4):257-263.

5. Speer BL. Chronic partial proventricular obstruction caused by multiple gastrointestinal foreign bodies in a juvenile umbrella cockatoo (*Cacatua alba*). *J Avian Med Surg.* 1998;12(4): 271-275.

Extensive Eye and Face Trauma Treated Successfully with Hyperbaric Oxygen Therapy Post-Enucleation

April Romagnano, PhD, DVM, DABVP (Avian Practice),
Carolyn Cray, PhD,
Lesley Phillips, DVM

Session #2102

Affiliation: From Avian & Exotic Clinic of Palm City, Palm City, FL 34990, USA (Romagnano), Division of Comparative Pathology, Department of Pathology and Laboratory Medicine, University of Miami, Miami, FL 33101, USA (Cray), Animal Emergency & Referral Center, Fort Pierce, FL 34982 USA (Phillips).

An 11-year-old Quaker parrot was attacked by a Congo African grey parrot and incurred massive blood loss. The bird presented moribund with a proptosed ruptured left eye, periocular and facial skin lacerations, and a fractured skull. The bird was stabilized in heat-oxygen and treated. Blood and cloacal samples were collected 48 hours prior to surgery. Injuries precluded mastication, which necessitated parenteral treatments and gavage-feedings. Enucleation and laceration repairs were performed under isoflurane. Despite hemoclipped orbital vessels (Hemoclips, Rica Surgical Products, Inc), excessive bleeding occurred. Gelfoam (Gelfoam sponges, Pfizer, Inc) was placed in the orbital socket, which was covered with Duoderm (DuoDERM dressing, ConvaTec, Inc) and sutured in place to achieve hemostasis. Due to massive trauma and blood loss, hyperbaric oxygen treatments (HBOT) were performed 4 days post surgery (rate set: 1.5 PSI, pressure set: 1.5 ATA, time: 40 min q12h, monitoring: q10min). During HBOT the bird did well standing in the chamber unsedated in a 10-gallon fish-tank. The bird became active and resumed eating almost normally after the first treatment. Tube feedings were discontinued and HBOT continued for 8 more treatments. To the authors' knowledge, HBOT hasn't been previously described in the literature for therapeutic use in avian species. HBOT delivers oxygen to the alveoli under increased atmospheric pressure resulting in large increases in dissolved oxygen (O_2-saturation > 110% of normal for 3 hours post treatment) increasing cerebral oxygen content 10-fold.[1,2] This hastens repair of central nervous system damage, increases wound healing, and decreases pain. It also potentiates fluoroquinolones often used in birds. HBOT hastened this bird's recovery. Activity, wound healing, and skin re-growth increased. Depression decreased and pain abated, which enabled normal chewing, drinking, playing, and vocalizing.

References

1. Edwards ML. Hyperbaric oxygen therapy, history, principles, applications in disease. *J Vet Emerg Crit Care.* 2010;20(3):289-298.

2. Lyman R. HBOT...what hyperbaric oxygen can do for your patients. Available at: www.animalemergency.net Accessed May 30, 2017.

Four Cases of Lymphoma in Unrelated Lineolated Parakeets

Bianca Murphy, DVM,
Brenna Fitzgerald, DVM, DABVP (Avian Practice),
Brian Speer, DVM, DABVP (Avian Practice), DipECZM (Avian),
Geoff Olsen, DVM, DABVP (Avian Practice)

Session #2065

Affiliation: From The Medical Center For Birds, 3805 Main Street Oakley, CA 94561, USA.

Since 2008, 87 lineolated parakeets (*Bolborhynchus lineola*) were presented for examination at Medical Center For Birds. Over that time, 16 of these birds died and were evaluated postmortem. In 4 of these birds, lymphoma was identified as the cause of death; cloacal involvement was a feature in 3 of the 4 cases. Presenting concerns in each case were tenesmus, hematochezia, and subtle changes in the appearance of the vent. Differential diagnoses at the time of presentation included infectious cloacitis, cloacallithiasis, injury from excessive masturbatory behavior, and neoplasia. In only one case was lymphoma identified premortem. Physical examination findings included slight protrusion of the vent, along with hyperemia of the proctodeal mucosa with palpable thickening of the cloacal wall. Presumptive diagnosis was made by cytology of mucosal cells collected by insertion of a sterile microtipped culturette into the cloaca. Based on these findings, prednisolone therapy was initiated (1 mg/kg PO q24h; PrednisoLONE syrup, Qualitest Pharmaceuticals, Huntsville, AL, USA). Clinical remission was achieved for a period of 5 months. That patient died following recurrence of hematochezia in spite of changes in the treatment protocol. Histopathology identified neoplastic lymphocytes effacing all structures of the vent, cloaca, and tail; these cells were also present in the spleen, bone marrow, and lymphatics. The authors conclude that lymphoma should be considered as a differential diagnosis for cloacal disease in this species. Early identification along with timely initiation of therapy has the potential to improve patient longevity and quality of life.

References

1. Speer B. *Current Therapy in Avian Medicine and Surgery.* St. Louis, MO: Elsevier Saunders;2016.

2. Schmidt RE, Reavill DR, Phalen DN. *Pathology of Pet and Aviary Birds.* 2nd ed. Ames, IA: John Wiley & Sons, Inc;2015.

3. Carpenter JW. *Exotic Animal Formulary.* 4th ed. St. Louis, MO: Elsevier Saunders;2013.

Design and Application of a Prosthetic Foot in a Rooster (*Gallus gallus*)

Len Donato, VMD, DABVP (Avian Practice), and
Douglas Kondro, BSc

Session #2024

Affiliation: From Radnor Veterinary Hospital 107 North Aberdeen Avenue Wayne, PA 19087, USA (Donato) and the Department of Biomedical Engineering at the University of Calgary 2500 University Drive Calgary, AB T2N 1N4, Canada (Kondro).

Abstract: A 3-year-old barred Plymouth Rock rooster (*Gallus gallus*) was presented with severe pododermatitis and osteomyelitis of the left foot. The digits and distal tibiotarsus were nonrecoverable and amputation was elected by the owner. A custom plastic 3D printed prosthetic was produced using open sourced and freely available software. In the postoperative period, between the initial mold formation for the 3D printer and the production of the prosthetic, post surgical changes to the amputation site occurred from trauma. The tissue had increased in size by a few millimeters. Initial application of the prosthetic was snug and used by the rooster well for 3 weeks. The owner was unable to place the prosthetic successfully after that time. This report demonstrates the use of a prosthetic foot in a gallinaceous bird and potential complications for practitioners to monitor.

Clinical Report

A 3-year-old barred Plymouth Rock rooster was referred for a several-month-history of severe pododermatitis of the left foot. Radiographs were taken of the foot revealing severe lysis of P1s and distal tarsometatarsus, and multiple fractures of the digits. Amputation, or euthanasia, were discussed with the owner who elected amputation and consideration for development of a prosthetic limb. Engineers were identified willing to develop the prosthetic while the amputation was performed. Complete healing of the amputation site had to occur before a mold could be made for the 3D printer.

A mid-tarsometatarsal amputation was elected to leave as much healthy tissue for the prosthetic, while minimizing the risk of leaving infected bone. There was a small amount of muscle to cover the bone and normal skin closure was achieved. Healing was uneventful and the sutures were removed by the referring veterinarian. The owner was instructed to keep a bandage on the end of the stump to prevent trauma. Unfortunately, the patient was unable to return for two months and the owner did not use the bandage regularly. During that time, the rooster used the amputation site regularly for balance and walking resulting in enlargement of the stump compared to immediate post-operative examination. Despite the enlargement the stump appeared to have healed adequately and a mold was taken.

In order to generate a custom prosthetic for the rooster, a mold was taken of the remaining distal tarsometatarsus to capture the unique geometry of the amputated limb. The mold was taken in a soft modeling clay to minimize discomfort for the animal which was post cured using heat. A negative of the mold was produced by pressing plasticine into the mold which was subsequently 3D scanned using Autodesk 123D Catch. This generated a 3D model in a StereoLithography (STL) format.

Design of the prosthetic was completed on the free software: Autodesk Meshmixer, which allows for modifications of STL's. The socket for the prosthetic was generated by expanding the residual limb model using an offset distance of 5 mm and subtracting the model of the residual limb, which had also been expanded by an offset of 2 mm. The resulting object was a socket with 3-mm thick walls and a 2-mm offset from the original model. Cut planes where used to shape and introduce an entrance to the socket.

Instead of using an anatomically accurate model for the prosthetic foot, a blade was selected which is better able to accommodate the different angles seen in the strike and push-off stance of the roosters gait. Blades were added to the socket and were produced using a semicircle of fifty millimeters in diameter that allowed for a 10-mm compression before becoming in contact with the opposite end of the socket. The attachment point of the blade was reinforced with a rib.

Once the model was completed the prosthetic was printed on a Lulzbot Taz 6 printer in acrylonitrile butadiene styrene (ABS) at a nozzle temperature of 240 °C and a bed temperature of 110 °C. Before insertion into the socket the residual limb of the rooster was lined with a polyurethane foam wrap to provide cushioning and breathability. The prosthetic was secured using circumferential Velcro straps.

The rooster returned one month later for the prosthetic to be placed, but the continued use of the amputated stump caused more damage to the stump. The stump required additional healing time and we encouraged the use of a bandage and pad to prevent further trauma. The owner returned an additional two months later after the stump healed properly and the prosthetic was placed. The fit was tight as the skin at the most distal part of the stump was slightly larger than when the mold was originally made. We tested the rooster's movement and after two steps the prosthetic appeared to function appropriately.

The distal part of the stump did increase in size and the prosthetic did not fit after a few weeks. The use of the stump after the initial surgery increased swelling and scar tissue, which eventually led to an improperly fitting prosthetic. While the use of this prosthetic was ultimately unsuccessful, continued development for this rooster continues. A mold has been retaken as of this writing to create a new prosthetic. This report demonstrates how a prosthetic can be used for heavy bodied birds undergoing similar procedures. Additional research is needed to minimize trauma and tissue changes to the amputation site post surgery.

3D Printed Prosthetic Limb in a Chicken (*Gallus domesticus*)

Stephanie Lamb, DVM, DABVP (Avian Practice)

Session #2043

Affiliation: From Arizona Exotic Animal Hospital, 744 N. Center St. Suite #101, Mesa, AZ 85201, USA.

Abstract: An approximately 2-year-old female domestic chicken presented after being attacked by a coyote. The bird sustained numerous injuries and ultimately had to have the left pelvic limb amputated at the level of the mid-tibiotarsus. After attempting a few different designs for a prosthetic limb made of conventional and home-made materials, a design for a 3D printed limb was created. The design was altered a few times to create an appropriately fitting structure for the individual.

Introduction

The use of prosthetics and orthotics are common therapeutic modalities in human medicine, which are emerging in the veterinary medical world.[1,2] Although there is more information on their use in canine patients, there are fewer reports on their use in birds. Reports that exist for birds include prosthetics applied to beaks and implant prosthetics in limbs.[3-6] The internet has several anecdotal reports posted to news and social media websites about their use in birds.[7-9] This report details the use of a prosthetic limb in a chicken after it lost its leg and the different types of limbs that were made for its use. This trial-and-error process is one that should be evaluated, along with other cases of prosthetics used in birds, to provide avian veterinarians a better understanding of their uses and how they may be able to help their patients.

Clinical Report

An approximately 2-year-old female domestic chicken presented to the veterinary hospital after having been attacked by a coyote three days prior. The patient suffered numerous lacerations, was anemic, and had maggots of various sizes within its wounds. The bird was managed with antibiotics, pain medications, topical cleansers, ointments and bandages. A blood transfusion was performed and wounds healed by second intension.

The patient was managed over several weeks but eventually it became apparent that the left limb had suffered unrepairable damage to the tendon of the long digital extensor. An amputation was performed with the intention of having a prosthetic limb placed afterwards.

The initial prosthetic limbs were made from various clinical products and altered children's toys. A few varieties were attempted but none seemed to work well for the patient due to various issues such as instability problems or being too heavy. Attempts were made to alter the form of the prosthetic but the patient did not take to any form well.

After 2 years the owner found a group willing to make a prosthetic limb for the patient out of a 3D printed plastic. The material used was polyethylene terephthalate glycol-modified (PETG) and it was printed on a Prusa i3 3D printer. The limb was altered a few times in an attempt to get the correct height and fit for the patient. The presentation will further elaborate upon the designs that were attempted for this patient.

References

1. Mich PM. The emerging role of veterinary orthotics and prosthetics (V-OP) in small animal rehabilitation and pain management. *Top Companion Anim Med.* 2014;29(1):10-19.

2. Millis DL, Levine D. *Canine Rehabilitation and Physical Therapy.* Philadelphia, PA: Elsevier Saunders;2014.

3. Bennett RA. Surgery of the avian beak. *Proc Annu Conf Assoc Avian Vets.* 2010;95-98.

4. Speer B. Surgical procedure of the psittacine skull. *Proc Annu Conf Assoc Avian Vet.* 2012;181-191.

5. Clipsham R. Beak repair, rhamphorthotics. In: Altman RB, Clubb SL, Dorrestein GM, Quesenberry K, eds: *Avian Medicine and Surgery.* Philadelphia, PA: Saunders;1997:773-786.

6. Rush EM, Turner TM, Montgomery E, et al. Implantation of a titanium partial limb prosthesis in a white-naped crane (*Grus vipio*). *J Avian Med Surg.* 2012;26(3):167-175.

7. Daily Mail. Bird amputee gets prosthetic wing to help it balance and protect its fragile stump. Available at: http://www.dailymail.co.uk/news/article-2824849/Bird-amputee-gets-prosthetic-wing-help-balance-s-fragile-stump.html Accessed May 30, 2017.

8. WBTV. Goose fitted with prosthetic foot. Available at: http://www.wbtv.com/story/24485119/university-makes-prosthetic-leg-for-goose Accessed May 30, 2017.

9. Pet Place. Artificial legs help maimed bird walk again. Available at: http://www.petplace.com/article/birds/general/enjoying-your-bird/artificial-legs-help-maimed-bird-walk-again Accessed May 30, 2017.

Bile Duct Carcinoma in a Young Hahn's Macaw

Louden Wright, DVM,
James Carpenter, MS, DVM, Dipl ACZM,
Pankaj Kumar, DVM

Session #2091

Affiliation: From the Departments of Clinical Sciences (Wright, Carpenter), and Diagnostic Medicine/Pathobiology (Kumar), College of Veterinary Medicine, Kansas State University, Manhattan, KS 66506, USA.

A 4-year-old Hahn's macaw (*Diopsittaca nobilis nobilis*) presented for a progressive lethargy and inappetance. Three years prior, the bird had been diagnosed with *Chlamydia psittaci* and treated with doxycycline. Despite resolution of clinical symptoms, elevated aspartate transaminase (AST), alkaline phosphatase (ALP), and bile acid levels persisted. On presentation, the bird was thin and depressed. A plasma biochemistry panel showed elevated AST, ALP, and bile acid levels. Radiographs identified generalized enlargement of the hepatic silhouette. The bird died after radiographs were obtained and was submitted for gross necropsy and histopathology. About 60% of the liver parenchyma was effaced with large raised tan nodules which had formed adhesions to the viscera. Histopathology identified the mass as a bile duct carcinoma, with neoplastic cells frequently infiltrating remaining areas of normal hepatocytes. A PCR probe of the liver tissue was negative for *Chlamydia psittaci.* Bile duct carcinomas have been reported in young Amazon parrots (*Amazona* spp.). Several reports hypothesized that these tumors occur in younger Amazon parrots due to infection with psittacid herpesvirus-1 (PsHV1). Psittacid herpesvirus-1 is the etiologic agent of Pacheco's disease and also causes mucosal papillomatosis in psittacine birds. In this case, formalin-fixed liver tissue was submitted for PCR and found to be negative for PsHV-1. This is the first reported case of a bile duct tumor in a Hahn's macaw. Although PsHV-1 was not identified in the tissue, it should remain a differential in young psittacine birds with bile duct carcinomas, as formalin fixation can reduce the sensitivity of PCR tests.

References

1. Greer CE, Lund JK, Manos MM. PCR amplification from paraffin-embedded tissues: recommendations on fixatives for long-term storage and prospective studies. *PCR Methods Appl.* 1991;1(1):46-50.

2. Hillyer EV, Moroff S, Hoefer H, Quesenberry KE. Bile duct carcinoma in two out of ten Amazon parrots with cloacal papillomas. *J Assoc Avian Vet.* 1991;5(2):91-95.

3. Phalen D. Psittacid herpesviruses and associated diseases. In: Speer BL, ed. *Current Therapy in Avian Medicine and Surgery*. St. Louis, MO: Elsevier Saunders;2016:47-51.

4. Zehnder A, Graham J, Reavill DR, McLaughlin A. Neoplastic diseases in avian species. In: Speer BL, ed. *Current Therapy in Avian Medicine and Surgery*. St. Louis, MO: Elsevier Saunders;2016:107-141.

Removal of a Tracheal Myelolipoma in a Cockatiel (*Nymphicus hollandicus*) by Surgical Resection and Anastomosis

Maria E Passarelli, VMD,
Natalie Antinoff, DVM, DABVP (Avian Practice),
Caleb Hudson, DVM, Dipl ACVS-SA,
Michael Garner, DVM, DACVP

Session #2068

Affiliation: From Gulf Coast Veterinary Specialists, 1111 West Loop South, Houston, TX 77027, USA (Passarelli, Antinoff, Hudson), Northwest ZooPath, 654 West Main St, Monroe, WA 98272, USA (Garner).

A 17-year-old female cockatiel (*Nymphicus hollandicus*) presented for a 1-week history of respiratory difficulties. On physical examination the patient was open-mouth breathing, had cyanotic mucus membranes, and was unable to vocalize. There was increased inspiratory effort with an audible wheeze, and a clicking noise on expiration. These signs were suspicious for a tracheal obstruction, so an air sac cannula was placed. Radiographs revealed a round intraluminal soft tissue opacity in the trachea. Tracheoscopy was attempted with a 2.7-mm endoscope, but the scope was too large to pass through the glottis. The obstruction could be visualized via transtracheal illumination. Attempts to coax out the obstruction were unsuccessful. The obstruction was firmly adhered to the tracheal wall, and the trachea subsequently avulsed in the mid-cervical region. Tracheal resection and anastomosis was performed, and the obstruction was removed with the associated tracheal ring. The patient recovered uneventfully and was doing well at follow-up exams 9 days and 24 days post surgery. The histologic diagnosis was ossifying myelolipoma arising from bone marrow within the tracheal ring. Tracheal resection and anastomosis has been widely reported as a treatment for avian tracheal lesions, such as strictures or masses. While there are numerous reports of the procedure in larger birds, including a cockatoo,[1] a currasow,[2] a goose,[3] macaws,[4,5] a duck,[6] a red tailed hawk,[7] and an eagle,[8] this is the first report of a tracheal resection and anastomosis in a cockatiel. This is also the first report of a tracheal myelolipoma in any species.

References

1. Doneley RJT, Raidal SR. Tracheal resection and anastomosis in a red-tailed black cockatoo (*Calyptorhynchus banksii banksii*). *Aust Vet J.* 2010;88(11):451-454.

2. Evans A, Atkins A, Citino SB. Tracheal stenosis in a blue-billed currasow (*Crax alberti*). *J Zoo Wildl Med.* 2009;40(2):373-377.

3. Clippinger TL, Bennett RA. Successful treatment of a traumatic tracheal stenosis in a goose by surgical resection and anastomosis. *J Avian Med Surg.* 1998;12(4):243-247.

4. Jankowski G, Nevarez JG, Beaufrere H, et al. Multiple tracheal resections and anastomoses in a blue and gold macaw (*Ara ararauna*). *J Avian Med Surg.* 2010;24(4):322-329.

5. De Matos REC, Morrisey JK, Steffey M. Postintubation tracheal stenosis in a blue and gold macaw (*Ara ararauna*) resolved with tracheal resection and anastomosis. *J Avian Med Surg.* 2006;20(3):167-174.

6. Guzman DSM, Mitchell M, Hedlund CS, et al. Tracheal resection and anastomosis in a mallard duck (*Anas platyrhynchos*) with traumatic segmental tracheal collapse. *J Avian Med Surg.* 2007;21(2):150-157.

7. Monks DJ, Zsivanovits HP, Cooper JE, et al. Successful treatment of tracheal xanthogranulomatosis in a red-tailed hawk (*Buteo jamaicensis*) by tracheal resection and anastomosis. *J Avian Med Surg.* 2006;20(4):247-252.

8. Aguilar RF, Redig PT. What is your diagnosis? (Tracheal fungal granuloma removed by surgical resection and anastomosis). *J Avian Med Surg.* 1997;11(2):121-124.

Successful Control of Seizure Activity with Levetiracetam in an African Grey Parrot (*Psittacus erithacus*) After Eight Years of Inadequate Control

Stephanie Lamb, DVM, DABVP (Avian Practice)

Session #2045

Affiliation: From Arizona Exotic Animal Hospital, 744 N. Center St. Suite #101, Mesa, AZ 85201, USA.

Abstract: A 24-year-old Congo African grey parrot (*Psittacus erithacus*) was evaluated for tonic-clonic seizure activity that had been occurring for a duration of 8 years. Several anti-epileptic medications had previously been attempted with poor success. Levetiracetam (100 mg/kg PO q12hr) was initiated and seizure activity promptly ceased. Ten months after starting the medication, tonic-clonic seizure activity began to reoccur. Levetiracetam was increased to three times per day and the seizures again stopped. Seizures continue to be well controlled 2 years later.

Introduction

Levetiracetam is a newer antiepileptic drug that has been used in the treatment of seizure activity in various species such as humans, dogs, cats, rodents, and birds.[1-5] It is believed to work through binding to synaptic vesicle protein 2A (SV2A), modulating its action, and inhibiting excitatory neurotransmitter release.[2] Studies have shown that it suppresses the inhibitory effects of some negative allosteric modulators at GABA and glycine receptors and can selectively inhibit N-type calcium channels.[6,7]

In the event of seizure activity, phenobarbital is often the first-line treatment selected by many but it is not always effective. Other antiepileptic drugs may be needed to help gain further control, an example of which of is levetiracetam, which can be added on or tried by itself. It has been used to treat both partial and generalized seizures.[1,8,9] In a study in humans where it was used to treat refractory seizure activity, monotherapy with levetiracetam resulted in a 73.8% reduction in partial seizure activity from baseline and 18.4% of participants remained seizure free on levetiracetam alone.[9] A levetiracetam study was performed in idiopathic epileptic cats that were previously treated with phenobarbital and had poor control or evidence of adverse reactions to the medication. Levetiracetam was administered at 20 mg/kg PO q8h. Seven out of 10 cats responded by having a greater than 50% reduction in seizure activity.[4] A separate study was completed in dogs with idiopathic epilepsy that were considered to be pharmacoresistant to phenobarbitone. These patients received levetiracetam as an add-on treatment at 10 mg/kg PO q8h. Eight out of 14 dogs had their seizure activity reduce by greater than 50%. When the dose was increased to 20 mg/kg q8h for dogs that remained refractory, 1 additional patient responded. Interestingly, 6 out of 9 responders had seizure frequency increase after 4-8 months of therapy.[3]

Serum ranges of levetiracetam in humans that are considered to be effective are 5.5 μg/ml to 45 μg/ml.[5] A study in Hispaniolan amazon parrots dosed birds at 50 mg/kg PO q8h or 100 mg/kg PO q12h and found that plasma concentrations of the drug were > 5.5 μg/ml for 9.4 and 12 hours respectively and therefore in the effective range for humans.[5] In one case, report of an African grey parrot dosed at 50 mg/kg PO q12h, the serum levels of levetiracetam were 1 μg/ml when measured at 12 hours after the last dose was administered. The patient had still experienced seizure activity at this dose, therefore, levetiracetam was increased to 50 mg/kg PO q8h. Repeated evaluation of the serum levels found that trough levels were 6.3 μg/ml and therefore within the therapeutic range for humans. Seizure activity was still occurring in this patient and additional drugs were added to the treatment regimen. Eventually, the patient had the levetiracetam dose increased to 100 mg/kg PO q8h. Serum levels were still within the therapeutic ranges for humans at this dose. Seizure activity improved with this dose and the addition of other anti-epileptic drugs.[1]

Numerous other antiepileptic medications have been utilized in birds with case reports available for the use of phenobarbital, potassium bromide, gabapentin and zonisamide.[1,10] Phenobarbital is one of the most commonly prescribed anticonvulsant medications for seizure activity in veterinary small-animal patients. A report in African grey parrots that gave a single dose of phenobarbital at 17 mg/kg revealed a short half-life and plasma concentrations that were far below the therapeutic ranges that are reported for dogs and cats. The authors concluded that treating African grey parrots at the dose described may not produce therapeutic effects.[10]

Less conventional treatments for seizures have been described and include things like acupuncture, dimethylglycine and cannabidiol. Dimethylglycine had been studied in humans with seizure activity after it was found to improve seizures in a few individuals; however, controlled studies revealed no difference in seizure frequency between controls and treated patients.[11] Cannabidiol is the main nonpsychoactive agent found in cannabis and has been explored in both human subjects and animal models for its use in the treatment of seizure activity.[12] Studies are limited in birds but cannabinoid receptors have been identified at various locations in the brain of budgerigars.[13] There is little scientific literature available on either dimethylglycine or cannabidiol for use in avian patients in the treatment of seizures and both should be further investigated to see if they could be efficacious. This report details the successful management of a Congo African grey parrot with the anti-epileptic drug levetiracetam after multiple failed attempts at seizure control with other medications.

Clinical Report

A 24-year-old, DNA-sexed, female, Congo African grey parrot (*Psittacus erithacus*) presented to the clinic for evaluation of poorly controlled tonic-clonic seizure activity for an 8-year duration. The seizure activity reportedly started at 16 years of age and episodes occurred, on average, 1 day out of the week. On the days when seizures occurred, anywhere between 1-5 seizures transpired. The episodes were described as lasting for 5-20 seconds at a time. After the seizure, the post-ictal phase was absent or would last for 1-2 minutes. If the patient experienced a post-ictal phase, the owners described the bird as being quiet and subdued. Occasionally, limb weakness was observed for 1-2 minutes during this time but this was not consistent or common.

When seizure activity first began at 16 years of age, diagnostics were performed to look for an underlying cause. Diagnostics included a complete blood count, biochemistry profile, ionized calcium blood levels, whole body radiographs, blood pressure evaluation, lead and zinc levels, fecal gram stain, and a *Chlamydia* PCR and titer. Results were reviewed and were within normal limits. Throughout the years various tests, such as the CBC, biochemistry and ionized calcium levels, were repeated as deemed necessary and values were within normal limits. At 18 years of age, the patient was tested for avian bornavirus by PCR of whole blood. The result was positive. Diagnostics that had been recommended but never completed included an echocardiogram to evaluate for cardiovascular disease and an MRI to assess for structural changes to the brain.

Numerous medications had been tried on the patient over the years at various doses and lengths of time. Medications were discontinued after several weeks-to-months of poor success at controlling seizure activity. Some drugs were used at various dose ranges. Pharmacologic agents that had been tried included phenobarbital (started at 2.5 mg/kg PO q12h, increased to 20 mg/kg PO q12h), potassium bromide (started at 75 mg/kg PO q24h, increased to 130 mg/kg PO q12h), gabapentin (started at 10 mg/kg PO q12h, increased to q8h), zonisamide (started at

20 mg/kg PO q12h, increased to q8h), calcium gluconate (100 mg/kg IM once), and isoxsuprine (5 mg/kg PO q12h). At the time the bird was found to be positive for avian bornavirus, the non-steroidal anti-inflammatory drugs, meloxicam (0.5 mg/kg PO q12h) and celecoxib (10 mg/kg PO q12h) had both been utilized. Meloxicam therapy was attempted first for 6 months and there was no change observed in the seizure activity. Therapy with celecoxib was attempted next for 3 months and also did not have any effect on the seizure activity. Non-pharmacologic agents like acupuncture and exposure to an ultraviolet B (UVB) light for 10 hours a day were attempted with no changes in the seizure frequency. The owner had attempted the use of a nutraceuticals dimethylglycine (3 mg/kg PO q12h) with response initially shown as no seizure activity for 5 weeks, then seizure frequency returned to previous levels. The owner had used a commercial cannabidiol powdered supplement (8 mg/kg PO q12h) which was compounded at home and noted a reduction in severity and length of time of individual seizure episodes but overall it did not reduce the frequency. Some of the medications that had been tried on the patient were used in combination. The patient was gradually weaned off of drugs when they were deemed to be ineffective. When medications were discontinued the seizure frequency did not increase.

At the time of presentation, the patient had been off of medications for the preceding 7 months. The bird received a pelleted diet and fresh water, which was provided ad libitum. The patient had a body condition score of 5/9 and was neurologically appropriate. The owner reported that seizure activity at home occurred on average 1 day out of a week. A complete blood count and biochemical analysis were performed. Results were within the normal reference ranges. The patient was started on a dose of levetiracetam (100 mg/kg PO q12h; levetiracetam, Diamondback Drugs, Scottsdale, AZ, USA).[1,5] The bird was monitored for seizure activity at home and the owners were instructed to record episodes as they occurred.

The patient returned for follow up 3 weeks later. At that visit, the owners reported that no seizure activity had been observed since the start of the medication. On physical examination, the patient was neurologically appropriate. There were no changes from the previous physical exam. Evaluation of levetiracetam blood levels was recommended but declined. The dose of the levetiracetam was continued as originally prescribed.

Four months after starting the levetiracetam, the owners returned to the clinic for re-evaluation. The previous evening the patient had one tonic-clonic seizure episode. The episode had occurred about 6 hours after the patient was due for her dose of the medication which the owner had forgotten to administer. This was the first time the owners had forgotten to give the medication since starting levetiracetam. Upon physical examination, there were no changes from previous. The owners were instructed to start the patient back on levetiracetam at the previous dose and return in 2 weeks for measurement of serum levetiracetam levels, 6 hours after the morning dose. This time was chosen to determine serum levels when the patient was reported to have a seizure.

The patient returned for re-evaluation 6 months after starting the levetiracetam. The owners reported no seizure activity had been seen and they had not missed any further doses of the medication. The owners reported that the patient seemed to be notably more active in the last 3 months than it had been since seizure activity started. This activity was described as the patient walking around the house and interacting with toys more. Blood was collected at 6 hours after the morning dose of levetiracetam had been administered. Blood was collected for a CBC and biochemical analysis. Results of the CBC and biochemical analysis were within normal limits. Results of the serum levetiracetam dose 6 hours post administration was 18.4 µg/ml. The patient was discharged on the same dose regimen of levetiracetam.

Ten months after starting the levetiracetam the patient returned for re-evaluation because it started to have seizures occur at about 11-12 hours post administration of the levetiracetam. The seizures were occurring about every other week. Levetiracetam serum levels were measured at 6 and 12 hours after the morning administration of the medication to assess suspected peak and trough levels. The results were 13.4 µg/ml and 2.5 µg/ml respectively. The dose was increased to 150 mg/kg PO q12h; however, this did not stop the seizure pattern. Blood levels were checked after being on this dose for 1 month. At 6 hours the levetiracetam serum level was 6.6 µg/ml and at 12 hours it was 2.3 µg/ml. It was determined that the dose should be increased to every 8 hours. At this dosing regimen the patient was reported to be free of seizure activity once more. After 4 weeks of being seizure free, the dose was then reduced to 100 mg/kg PO q8h. At the time of this writing the patient continues to be well controlled with levetiracetam and is reported to only experience a seizure if the owners forgot to administer a dose of the medication.

References

1. Beaufrere H, Nevarez J, Gaschen L, et al. Diagnosis of presumed acute ischemic stroke and associated seizure management in a Congo African grey parrot. *J Am Vet Med Assoc.* 2011;239(1):122-128.

2. Lyseng-Williamson KA. Levetiracetam: a review of its use in the treatment of epilepsy. *Drugs.* 2011;71(4):489-514.

3. Volk HA, Matiasek LA, Lujan Feliu-Pascual A, et al. The efficacy and tolerability of levetiracetam in pharmacoresistant epileptic dogs. *Vet J.* 2008;176(3):310-319.

4. Bailey KS, Dewey CW, Boothe DM, et al. Levetiracetam as an adjunct to phenobarbital treatment in cats with suspected idiopathic epilepsy. *J Am Vet Med Assoc.* 2008;232(6):867-872.

5. Schnellbacher R, Beaufrere H, Arnold RD, et al. Pharmacokinetics of levetiracetam in healthy Hispaniolan Amazons parrots (*Amazona ventralis*) after oral administration of a single dose. *J Avian Med Surg.* 2014;28(3):193-200.

6. Rigo JM, Hans G, Nguyen L, et al. The anti-epileptic drug levetiracetam reverses the inhibition by negative allosteric modulators of neuronal GABA- and glycine-gated currents. *Br J Pharmacol.* 2002;136(5):659-672.

7. Lukyanetz EA, Shkryl VM, Kostyuk PG. Selective blockade of N-type calcium channels by levetiracetam. *Epilepsia.* 2001;43(1):9-18.

8. Brazil CW. New antiepileptic drugs. *Neurologist.* 2002;8(2):71-81.

9. Ben-Menachem E, Falter U. Efficacy and tolerability of levetiracetam 3000 mg/d in patients with refractory partial seizures: A multicenter, double-blind, responder-selected study evaluating monotherapy. *Epilepsia.* 2000;41(10):1276-1283.

10. Powers LV, Papich MG. Pharmacokinetics of orally administered phenobarbital in African grey parrots (*Psittacus erithacus erithacus*). *J Vet Pharmacol Therap.* 2011;34(6):615-617.

11. Gascon G, Patterson B, Yearwood K, Slotnick H. N,N- Dimethylglycine and epilepsy. *Epilepsia.* 1989;30(1):90-93.

12. Devinksy O, Cilio MR, Cross H, et al. Cannabidiol: Pharmacology and potential therapeutic role in epilepsy and other neuropsychiatric disorders. *Epilepsia.* 2014;55(6):791-802.

13. Alonso-Ferrero ME, Paniagua MA, Mostany R, et al. Cannabinoid system in the budgerigar brain. *Brain Res.* 2006;1087(1):105-113.

Plasmodium sp. Infection in a Small Flock of Zebra Finches (*Taeniopygia guttata*)

Len Donato, VMD, DABVP (Avian Practice),
Allyson DeLozier, VMD,
Sherrill Davison, VMD, MS,
Linnea Tracy, BS

Session #2023

Affiliation: From Radnor Veterinary Hospital 107 North Aberdeen Avenue Wayne, PA 19087, USA (Donato, DeLozier) and the Laboratory of Avian Medicine and Pathology (Davison) and the New Bolton Center, University of Pennsylvania School of Veterinary Medicine, 382 West Street Road Kennett Square, PA 19348, USA (Tracy).

Abstract: An owner living in the Philadelphia area acquired 12 zebra finches (*Taeniopygia guttata*) from a breeder in Florida when she was there for the winter. The birds bred and five eggs hatched. The owner presented one of the juvenile birds at two months of age for necropsy. Three other birds also died. Necropsy revealed a yellow material throughout the coelom and marked hepatomegaly. Histopathology demonstrated hemozoin and protozoans consistent with *Plasmodium* sp., also known as avian malaria. This disease is uncommon in zebra finches and in the northeastern United States.

Introduction

Avian malaria is a blood-borne disease that has been documented in many species of birds. It is caused by parasites of the genera *Plasmodium. Plasmodium* species tend to have a severe health impact on birds. These parasites undergo both sexual and asexual life stages that necessitate a vertebrate and arthropod vector. For *Plasmodium* parasites, the main vector responsible for avian disease transmission is the mosquito. Once transmitted, the parasites undergo asexual replication within erythrocytes, which eventually lyse, enabling *Plasmodium* to spread within the host. Signs of malaria in birds vary from subclinical to acute death and can include hypertension, anemia, cardiomegaly, and debilitation. Many high morbidity incidents are suspected to involve an infection of an aberrant host. Diagnosis usually is based on histopathology, blood films, and PCR. Medical treatment rarely results in clinical success, but some zoo populations have seen some mild benefit to prophylactic and aggressive treatment.

Clinical Report

A juvenile zebra finch (*Taeniopygia guttata*) presented for necropsy in early summer in the suburbs of Philadelphia, Pennsylvania. The owner has a home on the east coast of southern Florida, where she acquired twelve zebra finches from a local seller six months prior. During that time they appeared healthy and were successfully breeding. The birds were being fed a seed diet and had access to fresh water.

In the beginning of spring she drove the twelve original finches plus five newer hatchlings to her second home in the suburbs of Philadelphia. The birds continued to breed and multiply, but in the first two weeks of June she noted three sudden deaths of the younger birds. Just prior to the deaths, the owner only noted a mild increase in one of the finch's respiratory rates. The juvenile that was brought to our hospital was the fourth death she noted from her collection. By July the owner had counted at least eight juvenile deaths.

At necropsy, there were some unclassified maggot-like parasites on the skin and in the oral cavity. The bird was in good body condition and had a lot of seed in its crop. An increased distance between the most caudal part of the sternum and the pelvis was noted. The coelom was diffusely yellow in color and there was marked hepatomegaly. The rest of the organs had no obvious gross lesions. Samples of the heart, gastrointestinal tract, liver, lung, kidney, and spleen were removed, fixed in 10% neutral buffered formalin, and submitted for histopathology. Before histopathology results were available, another juvenile died. That juvenile's tissues also were submitted for histopathology. Results were nearly identical to the first bird.

Histology revealed that in the heart there were multifocal areas of lymphoplasmacytic infiltrates. Some of the blood and myocardial cells had fine basophilic 1-mm diameter spheres consistent with hemogregarine parasites within the cytoplasm. There also were fragmented myocardial cells with a hypereosinophilia and necrotic areas with more severe inflammation. The pancreas, intestines, ventriculus, and kidneys had no significant findings. The liver was moderately autolyzed which obscured some features, but some intracytoplasmic, slightly refractile golden-yellow pigment was present. This pigment was interpreted as hemozoin. There also was a lymphoplasmacytic thickening in the interstitium of the lungs and around the blood vessels. There were many histiocytic cells and some lymphoid cells with the previously described pigment present inside of them. Finally, the spleen contained areas of plasmacytoic histiocytic cell populations, also with the presence of the previously described pigment.

The histologic lesions, presence of hemozoin pigment, and protozoa present were consistent with a plasmodium (malaria) infection. Exoerythrocytic schizonts were present and indicative of a severe parasitemia. Recommendations related to treatment and management were discussed with the owner. The owner elected to maintain the flock since no other birds had died and medical treatment historically is not successful.

Discussion

When avian malaria causes morbidity and mortality in a flock, a number of factors usually are playing a role. The birds in this report were placed in a car and driven for a few days. There were a number of them put together for transport and after arrival. While successful breeding occurred, that only led to an increase in the number of birds in the same space. A seed diet was being fed as well. Poor diet, stress and, overcrowding likely played a role in the emergence of clinical malaria in this flock. Most well kept and nonstressed flocks will not develop malaria. While poor hygiene also will contribute to an outbreak, in this case the owner felt they were able to maintain a clean environment. A physical inspection of the home environment was not done.

Medical treatments for *Plasmodium* sp. that have been used are primaquine and chloroquine, along with supportive medications like antibiotics and antifungals. They have been found to be fairly ineffective. Pharmacokinetic studies have not been done for many of species of birds so doses are empirical. In addition, these medications usually must be compounded and may not have the same efficacy as the manufactured products. Treatment is focused on identifying and treating underlying disease and improving husbandry issues. The owner did not want to pursue further treatment for this flock because of challenges treating a flock, cost, and the uncertainty that the medication would cure the disease. The owner was satisfied because there were no further deaths or significant illness.

There was no veterinary inspection performed prior to travel, so it is not known if there were issues at the time of transport. Ideally this flock would have been inspected by a U.S. Department of Agriculture accredited veterinarian before leaving the state. It is suspected the flock carried *Plasmodium* with them on their journey, but ultimately that was not known. Our research has found that *Plasmodium* sp. has not been reported in the literature for zebra finches, as well as not commonly reported in the northeastern part of the United States.

References

1. Ricklefs R, Fallon S, Bermingham E. Evolutionary relationships, cospeciation, and host switching in avian malaria parasites. *Syst Biol.* 2004;53:111-119.

2. Schrenzel M, Maalouf G, Keener L, Gaffney P. Molecular characterization of malarial parasites in captive passerine birds. *J Parasitol.* 2003;89:1025-1033.

3. Ferrell S, Snowden K, Marlar A, et al. Fatal hemoprotozoal infections in multiple avian species in a zoological park. *J Zoo Wildl Med.* 2007;38:309-316.

4. Hartup B, Oberc A, Stott-Messick B, et al. Blood parasites of house finches (*Carpodacus mexicanus*) from Georgia and New York. *J Wildl Dis.* 2008;44:469-474.

5. Olias P, Wegelin M, Zenker W, et al. Avian malaria deaths in parrots, Europe. *J Emerg Infect Dis.* 2011;17:950-952.

6. Tavernier P, Sagesse M, Wettere A, Redig P. Malaria in an eastern screech owl (*Otus asio*). *Avian Dis.* 2005;49:433-435.

Section 8

Tuesday Master Classes

Avian Heat Stress and Heat Injuries: Clinical and Conservation Relevance

Shangzhe Xie, BSc, BVMS, MVS, and
Todd J. McWhorter, BSc, MSc, PhD

Session #2006

Affiliation: From University of Adelaide, School of Animal and Veterinary Sciences, Roseworthy Campus, Roseworthy, South Australia 5371, Australia.

Abstract: Columbiform birds appear to have several adaptive advantages allowing them to tolerate higher environmental temperatures compared to psittaciform and passeriform birds. These include a higher critical thermal maximum, appropriate corticosterone response to heat exposure, and expression of heat shock proteins. This increased tolerance to higher environmental temperatures likely gives columbids the ability to continue foraging and display other normal behaviors during the heat, which confers them a potential survival edge compared to other birds. Columbiformes are still susceptible to heat injuries including heat stroke when pushed over the edge, and the most consistent changes in the blood associated with heat stroke were hyperkalemia and aspartate aminotransferase. There were increases in creatinine kinase and glutamate dehydrogenase in some heat stroke birds, but these were not as consistent. There were species differences in the histopathologic changes post-heat exposure amongst birds, whether or not heatstroke was induced, albeit none that were pathognomonic. Given the differences in heat tolerance and adaptation strategies amongst bird species, the importance of thermal refuges differs amongst species and the loss of these refuges has varying impacts on different species of birds, which can contribute to the restructuring of bird communities as a result of climate change.

Avian Heat Stress and Heat Injuries: Clinical and Conservation Relevance

Australian birds are under increasing threat from the effects of climate change and extreme events like heat waves. As temperatures continue to rise with climate change, sensitive taxa of birds will be challenged.[1]

Current Knowledge of Anatomic and Physiologic Adaptations

There are several anatomic and physiologic adaptations in desert birds that help them cope with the challenge of living in a hot desert. These adaptations are summarized according to organ systems below.

Kidney

Avian kidneys are elongated, flattened and closely fitted into the bony concavity formed by the synsacrum.[2] The kidneys contain both reptilian-type and mammalian-type nephrons, with species differences in the relative proportions of each nephron type.[2] Mammalian-type nephrons can be further differentiated into long loop and short loop mammalian-type nephrons.[2] When desert birds are subjected to severe dehydration or salt loads, they need to prevent the rise of plasma osmolality by secreting ions extrarenally, producing concentrated urine, reducing the glomerular filtration rate, or any combinations of these.[2] Desert quails (*Lophortyx*) can decrease the overall glomerular filtration rate by reducing the number of functioning reptilian-type nephrons to increase water conservation at the expense of waste excretion, whilst maintaining the functions of the mammalian-type nephrons to increase the urine-concentrating capability of the kidney.[2]

Desert mammals were found to have a larger medullary mass relative to total kidney mass compared to mesic mammals.[3,4] The vasa recta, loops of Henle, and collecting ducts in birds are encapsulated in the medullary cone, making the division between cortex and medulla less clear in bird kidneys.[5] Therefore, similar studies in birds to identify differences in kidney structure between desert and mesic birds will depend on measurements of relative medullary cone length instead of relative medullary mass. One such study by Goldstein and Braun found there was no association between relative medullary cone length and maximal ureteral urine concentration. Desert birds were found not to have superior urine-concentrating abilities to non-desert birds.[6] Williams and Tieleman commented that these results could be due to the small sample size, and the fact that two of the seven species studied were seabirds with salt glands and larger kidneys.[5] Goldstein and Braun suggested that smaller birds have better urine-concentrating abilities regardless of habitat affinity and that there was a negative correlation between the maximal ureteral urine concentration and the loop of Henle length.[6]

There were very few differences found between nephron ultrastructures of mesic and arid zone honeyeaters[7] and sparrows.[8] In the latter study, the Savannah sparrow (*Passerculus sandwichensis*) was found to have the renal morphology to produce more concentrated urine than the mesic House sparrow (*Passer domesticus*), but this was not reflected in its physiology.[7]

It has been suggested that the maximal ureteral urine concentration may correlate with the mass-adjusted field metabolic rate of a species because a lower metabolic rate means there is less metabolic waste for the kidneys to eliminate.[5] This could therefore partly explain the significance of a lower field metabolic rate in desert birds compared to non-desert birds.[5]

Gastrointestinal tract

The avian gastrointestinal tract has a very important role in the regulation of overall fluid and electrolyte balance. Birds are able to move urine from the cloaca by anti-peristalsis into the rectum,[9] where water is passively reabsorbed to reduce water loss in urine.[10] This likely explains the lower water content found in the lumen of the lower gastrointestinal tract of desert birds;[11] however, more studies are needed to evaluate this hypothesis.[5] Despite a lower renal urine concentrating ability compared to mammals,[12] this allows birds to be as good as mammals in conserving water during periods of high environmental temperature.[13-15]

Some birds can control cloacal evaporation as a means of thermoregulation at high ambient temperatures. This has been demonstrated in Inca doves (*Columbina inca*), whose cloacal evaporation was negligible at 30, 35, and 40 °C (86, 95 and 104 °F), but increased to 21.2% of total evaporative water loss at 42 °C.[16]

Skin

The balance between losing enough water to decrease body temperature and maintaining sufficient hydration is the key to survival in times of heat stress.[5] The avian skin has a well-vascularized dermal layer[17] and increased cutaneous water loss can be achieved by vasodilating the dermal capillary bed.[18] Conversely, the avian epidermis has the ability to secrete lipid-enriched organelles known as multigranular bodies during times of water deficit for rapid water-proofing.[19] Comparisons between the lipid composition of the epidermis of adult desert house sparrows and mesic house sparrows revealed that desert birds had a higher amount of ceramides and cerebrosides and a lower percentage of cholesterol within the stratum corneum.[20] A follow-up study revealed that desert house sparrow nestlings showed a greater degree of plasticity in cutaneous water loss and lipid composition of the stratum corneum compared to mesic nestlings, possibly due to the increased exposure to environmental stressors.[20] Similar differences were found between desert and mesic larks.[21] Desert sparrows had lower cutaneous water loss, which might be attributed to modifications in chain length and polarity of the sphingolipids that determine interactions among lipid molecules within the stratum corneum.[20] Namaqua doves (*Oena capensis*), laughing doves (*Spilopelia senegalensis*), Cape turtle doves (*Streptopelia capicola*) and crested pigeons (*Ocyphaps lophotes*) have been shown to be capable of dissipating heat very efficiently using cutaneous evaporation.[22] Passeriform birds have limited capacity for cutaneous evaporative cooling, which protects them from dehydration when temperatures are high and water sources are scarce.[23-26]

Respiratory tract

It has been suggested that desert animals could have more complex nasal turbinates to allow greater cooling of exhaled air, and therefore a larger reduction in respiratory evaporative water loss than non-desert animals.[27] Some birds are able to reduce respiratory evaporative water loss by recovery of water in the nose at moderate to low ambient temperatures, but this ability becomes insignificant at high ambient temperatures (eg 45 °C) for crested larks (*Galerida cristata*).[28] However, the results of that study did not support the hypothesis that desert birds have greater ability to recover water in the nasal passages compared to non-desert birds, because desert larks lack the ability to reduce respiratory evaporative water loss by recovery of water in the nose.[28] More studies directly comparing phylogenetically similar desert and non-desert birds are needed to confirm this finding.

Small passerine birds have high mass-specific metabolic rates and produce heat internally at a high rate.[29] Coupled with their high respiratory rate, this results in a high rate of pulmonary water loss.[29] Many other avian species including verdins (*Auriparus flaviceps*)[23] and nightjars,[30] have been shown to rely heavily on respiratory evaporative water loss to maintain body temperature when exposed to heat. While this helps to prevent body temperature from rising to lethal levels, it also increases the need for water. Maintaining hydration thus becomes a problem during extended periods of high environmental temperature.[30] Parrots can utilize gular fluttering in conjunction with panting to increase the efficiency of heat loss.[31-33]

It has been suggested, based on a three-dimensional model of the trachea and air sacs in the domestic fowl, that a long trachea could be helpful in controlling body temperature when metabolic rates and respiratory flow are elevated.[34]

Respiratory evaporative water loss is believed to be less efficient compared to cutaneous evaporative water loss because muscle contractions are required for the former, which results in heat production[29,35-37]; however, cutaneous evaporation appears to be affected by environmental humidity more than respiratory evaporation (panting).[37] Species of birds that are smaller, with low activity levels at high temperatures and rely on food as the main water source are more likely to utilize respiratory evaporative water loss as means of maintaining their body temperature.[26]

Gaps in Knowledge

Despite the high level of understanding of anatomic and physiologic adaptations of birds in coping with high temperatures, there are still significant gaps in the knowledge base. These gaps in knowledge are more important in the face of climate change, as global temperatures are starting to result in environmental temperatures that exceed the critical thermal maximum of most birds. For example, it is unknown if heat actually results in physiologic stress responses in Australian arid zone birds. There have been no studies on the changes in gene expression when these birds are exposed to high temperatures. There has not been any quantification of time budgets of different Australian arid zone birds to determine the difference in behavioral adaptations amongst them. There is currently no way of definitively diagnosing heat injuries and heat stroke ante- and post-mortem in birds. Instead, most avian veterinarians rely on human and mammalian literature to predict and extrapolate physiologic, histopathologic, and clinical pathologic changes that can be expected after birds have been exposed to extreme heat. We conducted a series of studies to fill these gaps in knowledge.

Stress responses

In one study, we investigated the corticosterone (CORT) and heterophil:lymphocyte (H:L) ratio responses of Australian arid zone birds to heat exposure. The CORT and H:L ratio responses of budgerigars (*Melopsittacus undulatus*), zebra finches (*Taeniopygia guttata*), and diamond doves (*Geopelia cuneata*) to heat exposures were different among the species. The birds were exposed to a temperature similar to what they experience during a typical summer day (35 °C) and a higher temperature (45 °C) similar to that experienced during a heat wave. There were no significant increases of CORT concentrations before and after heat exposure in zebra finches and budgerigars at 35 °C and 45 °C, but there was a significant increase in CORT in diamond doves after exposure to 45 °C.[38] The H:L ratios increased significantly after heat exposure in budgerigars at 35 °C and 45 °C, and diamond doves at 35 °C.[38] No significant correlation was found between the change in CORT and H:L ratios.[38] The data suggest that there are species differences in birds' stress responses to heat exposure that may reflect their ability to detect and adapt to high temperatures. There appears to be differences between the two types of stress measurements which may reflect differences in the timescales of these responses.

Gene expression

We studied the mRNA expression of 20 genes in the organs collected from the birds used in the stress response study described above. The results indicated that acute exposure of native Australian birds to high temperatures (45 °C) results in upregulation of heat shock protein (hsp) genes, but there was no significant upregulation of other genes with protective effects against cell damage, such as B-cell lymphoma 2 protein (BCL-2) and vascular endothelial growth factor A (VEGFA), nor genes associated with inflammation (interleukins). There was no downregulation of the genes involved in the coagulation pathway (eg fibrinogen) in these birds.

The genes for fibrinogen A, B and G were generally unchanged in this study. The genes for fibrinogen A and G were downregulated in the diamond dove gastrointestinal tract and diamond dove lung. There was no change in the expression of the fibrinogen genes in any other tissues or species. The general lack of downregulation across most tissue types indicates birds were not close to suffering from heat stroke under the heat exposure conditions in this study.

Interleukin 6 was only upregulated in the budgerigar liver, while interleukin 8 was upregulated in the diamond dove liver and lung. This is a possible indication that these particular organs in budgerigars and diamond doves were more susceptible to inflammation from the effects of heat exposure, and the immune-regulatory effects of interleukins 6 and 8 were important in protecting their organs.

Other non-hsp genes included in this study were albumin, BCL-2, and VEGFA. Albumin is involved in the binding of heat shock factor protein 1 to the heat shock element in the promoter to activate other hsps such as hsp70 and hsp90.[39] B-cell lymphoma 2 protein[40] and VEGFA[41] has anti-apoptosis effects that can be important in protecting against cell death during heat exposure. Similar to the genes for hsp27, there was generally little up- or down-regulation of these genes, with no clear pattern to their regulation. Albumin was upregulated only in the zebra finch gastrointestinal tract. Bcl-2 was only upregulated in the diamond dove liver. VEGFA was unchanged in all organs and species. This is likely an indication that the extent of heat exposure in this experiment was insufficient to induce cell apoptosis, and the anti-apoptotic effects of these proteins were less important.

The gastrointestinal tracts of all 3 bird species had the highest number of hsp genes upregulated, possibly indicating that this is the organ that requires the most protection to continue its function. Diamond dove organs had the highest number of hsp genes upregulated, possibly a reflection of their ability to protect their cells better during high temperatures.

Behavioral adaptations

Observations of 8 species of Australian arid zone birds at the Adelaide Zoo revealed that the proportion of time spent on stationary behaviors, feeding, in the sun and on the ground differed amongst species of birds, between mornings and afternoons, and weather. Wing-venting was used more frequently during the hottest observation periods. The smallest birds in the study utilized wing venting more than other species of birds, possibly because it is more important for them to conserve water.[42] Psittaciform birds spent less time feeding and more time resting in cooler microsites during hot periods.[42] Columbiform birds continued feeding and spent more time in the sun rather than resting in cooler microsites.[42] White-browed woodswallows (*Artamus superciliosus*) spent a significantly lower proportion of time on stationary behaviors and higher proportion of time feeding compared to the other species.[42] These results suggest that columbiform birds may have an advantage during heatwaves as they can continue feeding through high ambient temperatures, as long as there is adequate access to food and water.

The difference between columbiform birds in terms of their physiologic stress responses and behavioral responses to heat exposure corresponds to their higher critical thermal maximum (ie, the ambient temperature at which an animal loses its ability to regulate its body temperature) compared to birds belonging to other orders such as psittaciform and passeriform birds.[22,43-45] Arid-zone columbiform birds can utilize highly efficient cutaneous evaporative water loss during heat stress to remove heat produced or gained.[46] Wild pigeons living in arid habitats are able to withstand temperatures up to 60 °C (°F) for long periods without showing signs of stress.[46-48] Cutaneous evaporation is considered a more effective method of water loss than respiratory evaporative mechanisms as it prevents excess heat production from the increased respiratory muscle movement while preventing the influx of hot ambient air into the lungs of the bird.[22,46] The levels of cutaneous evaporation within pigeons change to accommodate changing body temperatures. There may be fewer restrictions on other activities important to their survival in the desert (eg, drinking, foraging and mating) during heatwaves, giving them an advantage over other birds.[49]

Clinical features of avian heat injuries

An inability of the body to cope with extreme temperatures will eventually lead to heat illness. Pathologic findings in animals that have died from heat illness are most often manifested in vascular and coagulation aberrations, such a congestion, hemorrhage and thrombosis. We histopathologically examined 8 each of captive-bred budgerigars, zebra finches and diamond doves post-heat exposure as model species for the bird orders Psittaciformes, Passeriformes and Columbiformes respectively.

While the exact pathophysiology of heat stroke in birds is not as well studied, the most common pathologic changes associated with heat stroke in mammals are damage to multiple organs including the jejunum, liver, spleen, lung and kidney.[50] In a case report, the pathologic changes of a bird that was suspected to have suffered heat stroke under general anesthesia revealed mild multifocal acute degeneration and contraction band necrosis of the biceps femoris muscle, as well as diffuse moderate acute congestion of the lungs.[51] There was a diffuse moderate chronic left ventricular hypertrophy, but that was suspected to be a coincidental finding that contributed to the bird's death rather than a result of the presumed heat stroke.[51]

There was mild to moderate congestion of the lungs of 28 of the 36 birds examined, including all the budgerigars and diamond doves. Eight of 15 zebra finches had no significant lung congestion. Interstitial and airway hemorrhage was in observed in one diamond dove. The most common finding in the livers was micro- and macrovesicular hepatocellular vacuolation, suggestive of lipidosis or glycogenosis; observed in in 4 zebra finches, 5 budgerigars and all 8 diamond doves. Hepatic lipidosis was more likely because the negative PAS stain, size of vacuoles, displacement of nuclei and other features were more consistent with lipid rather than glycogen.

Mild to moderate hepatic sinusoidal congestion was seen in two zebra finches and a budgerigar. There were no significant findings in the kidneys except for mild to moderate congestion in one zebra finch, two budgerigars and four diamond doves. Similarly, the gastrointestinal tract appeared normal except for mild to moderate congestion of submucosal vessels in one zebra finch and seven budgerigars. One diamond dove had focal infiltration of adipocytes within the myocardium and mild myocardial vacuolation. This is likely to be intracellular lipid reflective of a pathologic change associated with acute myocardial vacuolar swelling and degeneration as a result of acute ischemia. These findings reflect species differences in the histopathologic changes post-heat exposure amongst birds. The lack of significant histopathologic findings in the diamond doves may be another indication that columbiform birds are superior in coping with the effects of heat exposure.

In another experiment, we studied the heart rate, blood biochemistry, and histopathologic findings in galahs (*Eolophus roseicapilla*) and pigeons (*Columba livia*) after heat exposure under standardized conditions to induce heat stroke. We found that birds subjected to heat stroke had a sudden increase in heart rate and cloacal temperature, followed by a decrease in heart rate, when heat stroke has been induced. The most consistent changes in the blood associated with heat stroke were increases in potassium and AST. There were increases in creatinine kinase and glutamate dehydrogenase in some heat stroke birds, but these were not as consistent. The histopathologic changes in this study did not reveal any significant cardiac changes, but there were some acute congestion in the lungs. Some of the experimental pigeons had indications of necrosis in the pectoral muscle. There were significant hepatic changes in some experimental galahs, but not in the pigeons.

Heat injuries in birds remain difficult to diagnose both ante- and post-mortem due to the lack of pathognomonic features. However, our results have revealed some changes that may be seen in birds with heat injuries that may help guide avian veterinarians in making a diagnosis of heat injuries. These results form the basis for future research as they indicate that there may be species differences in the changes associated with heat injuries, and as such, further studies are needed for species that may be more susceptible.

Conservation Significance and Conclusion

The results of this project indicate that there are a variety of adaptations which birds can utilize to prevent themselves from succumbing to the effects of high environmental temperatures. Some groups of birds, such as psittacine and passerine birds, restrict their activities to shaded and cooler microsites at high temperatures, especially during the early afternoon when solar radiation levels are high. The loss of these microsites in the wild, as predicted with climate change,[52] will therefore affect these groups of birds more columbiform birds, that are less reliant on these microsites in order to continue displaying their normal behaviors during heatwaves. Without shaded microsites, birds cannot seek shelter during heatwaves, thus decreasing the amount of time birds can survive without water at high temperatures.[53] This will in turn decrease the amount of time birds have available to rest and "wait-out" a heatwave before they have to take the risk of flying to seek a water source.

Habitat, and consequently microhabitat, conservation is therefore crucial in ensuring the survival of these bird species as temperatures continue to rise, but unfortunately, this is also the most expensive aspect of conservation compared with other strategies such as captive breeding.[52] Failure to conserve thermal refuges for these more sensitive bird species will likely lead to a restructuring of bird populations in arid environments, which may have unpredictable ecological effects on the rest of the ecosystem.[53,54]

On the other hand, birds that can tolerate very high environmental temperatures (eg, columbiform birds) may hold secrets that can be uncovered with further research. These may include more advanced molecular genetic techniques such as RNAseq, with a focus on heat shock proteins. Further understanding of the genetic adaptations required to confer high tolerances to heat will allow better identification of vulnerable species of birds, so that appropriate resources can be allocated to helping them survive the effects of climate change proactively rather retrospectively.

Acknowledgments: We thank the Holsworth Wildlife Research Endowment and BirdLife Australia's Stuart Leslie Bird Research Award for providing funds for the research project. We also thank all the students and collaborators involved in the project.

References

1. Garnett S, Franklin D. *Climate Change Adaptation Plan for Australian Birds.* Victoria, Australia: CSIRO Publishing;2014.

2. Braun EJ, Dantzler WH. Function of mammalian-type and reptilian-type nephrons in kidney of desert quail. *Am J Physiol.* 1972;222:617-629.

3. Schmidt-Nielsen B, O'Dell R. Structure and concentrating mechanism in the mammalian kidney. *Am J Physiol.* 1961;200:1119-1124.

4. Heisinger J, Breitenbach RP. Renal structural characteristics as indexes of renal adaptation for water conservation in the genus *Sylvilagus*. *Physiol Zool.* 1969;42:160-172.

5. Williams JB, Tieleman BI. Physiological ecology and behavior of desert birds. In: Nolan VJ Jr, Thompson CF, eds. *Current Ornithology.* New York, NY: Kluwer Academics Plenum Publishers;2001:299-353.

6. Goldstein DL, Braun EJ. Structure and concentrating ability in the avian kidney. *Am J Physiol.* 1989;256: R501-R509.

7. Casotti G, Richardson K. 1993. A qualitative analysis of the kidney structure of Meliphagid honeyeaters from wet and arid environments. *J Anat.* 1993;182:239-247.

8. Casotti G, Braun EJ. Renal anatomy in sparrows from different environments. *J Morphol.* 2000;243: 283-291.

9. Brummermann M, Braun EJ. Effect of salt and water balance on colonic motility of white leghorn roosters. *Am J Physiol.* 1995;268:R690-R698.

10. Anderson GL, EJ Braun. Postrenal modification of urine in birds. *Am J Physiol.* 1985;248:R93-R98.

11. Amanova M. The ecological-physiological adaptations of water metabolism in desert birds. *J. Arid Environ.* 1984;7:199-200.

12. Braun E. An overview of avian renal function. *Proc 6th Int Symp Avian Endocrinol.* 1997;281-287.

13. Braun EJ. Regulation of renal and lower gastrointestinal function: role in fluid and electrolyte balance. *Comp Biochem Physiol A: Mol Integr Physiol.* 2003;136:499-505.

14. Laverty G, Skadhauge E. Physiological roles and regulation of transport activities in the avian lower intestine. *J Exp Zool.* 1999;283:480-494.

15. McWhorter TJ, Caviedes-Vidal E, Karasov WH. The integration of digestion and osmoregulation in the avian gut. *Biol Rev.* 2009;84:533-565.

16. Hoffman TC, Walsberg GE, DeNardo DF. Cloacal evaporation: an important and previously undescribed mechanism for avian thermoregulation. *J Exp Biol.* 2007;210:741-749.

17. Lucas AM, Stettenheim PR. *Avian anatomy: Integument.* Washington, DC: US Government Printing Office;1972.

18. Peltonen L, Arieli Y, Harjula R, Pyörnilä A, Marder J. Local cutaneous water barrier in cold-and heat-acclimated pigeons (*Columba livia*) in relation to cutaneous water evaporation. *J Morphol.* 2000;246:118-130.

19. Menon GK, Menon J. Avian epidermal lipids: functional considerations and relationship to feathering. *Am Zool.* 2000;40:540-552.

20. Muñoz-Garcia A, Williams JB. Cutaneous water loss and lipids of the stratum corneum in house sparrows *Passer domesticus* from arid and mesic environments. *J Exp Biol.* 2005;208:3689-3700.

21. Haugen M, Williams JB, Wertz P, Tieleman BI. Lipids of the stratum corneum vary with cutaneous water loss among larks along a temperature-moisture gradient. *Physiol Biochem Zool.* 2003;76:907-917.

22. McKechnie AE, Whitfield MC, Smit B, et al. Avian thermoregulation in the heat: efficient evaporative cooling allows for extreme heat tolerance in four southern Hemisphere columbids. *J Exp Biol.* 2016;219:2145-2155.

23. Wolf B, Walsberg G. 1996. Respiratory and cutaneous evaporative water loss at high environmental temperatures in a small bird. *J Exp Biol.* 1996;199:451-457.

24. Tieleman BI, Williams JB. Cutaneous and respiratory water loss in larks from arid and mesic environments. *Physiol Biochem Zool.* 2002;75:590-599.

25. Ro J, Williams JB. Respiratory and cutaneous water loss of temperate-zone passerine birds. *Comp Biochem Physiol A: Mol Integr Physiol.* 2010;156:237-246.

26. Smit B, Zietsman G, Martin R, et al. Behavioural responses to heat in desert birds: implications for predicting vulnerability to climate warming. *Clim Chang Responses.* 2016;3:9.

27. Schmidt-Nielsen K, Schroter R, Shkolnik A. 1981. Desaturation of exhaled air in camels. *Proc R Soc Lond B Biol Sci.* 1981;211:305-319.

28. Tieleman BI, Williams JB, Michaeli G, Pinshow B. The role of the nasal passages in the water economy of crested larks and desert larks. *Physiol Biochem Zool.* 1999;72:219-226.

29. Dawson WR. Evaporative losses of water by birds. *Comp Biochem Physiol A.* 1982;71:495-509.

30. O'Connor RS, Wolf BO, Brigham RM, McKechnie AE. Avian thermoregulation in the heat: efficient evaporative cooling in two southern African nightjars. *J Comp Physiol B.* 2017;187:477-491.

31. Bucher TL. Oxygen consumption, ventilation and respiratory heat loss in a parrot, Bolborhynchus lineola, in relation to ambient temperature. *J Comp Physiol B.* 1981;142:479-488.

32. Weathers W, Schoenbaechler D. Regulation of body temperature in the Budherygah, Melopsittacus undulatus. *Aust J Zool.* 1976;24:39-47.

33. Weathers WW, Caccamise DF. Temperature regulation and water requirements of the monk parakeet, *Myiopsitta monachus. Oecologia.* 1975;18:329-342.

34. Sverdlova N, Arkali F, Witzel U, Perry S. Computational fluid dynamics model of avian tracheal temperature control as a model for extant and extinct animals. *Respir Physiol Neurobiol.* 2013;189:67-75.

35. Calder W, Schmidt-Nielsen K. Panting and blood carbon dioxide in birds. *Am J Physiol.* 1968;215:477-482.

36. McKechnie AE, Wolf BO. Partitioning of evaporative water loss in white-winged doves: plasticity in response to short-term thermal acclimation. *J Exp Biol.* 2004;207:203-210.

37. Gerson AR, Smith EK, Smit B, et al. The impact of humidity on evaporative cooling in small desert birds exposed to high air temperatures. *Physiol Biochem Zool.* 2014;87:782-795.

38. Xie S, Romero LM, Htut ZW, McWhorter TJ. Stress responses to heat exposure in three species of Australian desert birds. *Physiol Biochem Zool.* 2017;90:348-358.

39. Yiangou M, Paraskeva E, Hsieh C-C, et al. Induction of a subgroup of acute phase protein genes in mouse liver by hyperthermia. *Biochim Biophys Acta.* 1998;1396:191-206.

40. Balcer-Kubiczek EK, Harrison G, Davis CC, et al. Expression analysis of human HL60 cells exposed to 60 Hz square-or sine-wave magnetic fields. *Radiat Res.* 2000;153:670-678.

41. Gerber H-P, Dixit V, Ferrara N. Vascular endothelial growth factor induces expression of the antiapoptotic proteins Bcl-2 and A1 in vascular endothelial cells. *J Biol Chem.* 1998;273:13313-13316.

42. Xie S, Turrell EJ, McWhorter TJ. Behavioural responses to heat in captive native Australian birds. *Emu-Austral Ornithology.* 2017;1-17.

43. Smith EK, O'Neill J, Gerson AR, Wolf BO. Avian thermoregulation in the heat: resting metabolism, evaporative cooling and heat tolerance in Sonoran Desert doves and quail. *J Exp Biol.* 2015;218:3636-3646.

44. Wolf B, McKechnie A, Gerson A, et al. Avian thermoregulation in the heat: tolerance to heat stress varies greatly among species. *Physiologist.* 2015;58:56.

45. Whitfield MC, Smit B, McKechnie AE, Wolf BO. 2015. Avian thermoregulation in the heat: scaling of heat tolerance and evaporative cooling capacity in three southern African arid-zone passerines. *J Exp Biol.* 2015;218:1705-1714.

46. Marder J, Gavrieli-Levin I. The heat-acclimated pigeon: an ideal physiological model for a desert bird. *J Appl Physiol.* 1987;62:952-958.

47. Marder J. Cutaneous water evaporation—II. Survival of birds under extreme thermal stress. *Comp Biochem Physiol A: Physiol.* 1983;75:433-439.

48. Marder J, Arieli Y. Heat balance of acclimated pigeons (*Columba livia*) exposed to temperatures up to 60° C Ta. *Comp Biochem Physiol A: Physiol.* 1988;91:165-170.

49. Walsberg GE. Thermal consequences of diurnal microhabitat selection in a small bird. *Ornis Scandinavica* 1993;174-182.

50. Roberts GT, Ghebeh H, Chishti MA, et al. Microvascular injury, thrombosis, inflammation, and apoptosis in the pathogenesis of heatstroke: a study in baboon model. *Arterioscler Thromb Vasc Biol.* 2008;28: 1130-1136.

51. Hofmeister EH. Anesthesia case of the month. *J Am Vet Med Assoc.* 2005;227:718-720.

52. Garnett S, Franklin D, Ehmke G, et al. *Climate Change Adaptation Strategies for Australian Birds: Final Report.* National Climate Change Adaptation Research Facility: Gold Coast; 2013:109.

53. Wolf B. Global warming and avian occupancy of hot deserts; a physiological and behavioral perspective. *Revista Chilena de Historia Natural.* 2000;73:395-400.

54. McKechnie AE, Hockey PA, Wolf BO. Feeling the heat: Australian landbirds and climate change. *Emu.* 2012;112:i-vii.

Medical Presentations in Free-flight Shows

Terri Parrott, DVM

Session #2011

Affiliation: From St. Charles Veterinary Hospital, Davenport, FL 33837, USA.

Abstract: Free-flight bird shows that encompass multiple species with different flight patterns can be medically challenging to the clinician. Traumatic events due to outside forces are common. Foreign body ingestion, toxin exposure, and behavioral changes in the birds are always in the subjective presentation and plan when working these athletes up in the clinical setting. Physical exam and work-up plans are discussed as well as treatments and outcomes.

Introduction

Free-flight bird shows and exhibits allow the birds to exhibit normal flight biomechanics and behavior patterns. Exhibits and stage shows are short flight patterns where as long range patterns (one or more miles) incorporate even more species specific behaviors and allow more "free-thinking" behavior for the animals. Traumatic injuries and ingestion of foreign and toxic substances can be real problems when these birds are allowed to take flight. The clinician must act quickly to triage these birds when presented. The goals in these programs are to diagnose, treat and return the birds back to their group as quickly as possible.

Trauma

Traumatic injuries vary greatly. Trainers will usually find blood in the cage or on the bird, feather or skin abnormalities and/or a depressed and isolated individual. The worst scenario is a bird that does not return and is found severely injured. The first call from the training staff is an indicator that an event has occurred. If possible have them look at the bird and give you a quick assessment on the phone while they are gathering a crate or cage for transport. If the bleeding is ongoing, have them isolate the area and apply pressure or give/pack the area with Yunnan Baiyao (Chinese herb). Sometimes the trauma is very evident when the physical exam is performed such as a wing droop or a displaced digit or nail. The trainer should be able to wrap or bandage the area for transport. When the bird presents at the hospital, a quick assessment and history can be taken and the patient should be anesthetized and thoroughly examined. CBC and chemistries are taken at this time as well as whole body radiographs. Oral and eye lacerations and open tissue with muscle, tendon or bone exposure should be addressed promptly. Cautery should be available and fractured nails and small vessel bleeding treated immediately. Small skin lacerations should be cleaned and sutured. Corneal integrity should be assessed promptly so a further workup plan can be formed. Intravenous catheters can be placed while under anesthesia with sites of placement depending on the size of the bird and area of injury. The ulnar vein is a preferred site as there is less problems with biting or chewing the catheter compared to the medial metatarsal vein.

Tongue trauma

Traumatic injuries to the tongue are over represented in psittacines with macaws taking a higher percentage in lacerations and injuries. The skin on the tongue is friable and usually more than one laceration is evident on exam.

Ophthalmic suture (5-0 to 7-0) is used on a tapered needle to repair these lesions. Historically, if more than 50% of the tongue is lost due to vascular supply or amputation, the bird's prognosis is poor for long-term survival.

Fractures

Skeletal injuries are a serious complication. Stabilization with a treatment plan must be discussed with trainers, as the ultimate goal is to return these birds to free flight. Injuries to the patagium carry a guarded prognosis for return to full flight as the area is hard to suture. Techniques to stretch the healing tissue have been described.[1] Traumatic injuries of the shoulder girdle should heal with conservative management. Fractures and large subluxations of the proximal coracoid from the sternum must be surgically reduced. The basics of orthopedic repair will not be discussed in this paper. Load bearing on legs and mobility of the wings as soon as possible is the goal of the surgeon. In birds, periosteal vascular integrity is very important for the healing process as there is minimal intramedullary blood supply into the periosteum.[1] External fixations with IM pins (titanium or steel) work well on many of the wing and leg fractures. In smaller bones (carpal and metacarpal bones) external fixation alone has been more successful when returning the bird to flight. Use of methacrylate based composite resin, Acrylx (Imex Veterinary, Inc, 1001 McKensson Drive, Longview, TX, USA), is ideal due to its ease of use, strength and light weight. After 24 hours of a pressure or coaptive bandage, the patient should be re-evaluated for swelling and any hemorrhage. Removal of the bandage in 2-3 days is the goal for increased range of movement in the thoracic limbs.[1] It is important to make sure these birds receive adequate pain control with a mix of both non-steroidal anti-inflammatories and opioids.

Head trauma

Any trauma involving the head is an immediate emergency and a complete neurologic exam should be performed with inclusion of skull radiographs under anesthesia. If the bird is actively seizing, midazolam HCl 5 mg/ml injectable can be given cloacally (authors preference) or IM at the dose of 0.1 ml/300 g to effect.[2] The bird may have to be stabilized with oxygen before any further diagnostics. In recent trauma, a fast-acting corticosteroid can be administered intravenously or intramuscularly.[3]

Air sac rupture

This condition frequently presents with a balloon-like deformity of the skin, many times around the head and neck, and sometimes involving the whole bird. The cause of the rupture may never be found, and the bird is uncomfortable and sometimes depressed. Giving a non-steroidal anti-inflammatory medication will help the bird become more comfortable and reduce traumatic inflammation. Radiographs under anesthesia and complete CBC and chemistries are important to rule out any internal injury. Making a percutaneous opening into the air sac and inserting a cannula for a few days will allow the area to heal and keep the bird "deflated".

Pododermatitis

Inflammation of the feet encompasses many etiologies. Traditionally, environmental factors such as perching and enclosure substrate are evaluated. In free-flighted birds landing areas also are investigated. Rope perches and swings, while being a soft landing platform, are abrasive and have been incriminated in our larger birds. Staging of the disease is important and has been described.[4] Arthritic changes in some of the birds have resulted in pressure sores on the plantar surface of the feet due to shifting leg lameness. Early identification of the problem with rapid topical treatment, therapy laser treatments, nutritional supplements and use of non- steroidal anti-inflammatories has helped many of these conditions improve. Do NOT remove the birds from flight as this will only cause the condition to worsen.

Beak trauma

Beak injury presents in many forms. In hornbills, traumatic tip amputation is common due to the bird sticking the end of the beak thru exhibit caging. Bleeding should be controlled with pressure or any clotting powder available and the bird transported to the clinic immediately. In many cases just the upper (maxillary) portion of the beak has been removed leaving a longer mandibular portion. Because the bird uses the beak for everyday prehension, it is important to even out the defect as soon as possible. Crushing defects are common in psittacines due to cage mate aggression and perforating wounds. These beak wounds can be cleaned, radiographed, and usually repaired with absorbable monofilament mesh (Bard- Davol, Providence, RI, USA) to fill the defect and acrylics. These birds may be returned to flight; however, they must be monitored and the repair re-evaluated as the acrylics and mesh may need re-application. Larger beak defects will have to be cleaned, radiographed, and evaluated for repair with prosthetic devices and/or acrylics. Species differences, injury location, as well as long-term care must be discussed as "return to flight" and "quality of life" are the ultimate goals.

Foreign Bodies

Inhalation

Nasal aspiration of plant material and seeds are manifested by sneezing and digging at the area by the bird. The trainer may notice a closed nostril and sometimes blood around the nares. These birds should be examined under anesthesia as many times manipulation while awake may further traumatize the bird and push the foreign body further into the cavity. Ophthalmic forceps can be used to grab the material. Radiographs of the skull and sinus should be taken to further evaluate the airway. Tracheal foreign bodies are a critical emergency. The call will come in that the bird is open mouth gasping or has collapsed in flight. Trainers should immobilize the bird and oxygen (if available) should be administered in transport. Anesthesia is important; however, if the trachea is occluded an abdominal air sac cannula should be placed immediately and oxygen and anesthesia administered. Quickly start a visual exam from the oral cavity to locate the foreign body. If located in the upper section of the trachea, a tracheostomy below the foreign body should be made and a small cannula or red rubber tube can be inserted caudal to the obstruction and the foreign body pushed out the oral cavity or blown out by air in a 6-12 ml syringe.

Ingestion

If the ingestion is acute and witnessed by the trainer immediate removal from the upper GI is imperative. Anesthesia and radiographs will locate most of these foreign bodies. Removal via endoscopy is usually non-complicated and minimally invasive. If the bird is discovered without a piece of equipment or regurgitation and anorexia with daily weight loss is observed, a foreign body should be suspected. Radiographs under general anesthesia may locate the problem. The birds species, size and habits will determine treatments. Species that "pellet" indigestible matter such as fur, bone and claws will be treated with lactulose and will continue to be fed their normal diets, if the foreign body is in the proventriculus or the ventriculus. Most shorebirds and hornbills will also pellet. If the object is further down the GI tract and the clinician is aware of the shape and size, lactulose will be given along with the bird's normal diet and, in some species, asparagus given to bind around the object. The trainers must monitor the isolated bird's enclosure until the object is found or the animal is re-radiographed and found to have passed the foreign body.

Toxin Exposure

Ingestion

The bird will frequently present with anorexia, weight loss or behavior changes such as depression, seizures, or exercise intolerance. The work up should include a complete blood count, comprehensive chemistry, gram stain and choanal and cloacal culture. If metallic foreign bodies are found, both zinc and lead levels should be measured. Treatments for these heavy metals have been described.[5] If elevated organ values are found on the blood screen, then further evaluation of plant and external toxins is warranted. Abnormal droppings should be collected and evaluated. Many times while in flight, plant material is ingested and some of these plants may be toxic. Trainers should be vigilant when watching the birds return or when out on perching areas as to what they have in their mouths or claws.

External exposure

Pesticides used in the flight area should be evaluated as to when and where the chemicals are used. In Florida, routine spraying and outside fumigation is common in the summer months and residual insecticide can be found in standing water and plants that the birds may have been exposed to. Herbicides and fungicides have not been incriminated in acute toxicity in any of our flighted birds. Birds showing acute neurologic signs or depression should have blood taken and cholinesterase levels checked for exposure to organophosphates. Extra label use of pralidoxime chloride (2-PAM) has been described.[2] Supportive care with fluids, oxygen, anti-seizure medications and parental nutrition is important while recovering.

Change in Physical Condition

Reproductive changes

Free-flight birds have intact reproductive organs that are maturing and changing. The gonadal changes occurring at times of the year will influence the bird's physical condition, attention, and flight behavior. Some of these birds will manifest self-mutilation of their feathers. Many will become anorexic or isolated and act depressed. Some species will exhibit "nesting" behaviors such as tearing up substrate in the enclosure, masturbation on objects, and/or regurgitation to the trainers, perches or other birds. Some birds may suddenly become aggressive. A complete work up of these individuals should include complete blood counts, serum chemistries, cultures, radiographs and parasite (external and internal) exams. Endoscopic evaluation of their gonads also can be performed. Treatments for reproductive behaviors should include manipulation of their light/dark cycle, hormonal therapy, and behavior modification techniques.

Age-related changes

Age-related physical changes are common in large groups of free-flighted bird teams. Having a balance of different age groups often is encouraged as the older birds will train acceptable flight behaviors into the younger members of the team. As older birds are flying they may be more prone to injuries, such as ligament and/or tendon strains and arthritic changes in the legs and feet due to repetitive landing behaviors. Larger birds, such as cranes and vultures, may exhibit shifting leg lameness with no external abnormalities. These birds should be radiographed and serum chemistries and complete blood counts performed. Elevated uric acid may be an indicator of gout, with or without radiographic changes, and should be addressed by diet changes and, if needed, supplemental medication.[4] Changing the landing areas on these birds also is critical. Use of non-steroidal anti-inflammatories may be indicated.

Cardiovascular changes in older birds may become evident with longer flights. The trainers will observe that the bird will not be able to complete the flight and will take short rest stops. The bird may be extremely "winded" when returning to the enclosure. These birds should have complete hematologic and radiographic workups as well as echocardiograms with Doppler flow measurements performed by a radiologist that has experience with bird cardiac measurements. Placing these birds on medication such as ace inhibitors can improve quality of life and exercise tolerance. Return to flight should be re-evaluated and shorter durations may be indicated.

Parasites

Free-flighted birds are exposed to other wildlife and are susceptible to both internal and external parasites. Feather loss warrants a thorough physical exam. An external parasite check should include cytology of the feathers. Quills may be sent off to a laboratory for a mite exam and identification. Browse that is brought in should also be cleaned and sprayed with a pyrethrin based bird spray. Both external spraying and monthly lufenuron can be used for parasite removal and prevention. Regular (monthly) fecals should be performed on groups of birds housed and flying together. If abnormal stools are observed or weight loss is documented isolated bird fecals should be examined. Internal parasite species that have been documented in our free-flight birds include, but are not limited to, *Capillaria*, *Isospora*, *Giardia*, *Syngamus*, ascarids, *Trichomonas*, cestodes, *Haemoproteus*, and *Sarcocystis*. Both direct, flotation and trichrome stains are routinely performed.

Outside Forces

Weather

Scheduled flights should be delayed or cancelled during inclement weather conditions. Lightning strikes and increased rain and wind gusts have caused disorientation and injury to birds that were allowed to continue flights.

Audience interference

With pre-show communication on rules about touching, using flash photography and using external devices to touch the birds, much of the audience interference has been curtailed. Some humans do not listen and bystander interference in flight happens. Once the bird's trajectory path has been changed, the bird becomes confused and frightened and many times will land before the "B" destination. Some birds will miss the "B" destination and fly off. Other birds see food sources held by small humans as an "easy-fix" and will deviate from the flight path. ALL these birds should be removed from the general flight schedule and "re-conditioned." Time frames on re-learning their flights vary with the species and age of the birds.

Group pressure

In a flock of free-flight birds there are personalities that range from followers to leaders. Some of the birds will stick together in every flight and others free fly on their own. Routines are to be followed but there are also "free thinkers" in some groups and there will be fly offs. One or two times in multiple flights should be expected. When the same perpetrators are discovered, these birds should be pulled from the flight group and re-trained.

Other species aggression

Encounters of birds in flight by birds of prey happen regularly. The free-flight captive bird may be physically and emotionally traumatized. A thorough exam should be performed when the bird returns and any injuries treated. The bird should be removed from the flight schedule for a few days to many weeks depending on the bird's injuries and behavior. Birds that deviate away from the flight route can also put themselves in danger from other exhibit animals. Birds that land in open enclosures of primates, large reptiles and carnivores will fall prey to these animals and may be found dead hours or days later.

References

1. Speer BL. *Current Therapy in Avian Medicine and Surgery*. St. Louis, MO: Elsevier Saunders;2016.
2. Carpenter JW. *Exotic Animal Formulary*. 4th ed. St. Louis, MO: Elsevier Saunders;2013.
3. Redig PT, Cooper JE. *Raptor Biomedicine*. U of Minnesota Press;1993.
4. Altman RB, Clubb S, Dorrestein G. *Avian Medicine and Surgery*. Philadelphia, PA: Saunders;1997.
5. Rosskopf WJ, Woerpel RW. *Diseases of Cage and Aviary Birds*. 3rd ed. Baltimore, MD: Lea and Febiger;1996.

Chronic Infectious Diseases and Treatment Options

Thomas N. Tully, Jr, DVM, MS, DABVP (Avian Practice), DipECZM (Avian)

Session #2014

Affiliation: From the Louisiana State University School of Veterinary Medicine, Baton Rouge, LA 70803, USA.

Abstract: Chronic infectious disease conditions are often difficult to both diagnose and treat. Patients are often presented in a debilitated state, are immunocompromised, and have secondary disease conditions. The secondary disease conditions may be the most life threating problems when the bird is first presented. It is incumbent upon the veterinarian to stabilize the patient and treat the most severe health problems first, while always considering the underlying cause of the initially diagnosed abnormalities. Although there are diagnostic tests available, unfortunately there or few that will provide a definitive diagnosis for chronic disease conditions apart from histopathological examination of affected tissue samples or bacterial isolation of the offending organism. Treatment options are available but generally require long term application and owner compliance in order to obtain an adequate treatment response. In some cases it is unlikely that one can guarantee complete clearance of the organism from the patient, even with appropriate treatment and compliance by the owner. A review of common chronic disease conditions, diagnostic tests available and current treatment options will be provided.

Avian Chlamydiosis

Parrot fever or psittacosis are two terms used to describe the *Chlamydia psittaci* organism when it infects humans. The terminology of this disease is important for veterinarians to know in order to communicate and inform owners of the importance of protecting themselves and birds against possible exposure. Since *C. psittaci* responds rapidly to treatment with the tetracycline group of antibiotics it is imperative that a diagnosis is made and treatment initiated as soon as possible. This non-motile, obligate, intracellular parasite has not been satisfactorily classified taxonomically but is generally considered a bacterial organism, has been one of the most common bacterial diseases that infect pet birds.[1] *C. psittaci* has been the subject of much research regarding avian species as well as humans and other animal groups.[1-5] The fact that this intracellular bacterium is a zoonotic agent has prompted the research community to improve diagnostic capabilities, treatment and prevention recommendations not only for the benefit of animal health, but human health as well. With the usage of antibiotic therapy and the development of better diagnostic testing techniques the pathogenic impact of *C. psittaci* has been reduced, but has not removed much of the mystery surrounding its diagnosis and treatment in animals, especially regarding avian species. Recent scientific studies have led to new findings on classification and diagnostic testing of the *C. psittaci* organism.[4-5] A protocol to control *C. psittaci* infection among humans and pet birds has been published by the Centers for Disease Control and Prevention (CDC).[6]

Even with the increased diagnostic testing capabilities, veterinarians who are the most successful in diagnosing and treating *C. psittaci* in avian species follow basic medicine principles incorporating the use of history, clinical signs, physical examination and finally, diagnostic test results. For an accurate avian workup, a routine list of procedures should be followed as they relate to the presenting complaint or clinical signs. Recommended procedures for a routine avian health examination include an external physical, fecal Gram-stain, direct fecal parasite check, appropriate cultures (choanal, crop, or cloaca), complete blood count (CBC), and testing for *C. psittaci.*[7] The same diagnostic protocol should be used for birds exhibiting clinical signs of diarrhea, respiratory distress, as well as for birds exhibiting nonspecific disease signs as a means to establish a minimum data base for that patient. When any abnormalities are noted on the primary workup, additional tests may be indicated (eg, serum chemistry panel, radiology, and/or other specific diagnostic tests).[7] Veterinarians should have a comprehensive understanding of the overt clinical signs associated with avian chlamydiosis, interpretation of CBC, serum chemistry panels, radiology and diagnostic test results.[7] *C. psittaci* is no different than most bacterial infections, the intensity of clinical signs depends on the pathogenicity of the organism and the avian species infected.[7] A bird becomes exposed and subsequently infected with avian chlamydiosis when the bacterium enters the host through the respiratory and/or gastrointestinal epithelium.[8] Depending on the specific biotype involved, dyspnea, nasal and ocular discharge, inflamed choanal and diarrhea are common signs observed in birds infected with avian chlamydiosis*)*.[9] In chronic infections cachexia, poor feather condition, depression and anorexia may be noted by the veterinarian.[8]

A misconception has existed for many years that "lime green" diarrhea is pathogenomic for a *C. psittaci* infection. This myth has been perpetuated because the infectious organisms are associated with liver infection/inflammation.[7] Many disease conditions cause or aid in the development of hepatitis that results in a yellow-to-green color stool.[7] A green-colored stool only indicated possible liver involvement; serum chemistry panel and liver bile acids should be reviewed to confirm the suspected diagnosis.[7]

Polymerase chain reaction (PCR) molecular diagnostic techniques are based on the detection of target DNA sequences, specifically of the pathogen in question.[5] Current tests that use PCR technology can test choanal and cloacal material harvested on culturette swabs or whole blood for DNA sequences of the *C. psittaci* organism. Using current commercially available diagnostic techniques, the target DNA sequences cannot determine if the material is from a viable organism.[5] Moreover, because of the sensitivity and specificity of molecular testing, false positives from environmental contamination, contamination by sample-to-sample DNA carryover or by DNA carryover from previous amplifications of the same target may occur.[10] Even with the disadvantages listed above, detection of *C. psittaci* genes by amplified and non-amplified nucleic acid-based detection methods is still a viable method to use in screening for this disease. PCR sample collections are easy, noninvasive with simple transport and storage requirements required by most diagnostic laboratories.[11] Other listed advantages of PCR testing include the minimal sampling, ease of testing large bird populations, rapid results and high sensitivity and specifity.[11] Current technology allows for further identification of PCR testing either to specifically identify the species or to confirm a positive test.[12] Serologic testing for *C. psittaci* in pet avian species has been available for a number of years.[13-15] These tests have often worked best when the patient was showing clinical signs associated with avian chlamydiosis. Currently the serologic tests that are available to the veterinary practitioner in North America are direct complement-fixation, elementary-body agglutination, micro immunofluorescence, and ELISA. Serologic testing may continue to be helpful in the future to confirm an overt clinical case of avian chlamydiosis or determine the efficacy of antibiotic treatment.[11] There are also some proponents of a multiparametric diagnostic approach, in which serology is part, that they consider the best avenue to achieve a *C. psittaci* diagnosis. The multiple testing panels are an avenue to overcome diagnostic test deficiencies when trying to definitively determine a patient's avian chlamydiosis status. The gold-standard for the definitive diagnosis of *C. psittaci* in an avian patient has been culture and isolation of the organism.[16-18] The problems have also been published on the difficulties of culturing this organism because of its obligate intracellular nature.[5,18,19] To have the best opportunity for *C. psittaci* culture and isolation on a live patient, serial fecal samples or combined choana/cloaca cultures should be collected for 3-5 consecutive days and pooled in transport media supplied by the diagnostic laboratory.[6] Liver and spleen samples are preferred from necropsied patients for isolation of the intracellular bacterium.

Medication protocols that have been used successfully in treating avian chlamydiosis depend on a number of factors including, owner administration of the antibiotic for 45 days, owner compliance of bringing the bird in for weekly injections, giving full dosage of the medication, reducing the amount of calcium chelation of doxycycline by restricting calcium supplementation at the time of treatment.[7] Owner compliance is often enhanced by the knowledge that avian chlamydiosis is a zoonotic disease. Whether the owner is treating at home with medicated pellets or giving weekly doxycycline injections the treatment should be under the direction and supervised before, during and after treatment by a veterinarian.

By maintaining the proper environment and nutritional requirements of a pet bird species, the owner has the ability to help preserve the health status of the bird by bolstering the immune system. Because of the intracellular nature of *C. psittaci*, even a bird that is properly managed can be infected without exhibiting clinical signs. It is usually through immunosuppressive conditions that the infection becomes a clinical entity. The ability of an infected bird to show no signs of illness emphasizes the need for a 30- to 60-day quarantine prior to placing the bird into an established avian facility or household. By educating the client, hopefully an understanding of the difficult nature of interpreting diagnostic test results will be conveyed, thereby clarifying that the bird may or may not be free of all diseases, particularly *C. psittaci.*[7] Medicated protocols that have been used successfully in treating avian chlamydiosis depend on a number of factors including, owner administration of the antibiotic for 45 days, owner compliance of bringing the bird in for weekly injections, giving full dosage of the medication, reducing the amount of calcium chelation of doxycycline by restricting calcium supplementation at the time of treatment.[6] Owner compliance is often enhanced by the knowledge that avian chlamydiosis is a zoonotic disease. Whether the owner is treating at home with medicated pellets or giving weekly doxycycline injections the treatment should be under the direction and supervised before, during and after treatment by a veterinarian.

Current published information lists the time between exposure to the *C. psittaci* organism to the onset of illness, in most cases from 3 days to several weeks but it may become an active clinical infection years after exposure.[6] There are three case definitions as established by the National Association of State Public Health Veterinarians (NASPHV) and the CDC.[6]

Case definition of avian chlamydiosis[6]

Confirmed case: is defined on the basis of at least one of the following laboratory results: a) isolation of *C. psittaci* from a clinical specimen, b) identification of chlamydial antigen by immunofluorescence (fluorescent antibody [FA]) of the bird's tissues, c) a greater than fourfold change in serologic titer in two specimens from the bird obtained at least 2 weeks apart and assayed simultaneously at the same laboratory, or d) identification of *C. psittaci* within macrophages in smears stained with Gimenez or Macchiavello stain or sections of the bird's tissues.

Probable case: is defined as compatible illness and at least one of the following laboratory results: a) a single high serologic titer in one or more specimens obtained after the onset of signs or b) the presence of *C. psittaci* antigen (identified by ELISA, PCR, or FA) in feces, a cloacal swab or respiratory or ocular exudates.

Suspected case: is defined as a) compatible illness that is epidemiologically linked to another case in a human or bird but that is not laboratory confirmed, b) a subclinical infection with a single high serologic titer or detection of chlamydial antigen, c) compatible illness with positive results from a nonstandardized test or a new investigational test, or d) compatible illness that is responsive to appropriate therapy.

These case definitions should help the clinician interpret the diagnostic test results as they relate to a patient that may have signs consistent with avian chlamydiosis or subclinical cases. To aid in diagnosis a complete patient history should be taken from the owner, an external physical examination performed and basic diagnostic tests submitted including a complete blood count.[19] If avian chlamydiosis is suspected, then treatment should be initiated until the diagnosis is confirmed, a clinical response is achieved or another diagnosis determined.

Treatment for avian chlamydiosis includes doxycycline formulation (Vibrovenos®, Pfizer Laboratories, London, UK). The dose recommendation is 75-100 mg/kg IM q5-7d for the first 4 weeks, then every 5 days for the remainder of the 45-day treatment period, doxycycline calcium syrup (Pfizer Inc, New York, NY) 25-50 mg/kg PO q24h, and doxycycline hyclate (50 and 100 mg capsules, Pfizer Inc, New York, NY) 400 mg/L of drinking water for cockatiels and 400-600 mg doxycycline hyclate/L of drinking water for parrots and macaws made fresh daily for 45 days.[20,21]

Aspergillus spp.

Upper respiratory aspergillosis usually presents as nasal granulomas that grow to deform the nasal anatomy. These fungal granulomas must be debrided and the underlying tissue cultured in order to obtain a definitive diagnosis.

Lower respiratory aspergillosis is diagnosed more often than the upper respiratory fungal granulomas. Both are however considered chronic infectious disease processes. Unfortunately lower respiratory aspergillosis is more difficult to diagnose than the overt nasal granuloma. The bird infected with aspergillosis involving the lungs and air sacs may be depressed and present with mild dyspnea. A complete blood count, plasma chemistry panel and radiographs are the basic diagnostic protocol for African grey parrots presenting with the clinical signs described above. Many times the complete blood count will show a severe, chronic inflammatory leukogram. The total white blood cell count can range as high as 70,000-80,000. Conversely the total white blood cell count can be in normal reference ranges or much lower. Radiographic evidence can range from "cloudy" air sacs, obvious radio opaque granulomas to no abnormalities noted, even in the most severe cases. At this time the best way to definitely diagnose aspergillosis in an African grey parrot is via ceolomic endoscopy. Serologic antigen and antibody testing has not been perfected to the point of offering reliable results for antemortem diagnosis.

Treatment

Once the diagnosis is confirmed or a differential diagnosis list is formulated treatment can begin. The proper treatment may be aided by specific methods that are effective treating the avian respiratory system.

Systemic therapy: There are several options for systemic therapy.[22]

Itraconazole 5-10 mg/kg. Caution should be used when contemplating the use of this drug in African grey parrots. Grey parrots are extremely sensitive to itraconazole and may become anorexic and develop a depressed state when treated with this drug. Alternative treatments should be considered.

Voriconazole 12-20 mg/kg.

Terbinafine 5-15 mg/kg. This is the systemic antifungal drug of choice for African grey parrots.

Sinus flush: Add 3.5 mg neomycin, 124.5 mg trypsin, and 10 mg amphoteracin B to 30 ml of a water soluble base. Sig: add 0.1-0.4 ml of the base solution to 10-20 ml of saline for direct flushing through the nares q24h. Flush until there is an easy uniform flow out of the choanal from both sides and with equal resistance.

Nebulization: Equipment for nebulization includes a nebulizer compressor (Sportneb, model 3050, Medical Industries America, Adel, IA), infant nebulizer, and a nebulizing chamber. Nebulize for about 15 minutes q12h, using a antimicrobial agent, saline and 3-5 drops mucomyst, amphotericin B (7 mg/ml saline), clotrimazole (1% × 30-60 min).

Macrorhabdus ornithogaster or Megabacteriosis

Macrorhabdus ornithogaster or megabacteriosis is a fungal organism that is found in the gastrointestinal tract of many avian species. This organism will cause severe disease by adversely affecting the ability of the gastrointestinal tract to properly function. This organism has been identified in budgerigars, parrotlets, lovebirds, canaries/finches, ostriches, and chickens. In severely infected birds a crop swab will capture the organism to place on a glass slide for identification using a Gram's stain. The gram-positive organism is then identified through microscopic examination of the stained slide. Unfortunately *M. ornithogaster* has a tendency to wash off the slide during the staining process and does not stain well even if present on the slide. Unstained slides of fecal material will often provide identification of the organism. The most common clinical signs associated with infection are weigh loss or "going light," and diarrhea. *M. ornithogaster* can be identified at necropsy by examining the gastrointestinal tract, especially at the isthmus between the proventriculus and ventriculus. Treatment is difficult and amphotericin B and acidification of the drinking water have been met with moderate success in reducing *M. ornithogaster* in birds that have been diagnosed with this fungal organism.

Mycobacterium spp.

Until recent advances in molecular identification, *Mycobacterium avium* subsp. *avium*, was thought to be the most common bacterium causing avian tuberculosis. However *Mycobacterium genavense* is now considered the more frequent pathogen in psittacine species. Other *Mycobacterium* spp. that have been associated with infection in avian species include *M. tuberculosis*, *M. intracellulare*, *M. gordonae*, *M. nonchromogenicum*, *M. fortuitum*, *M. avium hominissuis*, and *M. bovis*.[23] It is important for veterinarians to know the primary site of infection for *Mycobacterium* spp. is the gastrointestinal system since the fecal gram's stain is a very specific test for diagnosis and clinical signs in birds are related to this site of infection. Birds infected with *M. avium* often will present with weight loss and polyphagia; however, clinical signs can be nonspecific and diagnostic testing difficult. Birds that are infected often have hematologic abnormalities noted in their complete blood count results but again this is variable. Hematologic abnormalities often identified are a marked leukocytosis characterized by a heterophilia and monocytosis, along with a nonregenerative anemia of chronic disease.[23] Radiographic imaging abnormalities that may be observed are an enlarged cardiohepatic silhouette, thickening of bowel loops, and circumscribed medullary lucence of the long bones. If the fecal acid fast test is negative an exploratory endoscopic examination may be useful in order to establish a definitive diagnosis. The endoscopic examination will allow one to examine the major organ systems and take biopsy samples of suspect tissue or lesions. Although routine diagnostic testing is often nonspecific in determining a definitive diagnosis for avian mycobacteriosis there are tests will be more useful for disease confirmation. Histopathological and cytological evaluation of collected tissue and aspirate samples from suspect lesions and/or organs will provide a basis for identifying the organism. Unfortunately both histopathologic and cytologic identification of the *Mycobacterium* spp. organism will not allow for speciation. At this time PCR testing of feces and/or fresh tissue samples will provide an avenue to determine the species of *Mycobacterium* that is infecting the patient. A reminder, only *fresh* nonpreserved tissue samples can be used for PCR testing. If one suspects that a bird is infected with avian mycobacteriosis, fresh tissue must be frozen and saved in case PCR testing is necessary or requested by the owner. The author's practice uses the National Jewish Medical and Research Center located in Denver, Colorado (www.testmenu.com/NationalJewish) for PCR testing to determine the species of avian mycobacteriosis samples. Historically culture of suspect avian mycobacteriosis tissue and fecal samples has been used to identify the organism. The problem with culture of *Mycobacterium* spp. is that it requires an extended period of time to isolate and grow the organism, in some cases 6 months; therefore, this makes knowledge regarding sensitivity to antimicrobial agents difficult to achieve by the veterinarian who elects to treat a bird diagnosed with avian mycobacteriosis.

To quickly determine the species of *Mycobacterium* infecting the patient, PCR testing is recommended and for treatment options recommended drugs are often used without the benefit of sensitivity information.

M. avium has a slight zoonotic potential and owners need to be aware of this before the decision to treat is made. The author is unaware of any cases of avian mycobacteriosis being transmitted from an infected bird to a human but extremely immunocompromised humans are theoretically at risk for infection; however, prior to 1981, there were reported to be approximately 3000 cases of humans being infected with *M. avium* complex worldwide every year from exposure to sources (eg, water) other than birds.[24] Currently the CDC has no recommended restrictions for bird ownership to prevent exposure to *M. avium* complex to normal or immunocompromised owners.[24] In the author's practice the recommendation for all avian mycobacterium cases is euthanasia. There are treatment regimes investigated and used on psittacine species suffering from avian tuberculosis but the treatment period may be as long as 3 years with recheck visits scheduled every 2-3 months.

There have been many treatment protocols used in avian species that have been diagnosed with avian tuberculosis. The recommendation is to treat avian mycobacteriosis cases with combination drug therapy.[24] Often treatment may extend for years. The author had one case in which an Amazon parrot was treated for at least 4 years. After no evidence of disease by all diagnostic tests available, treatment was discontinued. The bird re-presented 10 years later with a recurrence of the disease. Therefore it is unknown if treatment will clear the organism from the patient's body. Treatment is required on a daily basis and best administered in a single dose of combined drugs. There are recommendations of placing the compounded drug on food but the author prefers a liquid formulation that can be administered orally to ensure treatment compliance. Drugs that have been prescribed for avian tuberculosis cases include isoniazid (30 mg/kg), rifampin (10-45 mg/kg), rifabutin (15-45 mg/kg), ethambutol (15-30 mg/kg), clofazimine (6 mg/kg), ciprofloxacin (80 mg/kg), clofazamine (6 mg/kg), streptomycin (20-40 mg/kg), and clarithromycin (60 mg/kg) all with a dose of q24h.[24] One protocol that may be curative as per reports of VanDerHeyden is a total daily dose of 85 mg/kg clarithromycin, 43 mg/kg azithromycin, 56 mg/kg rifabutin, 56-85 mg/kg ethambutol and 6-12 mg/kg clofazamine.[23]

Infectious Pododermatitis (Bumblefoot)

Pododermatitis is a general term for any inflammatory or degenerative condition of the avian foot and may range from very mild redness or swelling to chronic, deep-seated abscesses and bony changes. Instructors are encouraged to use specific scientific terms for diseases.

The constant pressure on the plantar aspect of the foot, on the metatarsal pad, and occasionally the digital pads, results in a thinning of the surface epithelium. The infection, most commonly caused by *Staphylococcus aureus*, is seeded by damage of the skin (devitalization, laceration, or puncture). Pressure necrosis and the swelling associated with the infection, reduces blood supply to the affected area. The reduced blood supply causes a failure to phagocyte the pathogens leading to chronic, granulomatous disease. In addition to reduced blood supply, the granuloma further isolates pathogens from systemic antibiotics and immune factors. The disease becomes progressive and degenerative. Rupture of the flexor tendons, osteomyelitis and septic arthritis of the tarsometatarsal phalangeal joints occur.

Classification of Lesions

Grade 1: Early insult or lesion of a prominent plantar area with no apparent underlying infection: hyperemia, early ischemia or hyperkeratotic reaction.
Grade 2: Infection of underlying tissues in direct contact with the surface lesion but no gross swelling; puncture (with localized infection), local ischemic necrosis.

Grade 3: Infection with gross pedal inflammatory swelling: serous, fibrotic or caseous.
Grade 4: Infection with swelling of underlying tissues involving deep vital structures; chronic, producing tendosy-novitis, arthritis and/or osteomyelitis.
Grade 5: Crippling deformity and loss of function.

Treatment options for birds infected with pododermatitis include Preparation H applied topically and massaging the area affected in mild cases in mild cases to improve blood supply. Dexamethasone, dimethyl sulfoxide (DMSO), and antibiotics, such as piperacillin or enrofloxacin, create an effective combination for topical application in acute cases with severe inflammation. Antibiotic-impregnated polymethyl methacrylate beads, following aggressive surgical debridment, offers and effective method for the delivery of antibiotics to an infected ischemic site. Combinations used are rifampicin-piperacillin, and rifampicin-pefloxacin.

Foot-pad toughening products applied topically are recommended during the advance stage of healing where it may strengthen the pink, tender skin has regenerated on the plantar surface.

Systemic antibiotics are recommended in grade 2 bumblefoot and higher.

Polymethyl methacrylate (PMMA) beads can be made with the following ingredients in respective antibiotic:cement polymer ratios.[25]

Simplex P formulation.

Bactericidal/water soluble/heat stabile antibiotics (eg, aminoglycosides, penicillins, fluoroquinolones, clindamycin) in powder form.
Aminoglycosides: mix cement polymer in a ratio of 1:14 (3 g/40 g packet bone cement).
Penicillin: mix cement polymer in a ratio of 1:5 (8 g/40 g packet bone cement).
Fluoroquinolones: mix cement polymer in a ratio of 1:6 (7 g/40 g packet bone cement).
Clindamycin: mix cement polymer in a ratio of 1:6 (6 g/40 g packet bone cement).

Steps to make 2- to 3-mm beads are listed below.

1. Mix antibiotic powder with cement polymer powder.
2. Mix by vigorously shaking for 2 minutes, then separate into 1-gram aliquots.
3. Transfer into an evaporation-proof containers and freeze to 0 °C.
4. Once chilled, quickly mix 0.7 ml of the chilled antibiotic/polymer mix with 1-gram aliquot of the powder compound to form a "loose batter" consistency and load into a 3-ml syringe.
5. Immediately expel the liquid dough in a line on a sterile surface approximately in 2- to 3-mm diameter beads.
6. A gloved assistant takes the soft beads and rolls them between their thumb and index finger making sure they are smooth and round.
7. The beads are then gas sterilized after which they must be allowed to dispel any gas through air exposure for 24 hours.

References

1. Everett KDE, Bush RM, Andersen AA. Emended description of the order Chlamydiales, proposal of Parachlamydiaceae fam. nov. and Simkaniaceae fam. nov. each containing one monotypic genus, revised taxonomy of the family Chlamydiaceae including a new genus, and five new species, and standards for the identification of organisms. *Int J Syst Bacteriol.* 1999;49:415-440.

2. Everett KDE, Andersen AA. The ribosomal intergenic spacer and domain I of the 23S rRNA gene are phylogenetic markers for *Chlamydia* spp. *Int. J Syst Bacteriol.* 1997;47:461-473.

3. Fukushi H, Hirai K. Restriction fragment length polymorphisms of rRNA as genetic markers to differentiate *Chlamydia* spp. *Int J Syst Bacteriol.* 1989;43:613-617.

4. Scieux C, Grimont F, Regnault B, Grimont PAD. DNA fingerprinting of Chlamydia trachomatis by use of ribosomal RNA, oligonucleotide and randomly cloned DNA probes. *Res Microbiol.* 1992;143:755-765.

5. Messmer T, Tully TN, Ritchie BW, Moroney JE. A tale of discrimination: differentiation of Chlamydiaceae by polymerase chain reaction. *Semin Avian Exotic Pet Med.* 2000;9:36-42.

6. Compendium of measures to control *Chlamydia psittaci* infection among humans (psittacosis) and pet birds (avian chlamydiosis), 2010. Available at: www.nasphv.org/Documents/Psittacosis.pdf. Accessed May 30, 2017.

7. Tully TN. Clinical aspects of companion bird chlamydial infections. *Semin Avian Exotic Pet Med.* 1993;2: 157-160.

8. Wyrick PB, Richmond SJ. Biology of chlamydia: reports from the symposium on avian chlamydiosis. *J Am Vet Med Assoc.* 1989;195(11):1507-1512.

9. Gerlach H, Chlamydia. In: Harrison GJ, Harrison LR, eds. *Clinical Avian Medicine and Surgery.* Philadelphia, PA: Saunders;1986:457-463.

10. Isaza R, Murray MJ, Vanrompay D, Yeary T, Tully TN, Fudge A. Round table discussion: use of PCR testing in diagnosing chlamydiosis. *J Avian Med Surg.* 2000;14(2):122-127.

11. Vanrompay D. Avian chlamydial diagnostics. In: Fudge AM, ed. *Laboratory Medicine: Avian and Exotic Pets.* Philadelphia, PA: Saunders;2000:99-110.

12. Everett KDE, Hornung LJ, Andersen AA. Rapid detection of the Chlamydiaceae and other families in the order Chlamydiales: three PCR tests. *J Clin Microbiol.* 1999;37:575-580.

13. Evans RT, Chalmers WSK, Woolcock PR, et al. An enzyme-linked immunosorbent assay (ELISA) for the detection of chlamydial antibody in duck sera. *Avian Pathol.* 1983;12:117-124.

14. Grimes JE. *Chlamydia psittaci* latex agglutination antigen and protocol improvement and psittacine bird anti-chlamydial immunoglobulin reactivity. *Avian Dis.* 1986;30:60-66.

15. Salinas J, Caro MR, Cuello F. Comparison of different serologic methods for the determination of antibodies to *Chlamydia psittaci* in pigeon sera. *J Vet Med B.* 1993;40:239-244.

16. Vanrompay D, Lublin A, Vanloock M, Mulenga S. Serology of *Chlamydiaceae psittaci* infection in psittaciformes by use of a recombinant enzyme-linked immunosorbent assay: recombinant DNA technology at work. *Semin Avian Exotic Pet Med.* 2000;9:43-49.

17. Fudge AM. A review of methods to detect *Chlamydia psittaci* in avian patients. *J Avian Med Surg.* 1997;11:153-165.

18. Vanrompay D, Ducatelle R, Haesebrouck F. *Chlamydia psittaci* infections: a review with emphasis on avian chlamydiosis. *Vet Microbiol.* 1995;45:93-119.

19. Flammer K. Chlamydia. In: Altman B, Clubb SL, Dorrestein GM, Quesenberry K, eds. *Avian Medicine and Surgery.* Philadelphia, PA: Saunders;1997:364-379.

20. Flammer K. Plasma concentrations of doxycycline in selected psittacine birds when administered in water for potential treatment of *Chlamydophila psittaci* infection. *J Avian Med Surg.* 2001;15(4):276-282.

21. Ritchie BW, Harrison GJ. Formulary. In: Ritchie BW, Harrison GJ, Harrison LR, eds. *Avian Medicine: Principles and Application.* Lake Worth, FL: Wingers Publishing Inc;1994:457-478.

22. Carpenter JW. *Exotic Animal Formulary.* 4th ed. St. Louis, MO: Elsevier Saunders;2013.

23. McRee AE, Higbie CT, Nevarez JG, et al. Mycobacterium in three companion psittacine species (*Eclectus roratus, Amazona aestivea, and Pionates melanocephala*): a case series. *J Zoo Wildl Med.* 2017. Accepted.

24. Lennox A. Mycobacteriosis. In: Speer BL, ed. *Current Therapy in Avian Medicine and Surgery.* St. Louis, MO: Elsevier Saunders;2016:94-97.

25. Remple JD. A multifaceted approach to the treatment of bumblefoot in raptors. *J Exot Pet Med.* 2006;15: 49-55.

Practice Pearls Potpourri

Anna Osofsky, DVM, DABVP (Avian Practice),
Jennifer Graham, DVM, DABVP (Avian Practice), DABVP (Exotic Companion Mammal Practice), Dipl ACZM,
Anneliese Strunk, DVM, DABVP (Avian Practice)

Session #2025

Affiliation: From Carrollton West Pet Hospital, 3729 Old Denton Rd, Carrollton, TX 75007, USA (Osofsky), Department of Clinical Sciences, Cummings School of Veterinary Medicine at Tufts University, 200 Westboro Rd, North Grafton, MA 01536, USA (Graham) and The Center for Bird and Exotic Animal Medicine, 11401 NE 195th St, Bothell WA, 98011, USA (Strunk).

Abstract: This master class reviews a variety of tips and tricks for the avian veterinarian. Topics include sedation and anesthesia techniques, collars and sock sweaters, microchips, venipuncture, wing/beak/nail trims, surgical tips, radiographs, incorporating low stress veterinary visit concepts into practice, and some additional diagnostic and therapeutic techniques.

Anesthesia

Humid-Vent

Avian patients are prone to transtracheal membrane formation post anesthetic events. Predisposing factors include length of the procedure, size and type of endotracheal (ET) tube, disinfectants inadvertently left on the tube, and whether intermittent positive pressure ventilation is performed (especially if a mechanical ventilator is used).[1] One trick the authors learned as residents was the use of a Humid-Vent in any procedure in which the bird is intubated and the procedure is going to last more than 30 minutes. The Humid-Vent is a heat and moisture exchanger placed between the anesthesia circuit and the ET tube; it is made of hygroscopic naturally absorbent microwell paper. These can be purchased at JorVet for an approximate cost of $58 for five units. Due to infectious disease risk, Humid-Vents should be considered single use only.

Doppler use in birds

The author (A.O.) uses a Doppler for all exotic pet anesthetic events. It usually can be placed quickly, especially in avian patients, and allows both the clinician and the anesthetist to listen to the patient's heart rate during a procedure. The easiest location for Doppler placement in avian patients is the superficial ulnar artery and vein located on the ventral wing near the cubital joint. The author uses two tongue depressors taped together at one end to hold the Doppler crystal in place. A very small amount of alcohol is used to moisten the skin and part the feathers but not so much that the patient is chilled or that the crystal has contact with the alcohol; Doppler gel is placed on the crystal, the crystal is applied to the artery/vein and then held in place with the non-taped end of the tongue depressors (one tongue depressor on the ventral side of the wing, one on the dorsal). The tightness of the tongue depressors can be adjusted for patient size. For small patients no tape is needed between the two tongue depressors; for larger ones, apply tape to the top of one tongue depressor for 3-4 revolutions prior to taping it to the second tongue depressor.

The Doppler can be used to measure blood pressure trends. Although studies indicate that there are no known reliable indirect means to measure blood pressure in avian patients,[2-4] measuring trends can be helpful. The blood pressure cuff can be placed on the humerus since there are no large flight feathers located here to interfere with placement.

Face masks

Creative use of face masks is often needed in exotic pet anesthesia. The author (A.O.) finds that avian patients often bite and irreparably damage the rubber diaphragms of face masks; in addition, the diaphragm opening is often not appropriately sized for these patients. The author uses self-adherent stretch bandage material to create custom diaphragms for each avian patient. Multiple layers are needed as this material is much more porous than rubber so exposure to inhalant gasses can occur more readily. Exam gloves also can be used to make a custom diaphragm that will not leak anesthetic gases so readily. Simply stretch the nonfinger portion of a glove over the mask opening and tape it to the sides of the mask so the glove material is taut; a small opening is cut into the center of the glove to pass the bird's head through. With this set-up, forced ventilation via the mask can often be achieved as long as the seal around the neck is good. To prevent inflation of the cervicocephalic air sacs during forced ventilation, a self-adherent bandage can be placed slightly loosely around the neck of the patient.

Emergency drugs

Because exotic pets are prone to destabilizing quickly while under anesthesia, the author (A.O.) prefers emergency drug and fluid doses be calculated and drawn-up prior to the anesthetic event. The author uses an Excel spread sheet divided by type of animal (eg, guinea pig, rat, rabbit, bird, snake, lizard) and drug (eg, atropine, epinephrine, flumazenil, colloids, crystalloids). The patient's name and weight are entered in the correct spaces and the appropriate doses are calculated. This sheet is also used to calculate the commonly used sedatives (eg, midazolam and butorphanol). If we anticipate the anesthesia will be performed using a face mask, the author always has an appropriately sized ET tube and laryngoscope ready in case of apnea. That said, nothing prevents anesthetic complications in exotic pets more than a vigilant, dedicated anesthetist; for exotic pet procedures, anesthetists should not be acting as concurrent surgical nurses.

Forced-air warming

Hypothermia under anesthesia is a major risk to small exotic patients with their high surface to volume ratios. Forced-air warmers have proven superior to infrared heat emitters and circulating water blankets at maintaining body temperature in parrots.[5] The author (A.O.) uses a forced-air warmer system for almost all exotic pet anesthetic events. The blanket is generally placed under rather than over the patient as monitoring respiratory rate and character is essential, even in procedures in which the body cavity does not need to be accessed. A rigid but porous surface (eg the top of a perforated sterilization tray) can be used under avian patients if more stability is needed to tape down wings etc. Reusable fabric blankets are available to keep down cost and allow washing between patients.

Buck Ear Curettes

Buck ear curettes are a great tool for avian and exotic pet practice. We have several different sizes we keep autoclaved and ready to use. In avian medicine these are most commonly used to carefully remove thick, caseous material from the oropharynx. The tip is blunt and unlikely to cause tissue trauma in gentle hands. However, for cleaning debris from the nares of birds, the author's (A.O.) favorite tool is a cotton tip applicator stick broken at an angle; the broken edge makes a perfect spatula for lifting material out of the nares. Care must be taken not to scrape the operculum or other soft tissue as this tool can cause bleeding in rough hands.

Avian Collars and Sock Sweaters

Restraint collars and various avian "outfits" are sometimes needed to manage avian patients manifesting mutilatory behaviors. Since these devices do nothing to treat the disease causing the behavior, it is essential that the underlying disease be addressed or the devices are simply unfair to the patient. The author generally reserves these devices for patients that are mutilating to the point of skin ulceration or active hemorrhage. Examples of potentially appropriate cases are cockatoos that mutilate their ventral keel/ingluvial region, self-mutilating Quaker parrots, and post-operative patients mutilating their surgical sites. For post-operative patients, the author almost always tries bandages prior to collars.

Avian patients are initially very stressed by collar placement; this can be extremely difficult for the owner to witness. Keeping the patient hospitalized for the day and sedating the patient (midazolam 0.5 mg/kg and butorphanol 1-2 mg/kg IM) prior to collar placement can facilitate acceptance by the owner and patient. The patient will usually roll around in the cage while getting used to the collar, though this is lessened by sedation. Keeping the patient for the day also allows the clinician to assess whether the collar is appropriately sized to be safe and achieve the desired goal.

While patients are adjusting to their collar, they often are clumsy and eat poorly. Thus, client education is essential. The cage may need to be changed while the patient is getting comfortable. For example, towels can be placed on the bottom of the cage in case of falling; higher perches can be removed or lowered; food and water dishes should be easily accessible. Rarely, patients will need to be placed in a smooth sided cage such as a glass aquarium while adjusting. To make sure the patient is eating acceptable amounts of food, the owners must have a digital gram scale to weigh the patient daily. The collars can interfere with normal eating behaviors. For example, the collars can prevent the patient from holding food in its foot and bringing the food to its mouth. If the patient loses more than 5-10% body weight then a recheck is needed and the collar likely needs to come off. The author (A.O.) has had one patient develop hepatic lipidosis secondary to anorexia after collar placement when the owner did not weigh the patient as promised. Have the owner offer food and water by hand several times a day during the adjustment period. Make sure owners know how to remove a collar in case of a crisis.

There are several types of collars available. The author most commonly uses the VSP collars in larger patients. Advantages of this collar are that it is fairly light, can be used with or without the disk, and is sturdy. Disadvantages of this collar are that it won't fit all birds, can irritate the carpi, and can be very restrictive to normal behavior especially with the disk in place. Recommendations for using this collar are to always start with just the collar and add the disk if needed after the bird adjusts to the collar. The disk should be as small as possible to be effective. Tape the collar on tightly rather than using the screws provided; it will be easier to remove if needed.

Another option is spherical collars. Advantages of this collar type are there are numerous sizes, they are fairly easy for the owner to remove, and a collar can be added under the disk (but not attached to it) if needed. The disadvantage of this collar type is that it is common for patients to reach wounds despite the collar.

For smaller birds, the author routinely uses foam insulation collars. Advantages of this collar type are that they are inexpensive, can be adjusted to almost any sized bird, and are light weight. The disadvantage of this collar type is that it is easily destroyed.

One of the authors (J.G.) has used surgical cotton tubular stockinette rolled on itself to create a soft E-collar for birds. The stretchy nature of the stockinette also makes it easy to place over the head of the bird. The tubular stockinette comes in a variety of sizes by multiple manufacturers. The primary drawback of this type of collar is the ability of the patient to chew the material, but it is easily replaced.

The author (A.O.) generally does not use Elizabethan type collars. These collars make it difficult for the patient to reach food and spread wings, depending on which direction the collar is placed.

When possible, the author (A.O.) prefers to use sock sweaters for ingluvial/keel mutilators in place of collars because they allow more normal behaviors than collars do. They must be changed monthly in most cases as patients will chew on them and they become soiled. The author has clients purchase an appropriately sized athletic sock and prewash it in a dye/fragrance free laundry detergent. When placing the sweater it is imperative to make sure that the sweater is not too tight in the shoulder region, will not restrict breathing and that the toes will not get stuck in the bottom of the sweater. Not all patients will tolerate a sweater. The author will illustrate how to make and apply a sock sweater in the lecture. As with collars, it is best to observe the patient for the day after the first sweater placement. There are also many commercially available "outfits" at this time, many of which have worked well for the author's patients; some of these will be reviewed in the lecture.

Disposable Heat Cautery Pens for Nail Trims

Cautery pens are intended for coagulating tissue or arresting bleeding from small vessels by heat conducted through a wire tip. Cautery pens are utilized for ophthalmic and general surgery and often disposed of in human hospitals after a single use. The tip heats in approximately 1 second so this provides rapid portable cautery at the touch of a button. Cautery pens are generally sold in boxes of 10, cost on average $10-15 (or less) per pen, and are available from multiple suppliers. As each pen can be used for weeks, these are a convenient and low-cost item to keep on hand for routine grooming. If needed, the pens can be gas autoclaved between uses (ie, if needed in a sterile setting).

Helpful hints

Do not use alcohol around the cautery as this is flammable. Do not depress the button until the pen is in place around the nail; remember, feathers/towels are flammable! Do not pull firmly when using the cautery pen as this can break the wire. Take off as much length as you would normally take with clippers; cauterizing the quick of the nail is painful, just as trimming a nail into the quick would be.

Sedation

Do you carry midazolam in your practice? If not, you need it! Sedation is useful for many aspects of avian medicine. Sedatives also decrease stress associated with induction but should be combined with analgesics for potentially painful procedures. Midazolam is a commonly used benzodiazepine that can be administered IM, SC, or intranasal (IN). Sometimes used as a sole sedative for nonpainful procedures, including radiographs, midazolam can be reversed with flumazenil. Midazolam is more potent and longer lasting than diazepam, and does not adversely affect mean arterial blood pressure or blood gases in select avian species.[6] Midazolam administration may be occasionally associated with regurgitation in larger macaws and cockatoos.[6] Routine use of flumazenil as a benzodiazepine reversal agent can be rarely associated with adverse reactions in human and canine patients.[7] While the author (J.G.) has witnessed sudden death in a bird following IM injection of flumazenil, no complications have been noted following IN administration by the author and others.[8]

Helpful hints

It is ideal for birds to be fasted for 2-3 hours prior to sedation although not mandatory (if the crop is overly full it is best to aspirate contents prior to sedation). Midazolam is helpful to reduce anxiety associated with patients in respiratory distress. Midazolam dosage recommendations range from 0.5 mg/kg to 5 mg/kg depending on species (consult Carpenter JW. *Exotic Animal Formulary.* 4th ed. St. Louis, MO: Elsevier Saunders;2013). Flumazenil reversal is generally 0.02 mg/kg to 0.1 mg/kg IV or IO but many use this drug IM and IN.

ET Tube to Facilitate Tracheal Wash

Patients are sedated and anesthetized for intubation. A sterile ET tube is placed. A red rubber feeding tube or polypropylene catheter is advanced to the level of the syrinx. Sterile saline is infused into the trachea and immediately aspirated back into the syringe. For safety reasons, keep equipment for an air sac cannula ready. However, if the bird has severe respiratory compromise, first place an air sac cannula.

Helpful hints

Volumes of 1-2 ml/kg of nonbacteriostatic 0.9% sterile saline are recommended. Samples can be submitted for cytology, culture (bacterial/fungal), and PCR analysis. As an alternative, a direct tracheal swab can be performed to sample the proximal trachea. A sterile culturette or mini culturette is used based on the size of the bird. Moisten the swab with sterile saline prior to collecting the sample.

Radiographs

Horizontal beam radiographs

Horizontal beam radiographs are routinely used with reptile patients but they can also be quite helpful for avian and small mammal patients in certain situations. Use of this technique requires a mobile x-ray tube head with a moveable imaging detector. Although you cannot evaluate avian films in the same manner as properly positioned films because of superimposition of wings/legs, you can use this technique in situations where regularly positioned films would be contraindicated. This may be particularly useful if you do not have access to ultrasound in your practice. If you have a bird with coelomic distention, a horizontal beam radiograph can determine if a bird has evidence of a mineralized egg or metallic densities. The major advantage of this approach can be in situations of ascites. For example, chickens are prone to development of ovarian neoplasia and concurrent ascites is common. The author (J.G.) has used this approach to evaluate hens with ascites for a mass effect in the region of the gonad/kidney prior to removal of coelomic fluid.

Helpful hints: Never lay a bird flat if you suspect it has coelomic fluid. You can remove fluid via aspiration prior to positioning for films to reduce the chance for fluid flooding the lungs. Ultrasound is helpful to determine location/extent of fluid.

Bird board

Many avian positioning devices are commercially available to obtain excellent quality radiographic images. For example: Bird Board, 8205 Alba Ct., Citrus Heights, CA, USA and VSP Avian Neck Restraint & VSP Avian Restraint Board Magazine, Veterinary Specialty Products, Shawnee, KS, USA. If devices are interposed between the bird and the film sensor, a small increase in kVp (2-4 kVp) may be needed to compensate.

Masking tape for radiographs

Tape can be helpful for positioning avian patients for radiographs but certain tape (for example, clear tape and cloth medical tape) may damage or remove feathers or scales. The authors find that masking tape can be placed on feathers for radiographic positioning and removed without causing damage to or removing feathers.

Vibravenos

Long-acting injectable doxycycline hyclate (Vibravenos; Psittavet) can be administered IM once every 5-7 days making this drug a nice option for bird owners that have difficulty giving oral medications; however, Vibravenos is not available in the United States so is generally ordered from the Netherlands. Contact dorresteingm@planet.nl for ordering information.

Vinegar for Papillomas

Routine avian examinations should include evaluating the vent for symmetry, integrity and hygiene. The cloaca should be everted circumferentially, checking for lacerations, swellings or papillomas. Vinegar (acetic acid) turns papillomas frosty white. Biospy and histopathology are needed to make a definitive diagnosis of a papilloma.

Nasal Flush (0.9% Saline versus Saline and Sterile Water)

A nasal flush can be helpful to obtain samples for diagnostics as well as a therapeutic treatment. Sterile non-bacteriostatic 0.9% saline (preferably warm) should be used if the sample is being submitted for diagnostic purposes. A 50:50 mixture of sterile 0.9% saline and sterile water can be used for therapeutic flushes since this mixture may be a bit less irritating to the bird than straight saline.

Restrain the bird firmly with the head positioned below the body (ie, head down, tail up, keel towards the floor). Warm flush is instilled into one naris and a collection device (sterile if obtaining material for culture) is held underneath the bird's head to collect fluid exiting from the choanal slit and out the mouth as well as out of the opposite naris. Note that the sinuses do not connect in passerine birds so fluid will not exit out the opposite naris in these species. The amount of fluid used can vary but flush volumes can range from 1-10 ml depending on the size of the patient. Fluid obtained in this fashion can be submitted for culture and cytologic examination.

Change in Recommendation for Figure-of-Eight Bandages

The figure-of-eight bandage can be useful to manage fractures, soft tissue injuries, luxations, or when the wing needs to be stabilized to hold a catheter in place. Cast padding is used for the first layer and this is covered by self-adherent bandage material. With the wing in a flexed position, start the bandage at the carpus; wrap the gauze around the carpus creating the top loop of the figure 8. The gauze is now directed toward the elbow and wrapping it around the area just proximal to the elbow helps prevent slippage and creates the lower loop of the figure 8. Care should be taken not to wrap too tightly or create too much bulk and avoid placing pressure over the patagium. A recent recommendation has been made against figure-of-eight bandages for humeral fractures.[9] In a captive/companion bird where flight is not critical and surgery is not an option, a body wrap and "cage rest," with the hope the fracture does not go to a non-union, can be attempted (J. Ponder, written communication, 2017).

Tape and Paper Clips for Oral Exam

Evaluation of the head should include examination of the external ear canals, nares, eyes and surrounding tissues, beak, and the oral cavity, including the choanal slit and glottis. Use of hard and metal speculums can cause trauma to the beak. Use of folded bandage tape is useful for oral examination of larger parrots while paper clips can be used for smaller birds including budgerigars, cockatiels, and lovebirds.

Microchips

Microchips can be placed in almost any species to help reunite pets with their families. The most important pearl is to *get in the habit of scanning **all** pets* (not just dogs and cats). The American Veterinary Medical Association (AVMA) policy is to microchip all companion animals and the AVMA promotes national "Check the Chip" day on August 15 every year.[10,11] There are promotional items available on the AVMA website.[10] The American Animal Hospital Association has a Universal Pet Microchip Lookup Tool available online.[12]

The International Standards Organization (ISO) developed microchip standards in 1996. This standard is a 134.2 kHz chip with a 15-digit numeric code including a 3-digit country code. These are generally required for pets traveling outside the United States. Non ISO-standard chips are currently the predominant chips in use within the United States. These chips contain 9 or 10 digits, without a country code. The primary frequency used in the United States is 125 kHz, although 128 kHz and ISO standard 134.2 kHz chips also are available. Most chips are made from bioglass and are 1/2 inch (12.7 mm) long. There are mini microchips available from MicroChip ID Systems (Covington, LA, USA) that are 0.25 inches (6.4 mm) long and inserted with a 15-gauge needle. HomeAgain (Madison, NJ, USA) also makes a slightly smaller chip than normal (HomeAgain XS) that is 10.8 mm long and inserted with a 15-gauge needle.[14]

Techniques for placement of microchips are similar between species, but location varies. The Association of Zoos and Aquariums (AZA) and the World Small Animal Veterinary Association (WSAVA) have established recommendations for a wide range of species.[15-17] The left side is preferred in most species.

The author (A.S.) prefers anesthetizing birds with isoflurane or sevoflurane for chip placement. This allows precise placement without risk of patient movement during the procedure, which might lead to trauma of surrounding tissue. The chip is placed in the left pectoral region (inserting the needle from cranial to caudal). Perform a brief sterile prep, mainly to ensure there are no feathers in the way. Use a small amount of tissue glue to seal the skin. Consider analgesia (butorphanol and/or meloxicam) for a short period post placement to help with any soreness. Complications are rare in birds.[18] Some practitioners have expressed concerns that intramuscularly placed microchips may affect flight ability in athletic (falconry) birds, and place them subcutaneously overlying the pectoral musculature parallel to the proximal keel (J. Samour, written communication, 2017).

Venipuncture[19,20]

Preparation!

The first step in any sampling of exotic pets is to be prepared for the process. Have in mind the volume needed for the tests desired and have syringes, alcohol, blood tubes, slides, and bandages (if needed for hemostasis) ready and available. Make sure slides are wiped clean. Remove needles before placing blood in tubes. Have large nail trimmers ready to cut off the needle end of the syringe if an insulin syringe with a permanent needle is being used for sampling in small patients. Our goal should be to minimize the time a patient is in hand in order to reduce stress.

Blood volume calculation

Bird blood volume is considered to be approximately 10% of body weight. The maximum safe volume of blood that can be obtained from healthy birds is 10% of the blood volume (this equates to a total sample size of approximately 1% of the body weight). Draw smaller samples in critically ill patients. A rule of thumb for ill patients is to draw a sample of no more than 0.5% of body weight. Contact your lab to confirm absolute minimum volumes required for testing (most lab guides overestimate the volumes needed). There is no reason to draw more than is absolutely necessary.

Alleviate stress during venipuncture[8,21]

Consider sedation with midazolam with or without butorphanol (IM or IN) or brief anesthesia with isoflurane or sevoflurane by mask. Midazolam is reversible with flumazenil, allowing for a very quick recovery time (see discussion above). Butorphanol should always be used concurrently in macaws to achieve adequate sedation. The author has had several cases where a bird has become more aggressive/excited after the administration of midazolam alone. These birds were given flumazenil and gas anesthesia was used for the venipuncture. Lowered blood pressure during sedation may help prevent hematoma formation, but might affect the ability to get an adequate sample from smaller vessels.

Locations

Medial metatarsal vein (author's preferred location): The pros of this location are easier restraint, ability to place a hemostatic bandage, and quick release of patients. Cons include limited volume in small patients; use only for < 0.1 ml draws in patients smaller than approximately 200 g. Consider placing a 25 ga needle into the vein and collecting blood as it wells into the hub. This provides enough blood for PCR testing, DNA sexing, or a CBC (blood smears and hematocrit). A second restraint often is required to remove the bandage.

Right jugular vein: The pros of this location are adequate sampling in all species (ie, ability to obtain large volumes quickly to minimize risk of sample clotting). The cons include the requirement to keep the patient in hand to gently apply pressure to venipuncture site, risk of hematomas (consider subcutaneous fluid if noted), and risk of more severe bleeding/death (including conure bleeding syndrome).

Ulnar vein: The pros of this location are good visibility and ability to collect adequate samples from most sizes of birds. Cons include severe risk of hematomas, requiring a relatively long period of pressure application and the risk of damaging the wing while restraining or applying pressure post blood draw.

Toenail clipping[22] (this method is discouraged): The pros of this method are the ability to collect blood from this location in all species. The cons include pain and a capillary sample that is more likely to be contaminated with feces or urates, which can lead to artifacts in test results (cleaning the toenail prior to sampling can minimize this risk).

Beak Trims[23]

Normal versus abnormal

The shape of beaks can vary between species. Review the normal appearance before starting any trim. It is also important to manipulate and evaluate the movement of the rhinotheca and gnathotheca in relation to one another to check for any mobility changes that could be contributing to the beak overgrowth. An important goal is to keep birds in hand for as short a time period as possible. Consider sedation or anesthesia in highly stressed birds or in cases of severe beak overgrowth, in which trimming is expected to take a prolonged period of time.[21]

Dremel versus rongeurs

The most common tool for beak trims is a grinding stone on a Dremel tool or other rotary device. Typically, the Dremel tool is used to form the keratin into a normal shape using the sides and tip of the bullet shaped Dremel bit. It is vital to use a light touch to prevent burns to the deeper tissue under the keratin.

For a very thickened or elongated beak, consider using tiny snips with a sharp tipped rongeur.[24] Cut only in the direction of the edge of the beak in very small "bites" and the keratin will break off in a natural fracture plane. The layering of keratin allows this technique be a useful method to aid in the beak trim process. This technique minimizes the amount of keratin that needs to be removed by Dremel, decreasing stress and risk of burns from the Dremel. Attempting to remove too much keratin at once with rongeurs can lead to beak fractures and capillary bed exposure. Once the majority of the keratin is removed with snips from rongeurs, it can be smoothed and shaped with the Dremel or nail file. Use water to wipe off keratin dust.

Treat all areas of beak overgrowth

If there is significant malocclusion with asymmetry, it is important to correct not only the obvious culmen and gonys changes, but also restore the tomia to a normal (and symmetrical) height. In addition, look in the roof of the mouth and make any corrections to allow normal lateral motion of the gnathotheca along the rhinotheca.

Disease prevention

Purchase Dremel bits to sell to clients and keep them for each individual bird. This will prevent cross contamination between birds, as Dremel stones are difficult to disinfect. We (A.S.) store ours in small labeled zip closure bags which are stored in a plastic craft box. The actual Dremel tools are also difficult to disinfect. For birds with confirmed positive infectious diseases (eg, *Circovirus*), consider having owners buy their own Dremel to bring in for grooming visits.

Wing Trims[25]

This is a hot topic in avian medicine and aviculture. There are some individuals who can refute any possible reason why an owner may elect to trim their bird's flight feathers. Allowing flight is an ideal way to provide exercise and enrichment for birds. However, there are a number of reasons why wing trimming may be beneficial for individual birds: decreasing the risk of injury or escape, reducing aggression, and facilitating training.

There are several techniques for trimming that allow varying amounts of flight ability. Only the distal 10 remiges (primary flight feathers) should be considered for trims. The number of feathers removed should be determined based on a bird's flight ability and body structure of the species, and may vary depending on the presence of blood feathers. The trim should be symmetrical, and the feathers distal to any blood feathers should be left longer to provide protection as the blood feathers are growing in. Record the number of feathers trimmed and the technique used in the medical record for future reference.

Birds should be able to fly down to the ground at about a 45 degree angle once a trim is performed. It is advisable to have owners work with their birds over a carpeted area while they are getting used to their trims to prevent injury if a fall occurs.

Updates

Some birds will gain strength and start to be able to fly better within the weeks after a trim and may require updates. Consider a policy allowing no charge updates to trims if a bird starts being able to fly well within the month after a trim.

Handout

Have a handout with a diagram of a wing available for owners to specify their preference for trim techniques. This may be kept in the chart for future reference, along with notes on how the bird responded to the trim.

History and Husbandry Forms

Collecting thorough husbandry information is vital for treating any bird case. Use a detailed form with clear questions. Consider having owners fill out the form online prior to the visit, allowing electronic transmission to the hospital. The form can be included in the electronic medical record and leads to shorter times in the exam room. The AAV website has a history form available for download to members.

Surgical Pearls

Removing feathers

Attach 4-5 long pieces of masking tape to the prep table (sticky side facing upwards). When plucking feathers in preparation for surgical procedures, place the plucked feathers on the tape to keep them secured (and prevent feathers from scattering or becoming airborne).

Eye protection

Using a forced air blanket during surgery leads to a higher risk of corneal drying. Use standard surgical eye lubricants multiple times to minimize this risk. For extra eye protection for the down eye when a bird is in lateral recumbency, cut a well in a small stack of 2" × 2" sterile gauze and fill it with eye lubricant, then place it under the bird's down eye.

Intraosseus catheter bandage

Intraosseus catheters are useful for fluid therapy in birds. Keeping birds from prematurely removing the catheter can be a challenge once they start feeling better. The wide ends of 35 and 60 ml syringe cases can be used to provide protection for the catheter and port as they exit the carpus.

After securing the IO catheter with tissue glue and a tape butterfly and sutures, wrap the wing in a figure-of-8 bandage with cast padding. Attach an injection plug or T-port to the catheter. Place the cut ends of the syringe cases over the end of the carpus, and secure them with thick elastic fabric tape, incorporating the syringe cases into the next layer of the figure-of-8 bandage. If intermittent IV fluids are being administered, the end of the syringe case can be covered with a flap of elastic fabric tape.

Low-Stress Handling[26-30]

The concept of low-stress handling is a hot topic in veterinary medicine. Certification programs such as Fear Free (www.fearfreepets.com) have been developed to promote the concept in dogs and cats, but are not yet available for exotic species. Birds have inherent stress associated with travel to the hospital and handling during the visit for examination, diagnostics, and treatment. Discussion of this could fill a master class of its very own; however, it is so important that the authors wanted to mention it in this lecture. Include behavioral and handling information in the medical record (what methods of handling and restraint worked best, how the patient responded to restraint). The author (A.S.) includes notes on preferred ways to get patients in hand as an alert.

Start early

It is ideal to start the process from the very first wellness baby bird visit. Discuss the benefits of reinforcement training to get birds used to veterinary related stimuli, such as towels, stepping up (on multiple peoples' hands as well as a perch, handling/touching all parts of the body, and taking liquid from syringes. Start with directing owners to try to find a consistent reinforcer (food item, touch, vocalization, etc.). There are multiple resources available to help owners with this process, including Barbara Heidenreich's videos and webinars, available at goodbirdinc.com.

Treat caddy

The author's (A.S.) hospital has placed a spice rack stocked with different types of pellets, seeds, chopped nuts, dried fruit, etc. in each exam room. These "treat caddies" provide a wide variety of reinforcement options to offer during visits and allow demonstration of pellets for clients unfamiliar with them.

Sedation

While in the hospital, consider sedation early in the visit if a bird is showing signs of fear or stress. Use midazolam and butorphanol IN or IM (see discussion on sedation).[8,21]

Follow up

Consider having the reception team call clients after a visit to conduct post visit interviews. The goal of the call is to see how the bird did after the visit and inquire about how the suggestions made during the visit are being incorporated into the bird's routine. The author has wellness plans with multiple visits, which allow clients to bring birds in for training/happy visits in addition to their twice yearly wellness visits.

References and Suggested Reading

1. Sykes JM, Neiffer D, Terrell S et al. Review of 23 cases of postintubation tracheal obstruction in birds. *J Zoo Wildl Med.* 2013;44:700-713.

2. Zehnder AM, Hawkins MG, Pascoe PJ, et al. Evaluation of indirect blood pressure monitoring in awake and anesthetized red-tailed hawks (*Buteo jamaicensis*): effects of cuff size, cuff placement, and monitoring equipment. *Vet Anaesth Analg.* 2009;36:464-479.

3. Acierno MJ, da Cunha A, Smith J, et al. Agreement between direct and indirect blood pressure measurements obtained from anesthetized Hispaniolan Amazon parrots. *J Am Vet Med Assoc.* 2008;233:1587-1590.

4. Johnston MS, Davidowski LA, Rao S, et al. Precision of repeated, Doppler-derived indirect blood pressure measurements in conscious psittacine birds. *J Avian Med Surg.* 2011;25:83-90.

5. Rembert MS, Smith JA, Hosgood G, et al. Comparison of traditional thermal support devices with the forced-air warmer system in anesthetized Hispaniolan Amazon parrots (*Amazona ventralis*). *J Avian Med Surg.* 2001;15:187-193.

6. Hawkins MG, Pascoe PJ. Cagebirds. In: West G, Heard D, Caulkett N, ed. *Zoo Animal and Wildlife Immobilization and Anesthesia.* Ames, IA: Blackwell Publishing;2007:269-297.

7. Plumb DC. *Plumb's Veterinary Drug Handbook.* 8th ed. Ames, IA: Wiley-Blackwell Publishing;2015.

8. Mans C, Sanchez-Migallon D, Lahner L, et al. Sedation and physiologic response to manual restraint after intranasal administration of midazolam in Hispaniolan Amazon parrots (*Amazona ventralis*). *J Avian Med Surg.* 2012;26(3):130-139.

9. Ponder JB, Redig PT. Avian orthodpedics: a game plan for success. ExoticsCon. 2016;185-193.

10. American Veterinary Medical Association. Microchips: The Objectives and Key Elements Needed for Effective Electronic Identification of Companion Dogs, Cats, Other Small Mammals, Birds, Fish, Reptiles, Amphibians and Equids. Available at: https://www.avma.org/KB/Policies/Pages/Electronic-Identification-of-Companion-Animals-Birds-and-Equids.aspx Accessed April 2, 2017.

11. American Veterinary Medical Association. Microchipping of Animals. Available at: https://www.avma.org/KB/Resources/Reference/Pages/Microchipping-of-Animals-Backgrounder.aspx Accessed April 2, 2017.

12. American Animal Hospital Association. Universal Pet Microchip Lookup Tool. Available at: http://www.petmicrochiplookup.org Accessed April 2, 2017.

13. MicroChip ID Systems. Available at: https://www.microchipidsystems.com/product/pro-id-mini-chip/ Accessed April 2, 2017.

14. HomeAgain. Available at: http://microchip.homeagain.com/faqs.html Accessed April 2, 2017.

15. World Small Animal Veterinary Association. Microchip Identification Guidelines. Available at: http://www.wsava.org/guidelines/microchip-identification-guidelines Accessed April 2, 2017.

16. Association of Zoos and Aquariums. Guidelines for Transponder Placement and Recording. Available at: https://www.aza.org/assets/2332/guidelines_for_transponder_placement_and_recording_aza_branded1.pdf Accessed April 2, 2017.

17. World Small Animal Veterinary Association. Veterinary List of Recommended Microchip Implantation Sites (Updated October, 1999). Available at: http://www.wsava.org/sites/default/files/Veterinary%20List%20of%20Recommended%20Microchip%20Implantation%20Sites%20%20.pdf Accessed April 2, 2017.

18. Granzow E. Microchip placement for identification of birds. *Lab Anim.* 2008;37:21-22.

19. Powers LV. Venipuncture. In: Mayer J, Donnelly TM, eds. *Clinical Veterinary Advisor.* St. Louis, MO: Elsevier;2013:546-547

20. Powers L. Venipuncture in birds. LafeberVet. Available at: http://lafeber.com/vet/venipuncture/ Accessed April 2, 2017.

21. Mans C. Sedation of Pet Birds. *J Exot Pet Med.* 2014;23:152-157.

22. Bennett TD, Lejnieks DV, Koepke H, et al. Comparison of hematologic and biochemical test results in blood samples obtained by jugular venipuncture versus nail clip in Moluccan cockatoos (*Cacatua moluccensis*). *J Avian Med Surg.* 2015;29:303-312.

23. Speer B, Powers LV. Anatomy and disorders of the beak and oral cavity of birds. *Vet Clin North Am Exot Anim Pract.* 2016;19:707-736.

24. Lintner MS. The beak trim. *Proc Annu Conf Assoc Avian Vet.* 2013;189-193.

25. Darbo-McClellan H, Pollock C. Grooming companion birds: a review. LafeberVet . Available at: http://lafeber.com/vet/grooming-companion-birds-a-review/ Accessed April 2, 2017.

26. Heidenreich B. An introduction to positive reinforcement training and its benefits. *J Exot Pet Med.* 2007;16:19-23.

27. Daugette KF, Hoppes S, Tizard I, Brightsmith D. Positive reinforcement training facilitates the voluntary participation of laboratory macaws with veterinary procedures. *J Avian Med Surg.* 2012;26:248-254.

28. Speer B. Management of aggressive and frightened birds. *Annu Conf PacVet.* 2015.

29. Cook EK. Teaching avian patients and caregivers in the examination room. *Vet Clin North Am Exot Anim Pract.* 2012;15:513-522.

30. Mattison S. Training birds and small mammals for medical behaviors. *Vet Clin North Am Exot Anim Pract.* 2012;15:487-499.

Avian Ophthalmology: Principles and Application

Rüdiger Korbel, Prof, Dr med vet, DipECZM (Avian)

Session #2034

Affiliation: From the Clinic for Birds, Small Mammals, Reptiles and Ornamental Fish, Center for Clinical Veterinary Medicine, University of Munich, German, Sonnenstrasse 18, D-85764, Oberschleissheim, Germany.

Manuscript adapted from: Korbel R. Avian ophthalmology: principles and application. *ExoticsCon.* 2016;173-180 and Korbel R. Avian ophthalmology: principles and application. *Vet Med Falconry 21st Century.* 2014;33-38.

Abstract: Ocular disorders in birds are of great importance to vision as well as the fact that they commonly suggest numerous systemic disorders, which may be transferable to man (eg, salmonellosis, mycobacteriosis). This lab will provide the practitioner with a practical review of suitable instrumentation following a step-by-step tutorial through the complete avian ophthalmologic examination procedure as a reliable basis for diagnoses and therapy of ocular disorders in birds. In the practical lab, diurnal and nocturnal raptors, with healthy and diseased eyes, will be used to demonstrate both physiologic as well as pathologic situations. Certain avian-specific capacities, like flying, are closely bound to fully functional vision underlining the importance of avian ophthalmology. In addition, ocular lesions in birds are an expression of systemic disorders more than in mammals and therefore represent an important diagnostic criterion. The ocular symptomatology frequently enables specific conclusions to be drawn on suspected disorders or it may even be pathognomonic for a certain disease. Thus, the avian eye may be seen, to a much larger extent than in mammals, as a diagnostic window. On average, more than 30% of all traumatized birds (incidence is higher in raptors than in pet birds) suffer from ocular lesions, which are most often hidden within the inner structures of the eye with hemorrhages arising from the pectin oculi. Ophthalmoscopy (ie, examination of the posterior eye segment) is obligatory in traumatized birds. Avian ophthalmology is not a highly specific working field within avian medicine but should be an integral part of the general examination procedure. After a short review on anatomic and physiologic peculiarities, common and newly developed routine ophthalmologic examination procedures will be described, followed by a description of clinical pictures in the eye of various patient groups based on an etiologic schematization. Avian ophthalmology is becoming more important in terms of avian visual perception, which strikingly differs from that of man (including avian visual perception within the UV and the ability to resolve high frequency flicker rates). These aspects will be of great importance for animal welfare since commercially maintained poultry are commonly kept under artificial light sources that meet human visual perception requirements but not those of avian vision.

Introduction

In many birds the eye is the most important sensory organ. The capacities of the avian eye which exceed those of the mammalian eye in part are an adaptation to the specific way of life and habitats as well as physical activities that are closely bound to perfectly functioning vision (eg, flying). Even partial impairment of vision that can be caused by any one of many known eye diseases, always has far-reaching consequences because compensation by other senses (including olfactory and acoustic sensory perceptions) is usually insufficient, if possible at all. As a result, ornitho-ophthalmology occupies an important position in avian medicine.

Part I. Anatomic and Physiologic Basics

Basic principles in avian ophthalmology are similar to mammalian ophthalmology. A broad variety of ophthalmologic physiologic and anatomic peculiarities should be considered within the class Aves representing 8800 species and more than 28,000 subspecies. One of the most important characteristics include the striated, rather than smooth, intraocular musculature, making it necessary to use bird-specific techniques for inducing mydriasis. In addition, the fundus oculi includes numerous peculiarities such as the pectin; an anangiotic retina; none, one or bifoveate retina; lack of the retinal pigment epithelium (RPE); and more. This lab will include a section on avian visual capacities, which are of great importance when keeping pet birds, raptors, and commercial poultry under artificial light sources, and thus are of animal welfare importance.

Significance of visual capacities

Visual function in raptors is essential for flying, surviving in the wild, and reproduction. The eye as the main sense organ in birds and its visual capacities have no general superiority compared to mammals but show a high degree of specialization as an adaption to living conditions. Visual acuity is 2-8 times higher than mammals. Visual fields are up to 360°. Stereopsis ranges from 0-70°. Maximum spatial frequency (the ability to dissolve a certain movement into single frames) is up to 160 frames/s (10 frames/s to 15 frames/s in man) and minimum detection of movements is up to 15°/hour (very slow movements). Ultraviolet-perception (UV-perception between 320 nm and 680 nm) is an ability common in diurnal birds bound to special UV-sensitive rods within the retina and an aspect not investigated very well to date. UV-perception probably plays an important role in inter- and intra-specific communication based on plumage-UV-reflection, even in birds that appear monomorphic for the human eye, for the identification and assessment of fruit ripeness based on varying UV-reflection of fruit wax layers, for phenomena of camouflage, orientation, and others. On the other hand, even complete blindness is not a reason for euthanasia in pet birds. Canaries suffering from blindness due to cataract formation—a condition that occurs often in this species—act normal if the interior of the cage or aviary is not modified.

Anatomic peculiarities

Although there are numerous anatomic and physiologic differences, like striated rather than smooth intraocular musculature, the anangiotic fundus oculi, and the pecten oculi, basically the approach to avian ophthalmology is similar to that employed in mammalian ophthalmology. Some anatomic peculiarities with relevance for the ophthalmologist are:

Size and weight of the avian eye:

- Axial length of 8 mm (kiwi, *Apteryx* sp.) up to 50 mm (ostrich, *Struthio camelus*).
- Weight of oculus sinister (OS) and oculus dexter (OD) in man 1%, fowl 7% (adult), and 12% (juvenile) of overall head weight.

Shape:

- Determination by annulus ossicularis sclerae (~10 single bony plates).
- Flat type (diurnal birds with narrow heads, for example Columbiformes).
- Conical type (diurnal birds with broad heads, for example Falconiformes).
- Tubular type (nocturnal birds, for example Strigiformes).

Adnexal structures:

Eyelids.

- Palpebra inferior is larger than palpebra superior (exclusively owls).
- Lower eyelid with tarsus, no Meibomian glands.
- Membrana nictitans is highly mobile, regulation of precorneal tear film, protective function for cornea, snorkel mask effect (compensation for loss of refractive power during underwater vision).

Lacrimal glands.

- Small or absent glandulae lacrimaliae.
- Large glandula lacrimalis membrana nictitantis (vel. Harderian gland).
- Replacement by nasal salt gland (aquatic species).
- No musculus retractor bulbi.
- Complete decussation of chiasma opticus (no consensual pupillary reflex).

Anterior eye segment.

- Striated intraocular musculature (highly mobile iris corresponding to external stimuli, no sensitivity to parasympatholytics/sympathomimetics, no reliable pupillary reflex corresponding to light stimuli).
- Fundus diameter much larger than pupil diameter.
- Ligamentum pectinatum/ciliary cleft with species specific peculiarities.
- Lens with annulus pulvinus (Ringwulst), accommodation range 2 (owls) up to 80 (waterfowl) D.

Posterior eye segment.

- Avascular (anangiotic) retina.
- Afoveate, uni- or bifoveate retina.
- Rods and cones (inclusive special UV-cones) in functional units.
- UV-sensitive cones in most diurnal birds.
- Slight retinal pigmentation in nocturnal, heavily pigmentated retina in diurnal birds.
- No retinal tapetum lucidum.
- Pecten oculi.
- Chorioidal, heavily pigmented structure.
- Protruding into the vitreous.
- Obscuring papilla nervi optici.
- Pleated, vaned and conical type.
- Functional theories, 32: nutritive, thermo- and pressoregulative function most obvious.

Part II. Instrumentation

This section on avian ophthalmology will provide the practitioner with detailed information on specific instrumentation needed in a routine ophthalmologic examination of an exotic animal. Specified instrumentation includes a Finoff transilluminator (which allows examination of the anterior and posterior eye segment using monocular direct ophthalmoscopy), a panoptic monocular ophthalmoscope (which meets specific needs for avian ophthalmology), or a combination of a head band ophthalmoscope in combination with each 30, 40, 78 and 90 diopter lenses. A review of luxmeter and spectral analytic devices will be provided to address measurement of suitable artificial light conditions.

General examination

Minimum requirements for basic and general purposes:

- Focused light source with magnification lens (Finoff transilluminator).
- Instrumentation for manipulation of the eye lids (Graefe hook).
- Lacrimal cannula (Anel).
- Topical anesthetics proxametacaine, oxybuprocaine (duration of action ~7-8 min) or lidocaine (duration of action ~17 min).

Advanced equipment:

- Slit lamp (magnification x5 - x15, better x20).
- Monocular direct panoptic ophthalmoscope, better.
- Head band ophthalmoscope with 30 and 78 D lens (additional aspherical 40, 60, 90 D).

General examination procedures for adnexal structures and anterior eye segment:

- Without restraint.
  Assessment of vision via food intake, reluctance to fly, orientation.

- With restraint.
  Examination of the ear opening.
  Pupillary reflex.
  Examination of the anterior eye chamber with lateral illumination.
  Examination of the anterior eye chamber with lateral transillumination.
  Examination of the anterior eye chamber with retroillumination.

Equipment and procedures for specific examinations:

- Slit lamp biomicrography.

- Gonioscopy (Lovac lens). Examination of the angulus iridocornealis with the pectinate ligamentum; etiologic assessment of primary/secondary glaucoma status.

- Tonometry. Estimation of the intraocular pressure (IOP). Use electronic short-acting tonometer or Schioetz tonometer in raptors. Standard reference values measured with an electronic tonometer calibrated for avian eyes, range from 9-22 mg Hg and are available for 42 species from 7 Orders. Standard reference values for psittacines range from 12-15 mm Hg for nocturnal birds from 9-12 mm Hg, for various poultry species IOP values are as high as 22 mm Hg. Minimum corneal diameter for reliable value is 9 mm. Reference values for newly developed tonometers (Tonovet and Tonolab) will be available soon.

- Schirmer tear test (STT). Test for the estimation of the lacrimal function. Use standardized filter strips of 2-, 3- and 5-mm width. Standard reference values using filter strips of various width for 42 species from 7 Orders showing a wide range of interspecific variations are available. Strigiformes show conspicuously low values.

- Phenol red thread tear test (PRTT). As an alternative for the standard STT the PRTT test may be used for evaluation of tear production. In principle, a cotton thread is placed into the lower conjunctival sac while wetting of the thread is visualized by phenol red as an indicator. Advantages of the PRTT over the STT include shorter reading times (15 seconds instead of 1 minute) and thus minimized stress to the bird and the possibility, using this test in birds even with very small eyelid fissures and low tear production rates, such as small psittacines.[1] Standard reference values are available for various psittacine,[1,2] raptor,[3] and other bird species. Values found in the literature are differing and strongly dependent on bird species and specific examination protocols used that have not been standardized.

- Electroretinography. Measurement of retinal function by recording electrical potentials after light stimulation. This technique gives no information about vision, only retinal function. Basic principles of electroretinography for routine examination have been established. Indications are retinal disorders and dioptric apparatus opacities.

- Laboratory examinations include bacteriologic examination of the conjunctival flora. Physiologic bacterial flora contains gram-positive bacteria, while gram-negative bacteria are an indicator for pathologic conditions. Standard reference values have been established for 42 bird species from 8 orders.

Part III. Examination

Basic examination principles are similar to those in mammalian ophthalmology; however, the small size of avian eyes and their many anatomic and physiologic peculiarities require specific adaptations. These will be demonstrated on a comparative basis to techniques used in mammals, including the basics of patient restraint, examining adnexal structures and anterior eye segment, induction of mydriasis using curariform drugs, mono- and binocular ophthalmoscopic techniques, and visualizing the fundus. The lecture will include a review on most recent and sophisticated diagnostic imaging techniques including ultrasonography, optical coherence tomography, and digital scanning ophthalmoscopy.

Mydriasis and air sac perfusion technique

Induction of mydriasis is indispensable for the examination of the posterior eye segment (ophthalmoscopy). A major difference between the mammalian and the avian eye is that the commonly used mydriatics of atropine and tropicamide have little effect in the avian patient due to a striated rather than smooth intraocular musculature. The iris is partly under voluntary control. It is essential to have a dilated pupil (mydriasis) to perform an ophthalmoscopy, (ie, examination of the posterior eye segment including the vitreous, the fundus and the pecten oculi).

Neuromuscular blocking agents such as d-tubocurarine (3%, 0.01-0.03 ml) have been used in the past but necessitate intracameral application via paracentesis, as the drug penetrates the cornea insufficiently and has to be administered directly into the anterior chamber by paracentesis using a 27- to 30-gauge needle. This technique includes substantial risk for injury of intraocular structures causing hyphema, increasing intraocular pressure (IOP), transmission of conjunctival flora with consecutive uveitis and systemic side effects if larger doses than recommended are used. Therefore, use this technique just for therapeutic reasons (eg, prevention of posterior or anterior synechia resulting from uveitis, secondary miosis and adhesions) where long-lasting mydriasis

(12-18 hours) is necessary. More updated techniques to introduce mydriasis include the use of rocuronium. The method of choice, conjunctival application of rocuronium (12 µl/eye to 35 µl/eye), has been recently described,[4,5] resulting in a mydriasis with species-specific differences with regard to lag of action after administration (up to 90 minutes in common buzzards and African grey parrots) and completeness of mydriasis and duration of action (20-360 minutes) have to be considered and require further scientific studies.[6]

An alternative for routine induction of a mydriasis as well as for intraocular surgery and surgery in the head area is air sac perfusion anesthesia (APA).[7] APA consists of a retrograde perfusion of the lung-air sac-system through a perfusion catheter via the left caudal thoracic air sac. As a carrier gas 0-3 L/min/kg BW of O_2 is used. Effects of nitrous oxide application include low potentiation of isoflurane of approximately 11% and thus improvement of the circulatory situation and protection of the surgeon from isoflurane waste gases. Higher perfusion rates than recommended result in respiratory alkalosis due to a CO_2-wash-out-effect causing severe cardiac arrhythmias. Isoflurane maintenance concentrations vary, depending on the bird species, between 1.0 vol% to 2.4 vol% (*Columba livia* Gmel. 1789). Pulse oximetry is indispensable, as APA causes a reversible apnea due to reduced CO_2 partial pressure, which results in nonstimulation of the respiratory center. Advantages of APA include an ability to provide long-period anesthesia, which is used for routine ophthalmoscopy, electroretinography and head surgery in birds, to free surgical access to the head for intraocular surgery, stable or decreased intraocular pressure, and reversible apnea with an absolute immobilization of the patient. Achievement of mydriasis for ophthalmoscopy may be optimized by systemic administration of 0.2 mg/kg BW of the muscle relaxant vecuronium, which allows a complete mydriasis and areflexia with a lag period of approximately 26 seconds and a duration of 256 minutes in pigeons (*Columba livia* Gmel. 1789) and a reduction of isoflurane consumption of approximately 25% at the same time. This technique allows examination even of the very lens periphery with the annular pad and the extreme fundus periphery.

APA principles:

- Anesthesia induced mydriasis and free surgical access to the head area.
- Perfusion of lung-air sac-system with oxygen-nitrous oxide-isoflurane or sevoflurane mixture.
- Retrograde perfusion via left caudal thoracic air sac using a specific air sac catheter.
- Induction of a reversible apnea due to lowering of the CO_2-partial pressure with subphysiologic values (below 48 mm Hg; no stimulation of respiratory center).
- Modified anesthetic machine with low flow flowmeter (0.01 L/min to 0.5 L/min).

APA performance:

- Carrier gas O_2.
- Induction via head chamber.
- Maintenance via air sac catheter.
- Isoflurane concentration of 1.0 vol% to 2.7 vol%.
- Monitoring via pulse oximetry (no respiratory movements), reflex score, blood gases, body temperature.

APA advantages:

- Routine method for ophthalmology and head surgery.
- Benefits of inhalation anesthesia and long period anesthesia.
- Free surgical access to the head.
- Stable or decreasing intraocular pressure.
- Reversible apnea, absolute immobilization of the patient.
- Achievement of mydriasis for ophthalmology.

Ophthalmoscopy, a technique to examine the fundus oculi (a clinical term, describing ocular structures, which are situated behind the lens) using a focused light beam reflected from the fundus, can be carried out by both monocular and binocular techniques and direct or indirect ophthalmoscopy in combination with double aspherical ophthalmoscopic lenses (at 30, 40, 60, 78 and 90 diopters [D, Volk Bio II]) refractive power (Figs 1 and 2). In all cases, indirect binocular ophthalmoscopy using a head ophthalmoscope is advisable. A 30 D lens is used in birds with larger pupil diameters (nocturnal and diurnal raptors), ophthalmoscopy of pigeons and larger psittacine species requires a 78 D lens, those in small birds (canaries, budgerigars) the use of a 90 D lens. Alternatively, monocular indirect ophthalmoscopy may easily be performed in all birds, especially in smaller species. A long-term survey including more than 32% of traumatized birds showed hemorrhage in the vitreous body originating from lesions of the pecten oculi. Thus, ophthalmoscopy is obligatory in traumatized birds.

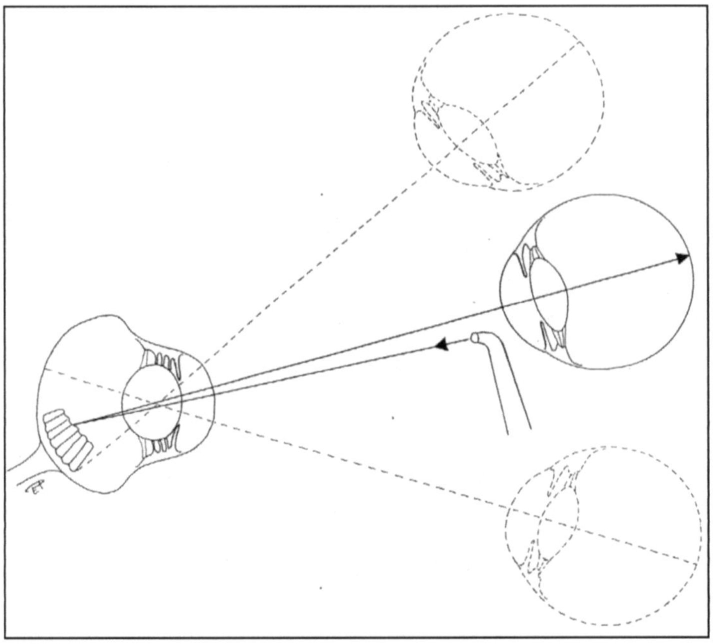

Figure 1. Optical principles of the (monocular direct) ophthalmoscopic examination of avian eyes. Ophthalmoscopy is the examination of ocular structures situated behind the lens, using a focused light beam directed through the pupil and performed using light, reflected from the fundus oculi. As a rule, a nealy coaxial direction of the light beam directed through the (dilated) pupil (mydriasis) of the patient's eye (left) on the one hand and the reflected light beam from the patient's fundus to the examiner's eye (right; viewing axis) on the other hand is indispensable in order to project an image on the examiner's retina (Kern TJ. Exotic animal ophthalmology. In: Gelatt KN, Gilger BC, Kern TJ, eds. *Veterinary Ophthalmology*. Ames, IA: Wiley-Blackwell;2007:1370-1405).

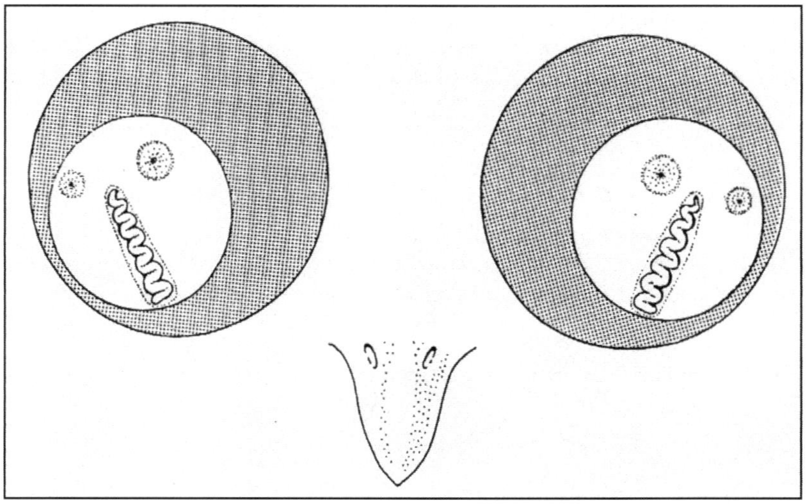

Figure 2. Topographic landmarks within the avian fundus, the pectin oculi, and the central and temporal fovea. Accessible viewing field (bright area) using monocular direct ophthalmoscopy (Kern TJ. Exotic animal ophthalmology. In: Gelatt KN, Gilger BC, Kern TJ, eds. *Veterinary Ophthalmology.* Ames, IA: Wiley-Blackwell;2007:1370-1405).

Part IV. Ocular Disorders

Fully functional vision is indispensable in raptors (falconry-kept) and wild birds and is of great importance in captive pet birds. Most often avian ocular diseases reflect ocular manifestations of systemic disorders, frequently representing zoonoses such as salmonellosis, mycobacteriosis, and avian influenza that are important to both the patient and the patient owner. In addition, many emerging diseases, such as borna virus infections, are of high relevance. This presentation will give a broad overview of avian ocular disorders and discuss diagnostic and therapeutic approaches on an interactive basis.

The ophthalmologist considers it logical to classify eye disorders based on the various ocular structures involved. A complete review of avian ophthalmology discords, based on morphology, will be provided in the presentation. For a literature review please refer to references 7-19. In a long-term investigation of ocular disorders in birds, an overall incidence of 7.6% was observed.[9] The highest incidence of eye disorders compared to other patient groups was found among wild birds (11.7%), among which most eye disorders were seen in Accipitriformes (26.1%), Strigiformes (20.0%), and Falconiformes (19.7%). In contrast, a lower incidence was found in falconry raptors with 7.3%, but again Accipitriformes (6.3%) were most frequently affected by eye disorders, followed by falcons (5.5%) and owls (5.0%). Among disorders of the posterior eye segment (fundus oculi) most often trauma-related hemorrhages (frequently arising from the pectin) are diagnosed. In general, chronic lesions caused by disorders located in the upper part of the fundus result in a poor prognosis because birds, especially raptors, are orienting themselves primarily using the upper part of the fundus (with the central and temporal fovea included).

References

1. Stadtbaeumer K, Korbel R. Untersuchungen zum phuenolrot fadentest (PRT) bei vöge ln. *Proc Conf Avian Dis.* 2002;17-25.

2. Rajaei SM, Ansari Mood M, Khorram H, et al. Measurement of tear production using the phenol red thread test in the common mynah. *J Avian Med Surg.* 2015;29(2):146-148.

3. Smith SP, Barbon AR, Forbes. Evaluation of the phenol red thread tear test in falconiformes. *J Avian Med Surg.* 2015;29(1):25-29.

4. Barsotti G, Briganti A, Spratte J, et al. Safety and efficacy of bilateral topical application of rocuronium bromide for mydriasis in European kestrels (*Falco tinnunculus*). *J Avian Med Surg.* 2012;26:1-5.

5. Petritz OA, Guzman DSM, Gustavsen KA, et al. Evaluation of the mydriatic effects of topical administration of rocuronium bromide in Hispaniolan Amazon parrots (*Amazona ventralis*). *Proc Annu Conf Assoc Avian Vet.* 2014:13-14.

6. Petritz OA, Guzman S-MD, Wiggans T, et al. Evaluation of the mydriatic effects of topical administration of rocuronium bromide in Hispaniolan Amazon parrots (*Amazona ventralis*) *J Am Vet Med Assoc.* 2016;248:67-71.

7. Kern TJ. Exotic animal ophthalmology. In: Gelatt KN, Gilger BC, Kern TJ, eds. *Veterinary Ophthalmology.* Ames, IA: Wiley-Blackwell;2007:1370-1405.

8. Koenig HE, Korbel R, Liebich HG. *Avian Anatomy: Textbook and Colour Atlas.* 2nd ed. Translated and revised by C. Klupiec. Sheffield, UK: 5M Publishing Ltd;2016:159-187.

9. Korbel RT. Disorders of the posterior eye segment in raptors-examination procedures and findings. In: *Raptor Biomedicine III.* Lake Worth, FL: Zoological Education Network;2000:179-194.

10. Korbel RT, Bohnet N. Ophthalmological examination of the avian eye: an interactive CD-ROM. Munich, Germany: 2007.

11. Korbel R, Schulze C, Velasco Gallego ML, Rinder M: Ocular coherence tomography (OCT) in the avian eye-clinical findings and evaluation protocols. *Proc Annu Conf Assoc Avian Vet.* 2013.

12. Korbel R, Rinder M. Ophthalmological findings in birds affected with PDD. *Proc Annu Conf Assoc Avian Vet.* 2011;215.

13. Korbel R. Gonioscopy in birds. *Proc Annu Conf Assoc Avian Vet.* 2011;218.

14. Korbel R, Sigg G. Central and peripheral avian corneal thickness measured with optical coherence tomography (OCT) in the normal eye. *ICARE.* 2015.

15. Korbel, R. Ocular 2D and 3D ultrasonography and ocular coherence tomography (OCT). *Proc Annu Conf Assoc Avian Vet.* 2012;27.

16. Korbel, R. Ophthalmological findings in chickens affected with the "popeye" syndrome. *Proc Annu Conf Assoc Avian Vet.* 2012;75.

17. Liepert A, Dorobek K, Hufen H, Korbel R. Application of 3D Ultrasonography in clinical avian ophthalmology. *Proc Annu Conf Assoc Avian Vet.* 2011:212-213.

18. Lierz M, Korbel R. Anaesthesia and analgesia in birds. *J Exot Pet Med.* 2011;21:44-58.

19. Murphy CJ. Raptor ophthalmology. *Compend Contin Educ Pract Vet.* 1987;9:241-260.

Avian Orthopedic Surgery: A Surgeon Specialized in Avian Trauma's Point Of View!

V.C.M. Ferraz, DVM, Ms, PhD, and
C.R.A. Ferrigno, DVM, Ms, PhD

Session #2036

Affiliation: From Strix Clínica Veterinária Especializada, Av. Prof. Vicente Rao, 1275, CEP: 04636-001, São Paulo, SP Brazil (Ferraz), From FMVZ USP, R. Prof. Orlando Marques de Paiva, 87, CEP: 05608-270, São Paulo, SP Brazil.

Abstract: Avian trauma is one of the most common reasons a bird is brought to a veterinary hospital; however, there are very few publications on orthopedics in birds. Understanding the forces that act on avian bones, as well as avian physiology and biology, and knowledge of available treatments are fundamental for good surgical outcomes. Intramedullary pins, external fixators, bone plates, and CRIF have been used successfully in avian patients. All will be discussed in this topic.

Introduction

There are hundreds of thousands of avian specimens in captivity, as pets and in conservation facilities. Every year many thousands of captive birds—as well as wildlife birds—suffer injuries, yet avian orthopedic surgery is still an unknown field by most veterinarians that work with this animal group. Scientific publications and text books are scarce on the subject and most of them are based on individual experiences, with very few case reports. A recent PubMed search using the words "birds" and "fractures" revealed only a few hundred articles, while "dogs" and "fractures" reveals almost 3000 articles. Knowledge of forces involved in avian bone fractures, how bones heal, which implants are adequate, and how environment management after treatment affect the results are still insipient.

Many concepts go unanswered, and many myths are still considered truth by many. For example, some believe the idea that birds can't withstand surgery or that their bones are more fragile, especially pneumatic bones, and thus more difficult to work with than medullary bones. Old orthopedic concepts are still used in birds, like the use of intramedullary pins as sole implants. New generation implants are seldom used, possibly due to their high cost, but also because of a lack of knowledge by practitioners.

This talk will provide basic concepts in avian bone fracture physiology based on prospective studies of histomorphometry and biomechanics under controlled circumstances. In addition, new concepts will be presented in the translation of orthopedic surgeries to avian surgery, the possibility of use of new-generation implants in avian patients, including a dynamic external fixator developed specifically for this group, with many different cases showing fractures, luxations, limb deformities in different bird species. These are important aspects to be understood by the avian veterinarian who attempts to perform orthopedic surgery in any avian species.

Review Article

Wings are the most important flight apparatus in birds and they involve complex cinematic and perfect aerodynamic control to make flight possible. Any condition that affects a specific part of the wing might lead to severe biomechanical and aerodynamic consequences during flight.[1]

Biomechanics

Bone primordial function is supporting loads of the body. According to the Wolff law, this tissue is capable of adapting to mechanical demands produced by individual weight and physical activities that cause deformation of the whole skeleton,[2] but it's important to remember that the anatomy of each animal species is forged following its own biomechanical needs for performing specific movements.[3] Pneumatic bones, as well as the fusion of different bones are an evolutionary adaptation for flight, decreasing skeleton weight. Also, an important characteristic of avian bones is that they present a higher deposit of phosphorus and calcium, resulting in a large cortical density. It is known that calcium density in chicken humerii increases during laying times, increasing its biomechanical resistance to fractures during this period.[4] In ongoing research, we have evaluated the maximum forces that different long bones of pigeons can withstand before fracturing, using biomechanical analysis (Fig 1). Our first hypothesis is that wing bones can withstand a larger load than leg bones in flying birds, independent of being cortical or pneumatic.

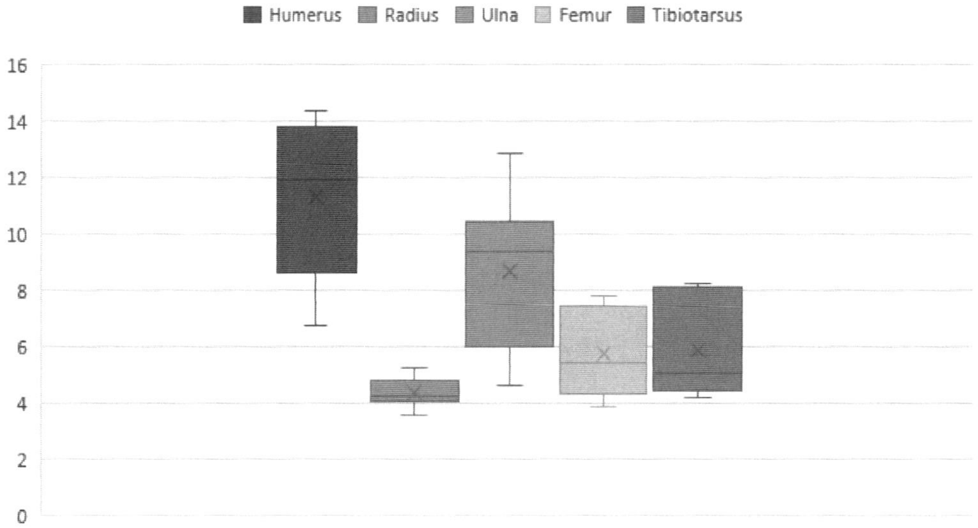

Figure 1. Maximum forces that pigeon bones (*Columba livia*) can withstand before fracture, measured in kgf.

Fractures

Birds bones have thin and brittle cortices, limiting effective purchase of orthopedic implants; Birds also have limited soft tissues on the limbs, and nerve and vascular damage commonly occur with fractures; and fractures are often open and many times the exposed segments are non-viable. Well stabilized and aligned avian fractures tend to consolidate faster than in mammals.[5]

Martin and Ritchie (1999) believe there are not enough studies about the consolidation process in avian bone fractures, but it is known that bone healing rate depends on fragment dislocation, blood supply damage, infectious agents and fragment mobility. The authors mention that fractures stabilized with bone plates, which translates into great stability, present a small bone callus, what is also observed in mammals.[6]

Diaphyseal fractures usually have good prognosis when there is minimal soft tissue damage and no nervous damage (radial nerve).[5]

Treatment

Defining the treatment for any orthopedics disease, including fractures, bone deformities and articular lesions, requires not only good knowledge of orthopedics, including bone physiology, forces acting on physiologically normal bones, and also forces that cause fractures and therefore, need to be neutralized, and characteristics of different implants, but also good knowledge of species biology, behavior and specific bone physiology.

Implants

Different implants have been tested on avian patients, but mostly described in only a few case reports for each species. No biomechanical analyses were made for the use orthopedic implants in birds, to the knowledge of the author.

Understanding each implant's mechanical forces, its features, materials and complexities is fundamental for defining which one to use in each case. We will discuss the main pros and cons of the main implants that have been used in avian patients.

Many stabilization methods have been used for the treatment of avian fractures such as external splinting,[5] intramedullary pins (metallic or nonmetallic),[7-9] external fixators,[10,11] and plates and screws.[12] Some of the recently developed implants propose to keep the vascular supply intact, by reducing contact area between bone and implant, which is consistent with the biologic osteossynthesis concept. Among these are locking compression plates and clamp rod internal fixator (CRIF).[13] In locking plates, the screw head is screwed to the plate and cause less periosteal vascular damage. The interface between screw and plate avoids movement between components, which means all screws within one segment must loosen for stability to be lost.[14]

Intramedullary pins

Intramedullary pins have been the most widely used implants in avian patients. Although they are known for neutralizing bending forces, which are the most important forces acting on wing bones, they do not neutralize compression or torsion forces that cause movement in the fracture site, even visible during surgery. They are an inexpensive implant and the learning curve for their use is lower. They can be used as a primary implant combined with a secondary implant (external fixator, bone plate or even a bandage) to increase its force, and in fact that is a very common technique used in other animals and humans. It is the opinion of the author that other fixation methods are usually strong enough for the forces withheld by avian bones and that the intramedullary pins usually cause more harm than good.

External fixators

External fixators are a good method for stabilizing fractures in medium to large birds. Advantages of using external fixation include shorter surgical and anesthetic times, ease of placement and removal, limited operative iatrogenic trauma, low infection rate, stable fixation, the ability to make alignment adjustments following surgery, it promotes an endosteal callus, which is beneficial, because a very developed periosteal bone callus may affect muscular activity, avoiding return to flight, and finally it also allows early return to function in some humeral fractures, not interfering with movement amplitude of the joints, therefore keeping wing function and bone length.[10,15]

It is the opinion of the author that external fixators are an excellent choice for avian patients because the learning curve is lower compared to more advanced implants. It is harder to cause an iatrogenic fracture using these implants, compared to other implants, and this is especially true when considering smaller patients. Birds usually adapt well, even though it is an external implant; psittacine birds rarely even bother with them. They are a versatile implant and their cost is usually lower than other implants. No special instruments are needed for their use.

As disadvantages, we can point to the fact that they are not as stable as bone plates or the CRIF system. Some animals do not withstand their use and animals cannot be let loose even in a controlled but large environment because of the risk of getting tangled, while the implant is not removed. We will present many cases with the use of external fixators for long bone fractures, articular luxations, and bone deformities.

Bone plates

Bone plates historically have been contraindicated for use in birds. With the evolution of technology for producing these implants, and with new and smaller plates and more precise instruments, bone plates (specially locking plates) have been used successfully in a variety of bird fractures.

Locking plates have many advantages compared to dynamic compression plates, since they do not need perfect reconstruction of the bone column.[16] The use of this kind of implant is advantageous because of greater stability and it allows the use of fewer screws, which minimizes the possibility of iatrogenic fractures during surgery. It also helps preserve blood supply as well as minimize soft tissue damage.[14]

Clamp rod internal fixation system

The CRIF system consists of clamps that slide in a rod and may be positioned around the rod in different positions. These clamps are compressed by means of tightening of the screws, which fix the rod internally to the bone.[17]

There aren't any published data on the use of the CRIF system in avian patients, even though it is a simple implant that was designed for veterinary use, focusing on ease of use and lower cost. We have used the CRIF system in one avian patient, combined with a locking plate in a female adult cape barren goose (*Cereopsis novaehollandiae*), with a complete humeral fracture after a fight with a cagemate. The patient underwent surgery for the stabilization of the fracture, which was fixed with a 2.7-mm locking compression plate (LCP), and a 2.0-mm CRIF. At 60 days, complete bone healing was seen. At 218 days post surgery, a remodeled bone callus was observed and the implants were removed. The reason we chose this combination of implants in this case was mainly the concern with the species behavior, which includes using the wings to fight for females and for territory. Even though just the locking plate would be sufficient for the fracture type and animal weight, we believed he could refracture it during one of those behavior displays if it didn't have a more stable fixation.

There is no need for removal of implants in captive birds, unless these are causing discomfort.[18] Whenever the patient is intended for release, all implants should be removed, and enough time should be allowed for filling the holes (from pins and screws) after implant removal, since these can cause fragility of the bone, predisposing the bird to new fractures.

Conclusions

Knowledge of bone biomechanics, forces acting on bones, and avian physiology and biology are all important for the treatment of avian trauma. Pins, external fixators, bone plates, and the CRIF system all have advantages and disadvantages when used in avian patients.

References

1. Beaufrere H. A review of biomechanic and aerodynamic considerations of the avian thoracic limb. *J Avian Med Surg*. 2009;23(3):173-185.

2. Gusmão CVB, Belangero WD. Como a célula óssea reconhece o Estímulo Mecânico? *Revista Brasileira de Ortopedia*. 2009;44(4):299-305.

3. Mota NL, Castro Júnior AF, Castro BK et al. Embriologia e histofisiologia do tecido ósseo: revisão de literatura e bases histofisiológiaos das principais doenças ósseas metabólicas. Boletim do centro de biologia da reprodução. *Juiz de Fora*. 2008;1:3.

4. Flemming RH, Fleming HA, McCormack, L, et al. Medullary bone and humeral breaking strength in laying hens. Res Vet Sci. 1998;64(1):63-67.

5. Maccoy DM. Treatment of fractures in avian species. *Vet Clin North Am Sm Anim Pract*. 1992;22(1): 225-238.

6. Martin H, Ritchie BW. Orthopedic surgical techniques. In: Ritchie BW, Harrison GJ, Harrison LR, eds. *Avian Medicine: Principles and Application*. Lake Worth, FL: Wingers Publishing, Inc;1999:1137-1169.

7. Alievi MM, Oliveira ANC, Ferreira PA, Traesel C, Guimarães LD, Flores F, Silva SF, Schossler JEW. Osteossíntese de úmero em pombos domésticos (*Columba livia*) associando-se pinos metálicos e polimetilmetacrilato intramedulares após osteotomia diafisária. *Arquivo Brasileiro de Medicina Veterinária e Zootecnia*. 2008;60(4):843-850.

8. Cunha O, Rios A, Taffarel MO, et al. Osteossíntese de úmero com êmbolo de seringa em pombos domésticos. *Medvep-Revista Científica de Medicina Veterinária-Pequenos Animais e Animais de Estimação*. 2004;2(6):87-90.

9. Schossler JEW, Garcia ÉFV. Osteossíntese de úmero por xenoenxerto ósseo preservado em glicerina 98% em pombos domésticos (*Columba livia*). *Medvep-Revista Científica de Medicina Veterinária-Pequenos Animais e Animais de Estimação*. 2011;9(29):302-307.

10. Ferraz VCM. Avaliação radiográfica e de função de vôo após fixação de osteotomias distais do úmero em pombas (*Columba livia*), com modelo de fixador externo articulado. *Pesquisa Veterinária Brasileira*. 2008;28(8):351-357.

11. Leotte AM, Raiser AG, Braga FA, et al. Fixação esquelética externa tipo Ia (unilateral-uniplanar) para osteossíntese diafisária de úmero em pombos domésticos (*Columa livia*). *Ciência Rural*. 2004;34(6): 1865-1870.

12. Kuzma AB, Hunter B. Osteotomy and derotation of the humerus in a turkey vulture using intramedullary polymethylmethacrylate and bone plate fixation. *Can Vet J.* 1989;30:900-901.

13. Zahn K, Matis U. The clamp rod internal fixator-application and results in 120 small animal fracture pacients. *Vet Comp Orthop Traumatol.* 2004;17:110-120.

14. Ferrigno CRA, Cunha O, Caquias DFI, et al. Resultados clínicos e radiográficos de placas ósseas bloqueadas em 13 casos. *Brazil J Vet Res Anim Sci.* 2011;48(6):512-518.

15. Bush M. External fixation of avian fractures. *J Am Vet Med Assoc.* 1977;171(1):943-946.

16. Maccoy DM. General principles of avian surgery. *Comp Cont Educ Pract Vet.* 1991; 13(6):989-992.

17. Zahn K, Frei R, Wunderle D, et al. Mechanical properties of 18 different AO bone plates and clamp-rod internal fixation system tested on a gap model construct. *Vet Comp Orthop and Traumatol.* 2008;21: 185-194.

18. Guzman DSM, Bubenik LJ, Lauer SK, et al. Repair of a coracoid luxation and a tibiotarsal fracture in a bald eagle (*Haliaeetus leucocephalus*). *J Avian Med Surg.* 2007;21(3):188-195.

Avian Metabolic Bone Diseases

M. Scott Echols, DVM, DABVP (Avian Practice)

Session #2041

Affiliation: From Echols Veterinary Services, Salt Lake City, UT 84121, USA, Medical Center for Birds, Oakley, CA 94561, USA, Parrish Creek Veterinary Clinic, Centerville, UT 84014, USA.

Abstract: Metabolic bone diseases relate to disorders of the bone and especially include osteoporosis. Until recently osteoporosis has been largely ignored in captive birds, especially adults. This discussion covers osteoporosis, diagnosis, management, and treatment options in birds.

Metabolic Bone Disease

Metabolic bone disease (MBD) is a non-specific phrase pertaining to diseases of the bone that have many causes. These 'diseases of the bone' are related to loss of trabecular and cortical bone mass. [1] Conditions such as osteoporosis (most commonly described), rickets and osteomalacia are considered MBDs. Most often MBD relates to degree of mineralization, bone formation and ultimately bone strength.

Rickets is generally due to vitamin D, calcium and/or phosphorus deficiency (usually from improper diet and sunlight [vitamin D]) or impaired metabolism in young animals. [2] As the animal grows the bone may fracture or deform. In birds it is most often seen as a curved keel or misshapen long bones. Some of these young birds go on to develop improved bone mineralization and strength but retain their deformed bones. Because this is more of a developmental disorder, rickets is not further discussed in this article.

Osteomalacia is softening of the bones in adult animals. Osteomalacia is simply the adult form of rickets; however, bone deformation is far more pronounced in young growing animals with rickets than in adults with osteomalacia (that started with normal bones). Certain diseases (hyperparathyroidism), drugs, inadequate sunlight and more are known causes of osteomalacia in humans. The same correlations have been proposed but inadequately proven in birds.

Osteoporosis is simply decreased bone strength and is associated with an increased risk of fractures. Osteoporosis is more accurately described as lower than normal peak bone mass for the animal. At least in humans, bone density may be classified as normal, osteopenic, osteoporotic and severely osteoporotic based on a T-score generated from dual energy X-ray absorptiometry (DXA scan). While DXA scans have been performed in research birds, such data are limited.[3] Of the definitions above, osteoporosis best encompasses the nature of discussion presented here.

Osteoporosis

While osteoporosis conjures up images of old people hunched over and at risk of breaking their back or hip, the disease has far wider implications. In humans, osteoporosis has been associated with loose teeth (similar parallels have been made in non-human mammals), musculoskeletal pain, nerve pain, joint disorders and more. The typical age related bone changes in humans with osteoporosis including increased bone marrow space, endosteal resorption of cortical bone and increasing bone tissue porosity, are also recognized in birds. [1]

Cage layer paralysis was first recognized in the literature in the 1950s shortly after the introduction of battery cages.[4] The disorder is essentially due to osteoporosis and weakened bones.[4] Riddell, et al stated the disease should better be termed cage layer osteoporosis.[5] Osteoporosis is simply the loss of structural bone that leads to bone fragility and susceptibility to fractures. Spinal degeneration and/or transient irregularities in intracellular calcium metabolism may lead to abnormal muscle activity and paralysis seen with this disease.

Common boney abnormalities of chickens with cage layer fatigue include an S-shaped keel, rib deformities (swollen vertebral-sternal junctions), and thin and easily broken bones including those of the cranium.[5] Additionally, vertebral fractures are common in poultry and narrowing of the 4th and 5th rib space adjacent to the spinal column due to collapse and fusion of the individual vertebrae have been described.[5] Overlying vertebral compression fractures have also been correlated with pulmonary hemorrhage and spinal cord degeneration and/or compression in poultry.[5] One study showed that by the time battery raised chickens had been fully processed, 98% of the carcasses had broken bones with an average of 6 fractured bones in each bird. [6] The authors noted that the bones most frequently broken are recognized to be relatively liable during periods of skeletal depletion.[6] Other studies have reported that up to 70% of laying hens have old healing fractures and the incidence varies based on husbandry and strain.[7] Osteoporosis is considered widespread in the poultry industry and is considered responsible for 30% of the fractures that occur during production and at depopulation.[4] Recognition of these problems has led to the serious welfare concerns of cage/battery raised poultry.

In chickens with cage layer fatigue, most soft tissues are histologically normal except the parathyroid glands, which are often hyperplastic.[5] Of course, secondary soft tissue pathology (eg, inflammation, spinal cord trauma) can be seen with osteoporosis.

Histologic changes in the bones with cage layer fatigue are far more pervasive. As expected, the cortex of many long bones tends to be thin and absorption cavities are prominent. Also common is a layer of subperiosteal bone overlying the cortex.[5] In many areas this new bone can also be seen communicating with the medullary bone through the absorption spaces underlying the thin cortex. Medullary bone tends to be fragmented and osteoid is commonly greater than calcified medullary bone. Some of the weight bearing bones, the femur in particular, may not display these characteristic changes in affected birds. Vertebral fractures and damage directly to the spinal cord are likely what results in the paralysis of cage layer paralysis.[5]

Based on micro computed tomography, review of radiographs through a consultation service and recent advanced X-ray technology taken of birds from across the US, the author feels that osteoporosis is common in the captive avian population representing numerous orders of birds.

Formation of Bone in Birds

Understanding basic bone biology helps researchers better understand bone pathology. In the appendicular skeleton, longitudinal growth of the long bones occurs via endochondrial ossification. Resting chondrocytes within the germinal layer of the epiphyseal growth plate become committed to differentiate into proliferative chondrocytes. Nourished by epiphyseal capillaries, a zone of proliferation forms as columns of packed chondrocytes within an extracellular matrix that contains high amounts of type II collagen.[4]

As more matrix is secreted the chondrocytes within the columns separate, hypertrophy, and start secreting type X collagen. This hypertrophic zone receives its blood supply from metaphyseal blood vessels. The development also creates a narrow avascular zone of pre-hypertrophic chondrocytes between the hypertrophic and proliferative zones. In addition to collagen, the chondrocytes secrete proteoglycans, growth factors and other matrix components which all regulate further development of the cells.

In the lower hypertrophic zone, chondroclasts absorb matrix while fully hypertrophied chondrocytes secrete alkaline phosphatase (ALP). ALP initiates formation of hydroxyapatite crystals (ie, the bone mineral).

Next, osteoblasts formed from precursor cells in the marrow resorb chondrocytes undergoing apoptosis. Osteoblasts produce bone matrix (out of type I collagen fibrils and increased Ca^{2+} and PO_4^{3-}) that forms hydroxyapatite crystals in and outside of the matrix. Ongoing osteoclastic resorption combined with osteoblastic formation results in the development of woven trabecular bone which is based on the irregular structure of collagen fibrils.

Bone elongation occurs by continued chondrocyte proliferation at the head of the growth plate and mineralization and hypertrophy at the rear. The trabecular network is mostly resorbed forming bone marrow.

The above process is different from that in mammals in that birds don't form secondary ossification centers and do tend to have much better vascularized growth plates.

Long bone widening involves intramembranous ossification. Within the perichondrium, osteoblasts develop and produce bone spicules. These bone spicules merge to form bone with cavities lined by osteoblasts. The osteoblasts fill in the cavities by secreting concentric layers of lamellar cortical bone.

Meanwhile osteoclasts absorb bone on the endosteal surface so that bone widens on the outer surface with an expanding ring (of bone) and inner resorption (of bone). During early growth the expanding bone rings grow too rapidly to fill the resulting cavities with bone. As growth slows, bone remodeling increases and bone cavity filling rises.

Towards the end of the growth period, secondary osteons are formed by osteoblasts cutting a tunnel through the bone followed by osteoblasts forming concentric layers of lamellar bone. During this period of osteon formation, osteoblasts become entombed within the bone and differentiate into osteocytes. Finally, osteocytes develop a network of interconnections and regulate bone remodeling based on biomechanical forces.[4]

Bone changes during reproduction

Bone biology changes dramatically at the onset of sexual maturity in hens. Osteoblasts stop forming lamellar bone and start producing woven or medullary bone. This medullary bone is laid down on the surfaces of structural bone and in spicules within the medullary cavity and is designed to provide a source of calcium for egg shell production. This is most dramatic in the leg bones but can also be found in wing and pneumatic bones. Medullary bone deposition is fastest during the early stages of lay and then tends to accumulate slowly throughout the laying period.[4]

In the mature hen, rising estrogen stimulates osteoblasts causing them to lay down medullary bone rather than structural bone. The reverse occurs when birds are out of lay and estrogens decline. In human post-menopausal osteoporosis, declining estrogen increases bone resorption and trabecular bone loss and suppresses structural bone formation which is notably different from what occurs in birds.[1,4,8]

During lay, structural bone formation may cease entirely to give way to medullary bone production; however, osteoclastic resorption continues resulting in declining structural bone content in the hen. During the early laying period, the trabecular bone loss in the spine and epiphyses of long bones can be marked. Cortical bone loss tends to be more gradual. Overall structural bone loss is noted throughout the skeleton especially in the long bones of the wing, leg, keel, vertebrae and others even though loss may be replaced by medullary bone. Through the lay period and with chronicity, the end result is a progressive weakening of the skeleton.[7]

Even though potentially more medullary bone is produced than structural bone is lost, accounting for stable or increased total bone content, the bones become weaker. Because medullary bone is a type of woven bone based on irregularly arranged collagen fibrils and is present in isolated spicules it is fundamentally not as strong as structural bone. The overall strength of the skeleton is weakened resulting in increased risk of fractures.

Diagnosis of Osteoporosis

Bloodwork

Unless the bird is suffering from advanced MBD, severe electrolyte disturbances (most commonly hypomagnesemia or hypermagnesemia resulting in hypocalcemia), is taking certain medications, has other disease states or competing elements in the diet, most lab work abnormalities will be non-specific in birds with osteoporosis and more likely related to secondary complications.

Radiographs

The classic radiograph of paper thin cortices and folding fractures related to MBD represents the extreme in disease. Radiographs truly are shades of grey and interpretation of bone loss is subjective. Additionally, the X-ray machine, settings, patient positioning, patient motion, developing process and interpretative software (if digital) all play a role in the image we view. Finally, post-imaging software (sharpening tools are commonly added to digital radiographs), viewing monitor and the viewer's anatomic understanding and interpretation also add to the subjective nature of reading radiographs.

Because medullary bone is as radiodense as structural bone, standard radiographs and bone ash measurements are not a good means to evaluate bone structure when medullary bone is present;[4] however, when medullary bone is at a minimum the cortex of long bones can be subjectively measured for relative thickness and density. It should also be noted that external bone diameter is not adequate when evaluating bone density. It is well described that in human aging osteoporotic bones, the external bone diameter increases along with the marrow space as the cortices thin. The same is believed to happen in avian bones.[7]

There are common areas on radiographs the author has recognized and correlated with CT and basic physical activity that may give veterinarians more insight into bone and animal health.

Birds that tend to not fly much have generally poor bone density along the pectoral girdle (coracoid, furcula and proximal humerus) and distal neck/proximal notarium vertebrae.

Birds that tend to not walk or climb much generally have poor bone density along the synsacrum, proximal femurs and sometimes extending down the legs. Birds that tend to not walk or fly much have the aforementioned deficits plus a more generalized radiolucency along the entire spine that can include the caudal vertebrae.

Specific non-invasive bone mineral density diagnostics

The currently available non-invasive bone density diagnostics, other than the gold standard DXA, include pDXA (peripheral dual energy x-ray absorptiometry), QUS (quantitative ultrasound), QCT (quantitative computed tomography) and pQCT (peripheral quantitative computed tomography). Abnormal pDXA, QUS, QCT and pQCT are screening tests that would ultimately lead to an additional test- central DXA as the preferred technique to measure bone mineral density. There have been limited published works describing these modalities being used to evaluate bone density in birds.[3,9-11]

While DXA scans are typically discussed for use in adult humans as the gold standard in measuring bone mineral density, their use has been explored in birds.[3,9] DXA scans use 2 X-ray beams with differing energy levels aimed at specific bones. The soft tissue is subtracted out and the bone mineral density is measured by the absorption of each beam by the bone studied. A 'T' score is generated and these values are compared against young normals (eg, a healthy 30-year-old) to determine if the bone mineral density (BMD) is low, normal or high for the bone studied. A 'Z' score is 'age matched' and designed to compare BMD scores to someone of similar (or same) age and body size.

DXA is not without fault as it cannot distinguish flat mineral dense versus thick mineral poor bone, shows ambiguity between elements of varying shape, depends on bone orientation during scanning and also depends on the animal's position and regions of interest.[3] Because x-ray absorption is unique at different energies, we can use the 2 readings to determine the thicknesses of 2 different tissues ASSUMING that the tissue absorptions are already known. This is an over-simplification (eg, x-ray can pass through more than 2 different tissues or the tissue might have different absorptions). DXA tries to solve an equation with 4 or more unknowns (absorption and thickness of muscle and bone with 2 measurements). The intention is to measure absorption (bone density); however, it assumes tissue *thickness* is known.

In addition, in the case of birds and some reptiles, reproductive activity can affect bone mineral density measurements via DXA (and really most density measurement tests); however, if the above factors are adjusted for, DXA can provide a precise T score for site specific measurements and can be used on animals as small as rats.[3]

While DXA scans have primarily been reserved for humans and research animals, there is the potential for use in avian clinical cases; however, this diagnostic tool needs additional study to better understand its role in avian medicine.

Computed tomography

Computed tomography or CT is considered the gold standard when measuring bone (or any tissue) density. As opposed to 2-dimensional radiographs, CT can give volumetric density in Hounsfield units (HU).

The HU scale is a quantitative scale for describing radiodensity. Air is given a value of -1000 HU, water 0 HU and the number increases with increasing tissue density. Using CT, one can get a point HU value over any tissue scanned.

As described above, medullary bone can be as radiodense as structural bone and the overall bone content can be the same or higher when medullary bone is deposited at a greater rate than structural bone. The end result may be weaker bones; however, CT measures the density represented over a given volume. The HU reading may be higher in bones with increased medullary bone over those with normal structural bone (given the same size of bone).

With the above limitation, CT still has advantage over radiographs when evaluating bone density in a few situations. First CT should give representative HU values with typical structural bone (not filled in with medullary bone). Second, using computer sectioning, CT HU can be measured just over the cortex. In this situation, the volume is just one cortex wall rather than the entire thickness of the bone. This is of course assuming the medullary bone has not invaded the cortex and the value is compared to cortex only readings from other comparable animals and bones. Third, CT can be used to visually see radiolucent holes in the bones. Once CT data is rendered the operator can peel away densities. By digitally removing densities, soft tissues can be 'stripped away' leaving only the denser tissues such as bone. If the density of the bone is the same or less than the stripped soft tissues, the bone will disappear. Visually the bone is either faintly visible or non-visible. It may also appear as if the bone has holes in it as density loss tends to be patchy rather than uniform.

Modification of Osteoporosis

Reproductive activity

Getting birds out of lay activity can be highly beneficial for those at risk of osteoporosis. Birds that are selected and genetically bred for high egg production tend to have shortened (or even absent) rest periods and are highly susceptible to osteoporosis.[4] Once lay activity has stopped, structural bone formation commences and medullary bone gradually disappears.

Interestingly, egg production is not directly correlated with osteoporosis. Studies in hens support that egg production has little or no effect on bone quality. After all, most egg laying hens take in more calcium than is put into shell production. Rather, it is likely the duration of increased estrogen (related to prolonged egg laying periods) that correlates with poor bone quality.

Bone calcium metabolism is highly complex and not completely understood in the laying hen. At least in domestic laying hens, eggshell calcification occurs at night after eating ceases.[12] It is believed that the majority of eggshell calcium must come from bone stores and not the gut. Even though intestinal calcium absorption is more intense during eggshell formation there is simply less calcium available in the gut (which would be mostly empty during night time hours); however, in a normally functioning laying hen, the bone calcium reserves would be replenished from intestinal sources while the egg is plumping and eggshell is not being laid. Simply put, intestinal (or dietary) availability of calcium is not sufficient to satisfy the high calcium required for eggshell formation.[12]

The remedy is to get pet birds to stop laying (or being broody). This can be accomplished by antecedent-behavior-consequence rearrangement (discussed elsewhere); behavior modification of bird-owner interactions (eg, stop petting, stroking, cuddling the bird); environmental, dietary and general activity changes (also discussed elsewhere); and select medicinal (GnRH agonists) use. Rarely, surgical modification (eg, removing cystic ovaries, neoplastic tissues and *questionably* ovariosalpingohysterectomies) can be used to manage excessive egg laying.

Genetics

As more breeds of various species of captive birds are created, genetics play an increasing role in the health of these animals. Some genetic variations, such as the English trumpeter pigeon with feathered feet and altered leg bone formation, prevent normal movement and likely affect weight loading and subsequent structural bone formation. Significant tibiotarsal breaking strength differences have been noted in certain lines of chickens.[4] The differences in some of these chicken strains have been primarily attributed to structural bone quantity.

Genetic selection has been shown to improve bone strength and resistance to osteoporosis in laying hens.[7] Genetic factors and appropriate breeding may also need to be considered for heavily inbred groups of captive birds.

Drugs

During lay, bisphosphonate (inhibits osteoclast activity) can alleviate structural bone loss.[4] Parathyroid hormones, other osteoclast inhibitors, estrogens and other drugs have been well described as means to alter the course of osteoporosis in humans. Even with the large array of drugs available, these medications are rarely effective as the only treatment of osteoporosis in humans.

It seems reasonable that by understanding bone biology, select drugs (some of which may already be available) could be beneficial in treating osteoporosis in birds; however, no such trials were found at the time of writing.

Diet

Deficiencies of phosphorus, calcium and cholecalciferol have been shown to result in osteomalacia (ie, softened bones) bone loss;[4] however, simply avoiding osteomalacia (with proper dietary phosphorus, calcium, and cholecalciferol) has not been shown to prevent osteoporosis in egg laying hens. Particulate forms of calcium supplementation have been shown to improve shell and bone quality over powdered supplements. The belief is that the particulate calcium remains in the gut longer (than powders) and is more consistently available through the day and/or night for egg shell production.

While various forms of calcium supplementation (eg, oystershell, limestone) may increase medullary bone and possibly reduce cancellous bone loss, calcium deficiency is currently not considered to be a primary cause of osteoporosis (at least in hens).[7] As is also true in postmenopausal women, dietary vitamin D and calcium supplementation are of limited benefit in osteoporotic chickens.[8]

Consumer reports show that 90% or greater of the captive pet parrot population is fed seed-based diets. These diets are generally considered to be nutritionally inadequate for long term health. How exactly seed-based, pellet-based, fruit and vegetable based, combination and more diets relate to bone development and, more interestingly, long term bone maintenance for captive birds is not exactly clear. Unfortunately, the general recommendation of feeding an appropriate diet (often consisting of a species-formulated pellet-based diet plus added vegetables and some fruits) for captive parrots (and other species) is currently the most common advice avian veterinarians give concerning captive parrots. The appropriate diet must be considered not only for the species in question, but also that animal's lifestyle, health status, and more. There is no one magic diet.

One study used peripheral quantitative computed tomography to study BMD and content of the left tibiotarsus in healthy male and female 2-year-old budgerigars (*Melopsittacus undulatus*) housed under identical conditions for 12 months.[10] One group received a commercial pelleted diet while the other a commercial seed diet. There was no difference in bone density and mineral content between the birds fed the different diets; however, the female birds of both groups had higher values than the males. The cause for the increased bone mineral content and density in the females was not proven; however, the authors postulated the increased values may represent hyperostosis and/or prophylactic storage of calcium during reproductively dormant periods.[10] This one study shows an interesting lack of effect of a pelleted diet compared to a seed diet fed for 1 year in budgerigars.

Omega-3 fatty acid (O-3 FA) supplementation has been shown to improve bone mineral density and/or reduce bone breakage in poultry species. The benefits of O-3 FA supplementation have been widely published in the avian literature and should be considered when managing osteoporosis.

As most commercial bird foods are made from corn and soybean components, these diets are typically high in omega- 6 fatty acids (O-6 FA). Fish oils are highly unstable when manufactured in foods. Flax seeds provide a more stable source of O-3 FA in manufactured diets. Flax seed oil provides a more concentrated source (than flax seeds) of O-3 FA and can be added to the diet. Finally, fish oil (eicosapentaenoic acid [EPA] and docosahexaenoic acid [DHA]) provides the most biologically potent form of O-3 FA and must be supplemented fresh due to its instability to heat, oxygen and physical mixing. Based on research in multiple animals, fish oil supplementation is likely the best means to rapidly increase body EPA and DHA levels, even for herbivorous species that would naturally consume more alpha linolenic acid (ALA).

Omega-3 FA contribute to several factors involving bone health. Four-week-old quail fed identical diets for 7 months except for fat content (soybean oil [SBO], hydrogenated soybean oil [HSBO], chicken fat [CF] or menhaden fish oil [FO] all at 50 g/kg of feed) had notable differences in bone parameters.[13] The ratio of O-6 FA:O-3 FA in the diets were as follows: SBO 12.55, HSBO 17.85, CF 18.47 and FO 0.66. As O-3 FA's reduce the concentration of arachidonic acid and subsequent production of PGE_2 (the opposite is true with O-6 FA), bone formation is increased. PGE_2's long-term goal is to stimulate bone resorption. As expected, quail fed FO (and HSBO) had markedly improved cortical bone thickness and density compared to the other groups. Also, quail fed FO had higher percentages of tibial ash, calcium and phosphorus. Last, quail fed FO or HSBO had increased bone shear force over the other groups. All the findings indicated bones from FO and HSBO groups were stronger. As a potential negative, HSBO quail had more trans fatty acids in studied tissues.[13]

A similar study used 16-week-old chickens fed diets of varying O-6 FA:O-3 FA (ranging from 48.7:1 to 4.7:1) for 42 weeks. While the study did find increasing cortical bone thickness as the O-3 FA concentration increased (up to the highest O-3 FA level), other parameters such as bone strength and mineral content were similar between the different diets.[14] One of many differences is that this second study used a much lower O-3 FA content diet than in the quail study above. These findings support a role of supplemental O-3 FA in improving bone quality in young Galliformes and may similarly affect other species.

Free-range hens supplemented with ALA had a significant (40-60%) reduction in keel fractures, reduced breakage severity, increased BMD and content, greater bone and trabecular volume, and finally enhanced bone turnover compared with control groups that ate a typical diet high in O-6 FA.[9] Supplementation began between 23 and 30 weeks of age and birds were evaluated by numerous means (DXA score, histomorphometry, biomechanic studies and more) at 30, 50 and 70 weeks of age. Keel fractures were even lowered by 40% at the end of lay compared to control groups.[9]

In general, feeding diets mostly devoid of calcium, containing excessively high phosphorus and other deviations from diets considered normal for the species puts the animal at risk of developing some form of MBD if fed too long. The exclusive red meat diet fed to raptors is a well-known risk factor for severe MBD including folding fractures;[15] however, MBD inducing diets are rarely this obvious or cause such severe nutritional disease.

For granivores, seed based diets are typically high in phosphorous and/or low in calcium and are likely the most common dietary cause of low grade MBD. In addition, research supports that diets either high in O-6 fatty acids and/or low in O-3 fatty acids may be another risk factor for osteoporosis and other forms of MBD.

Exercise

While studies are lacking relating exercise as a treatment for avian osteoporosis, it is known that bone strength is related to the degree of biomechanical load placed on the bone.[4] Load bearing is considered essential for normal skeletal development and turnover, is well documented and often termed adaptive bone remodeling.[7] Long periods of inactivity are well known to be detrimental to bone mass in humans and chickens.[7] General recommendations for prevention of osteoporosis in humans include building bone mass prior to older age and/or estrogen decline (post-menopause) and maintaining load bearing exercise.[8] Hens housed in battery cages have notably decreased bone strength compared to those housed in alternative systems that allow for exercise. Along the same lines, allowing hens (chickens) freedom of exercise improves bone strength.[7] All the above comments are expected and common sense based on our understanding of bone development and maintenance across numerous classes of animals. Based on the author's work with radiographs and CT, inactivity does seem to play a role in avian bone density.

Tatara, et al surgically denervated the right humerus (radialis and mediano-ulnaris neurectomy) in 22 day old turkeys and then compared this bone to the left (untouched) humerus and those from sham operated birds at 17 weeks of age.[1] Experimental group birds were not able to flap their right wing and experienced a moderately drooped wing. The denervated humeri had significantly decreased volumetric bone density, relative mean wall thickness, cross sectional area, overall strength and more that supported focal (one bone) cortical osteopenia resulting from disuse atrophy and possibly related to elimination of neuronal regulation of the bone.[1] This study supports that lack of physical exercise (in this case, likely due to neurectomy) can significantly and adversely affect bone mineral density in birds.

Lay activity in hens also has an important role in bone mineral density relative to exercise. In studied chickens, exercise does have a positive benefit on bone strength in the early lay period when osteoclasts and subsequently bone resorption are low;[7] however, later in lay, continuous exercise does not continue to suppress osteoclast recruitment and structural bone resorption. As a result, even continuous exercise in late lay birds does not benefit bone quality.[7] This brings us back to earlier comments that it is important to cease lay activity in effort to improve bone mineral density.

The obvious remedy is to increase bone loading activities such as flying, climbing and walking for pet birds. This can be accomplished by providing larger space, creating an environment that encourages movement (eg, oraging, outside play time, flight training) and even forced exercise (eg, moving the food and water bowls away from the bird, training the bird to flap or perform various activities). What is not clear is just how much activity (and for how long) is needed to improve structural bone formation.

Sunlight

Numerous studies have evaluated the role of either natural unfiltered sunlight or supplemental UVA/B exposure to growing and adult birds. The effects of UV radiation on growing chicks of precocial species generally are benefits related to bone health. These effects are more difficult to measure in adults as mature bone is more slowly modified. Also the addition of UV light (natural and artificial) can affect behavior and has been related to increased activity. It is not always clear if the increased activity (and subsequent bone loading) or the actual UV light is improving bone quality. Either way, exposure to natural sunlight for diurnal species seems both humane and makes common sense for improving the quality of a captive bird's life.

Cousquer et al studied metabolic bone disease in wild collared doves.[16] The authors found that 51.2% of juvenile and 9.6% of adult birds admitted to a hospital in the United Kingdom had clinical signs of metabolic bone disease (eg, bendy beaks, deformed bones). The histologic and blood work findings most supported vitamin D deficiency induced rickets in the birds. The incidence was highest between December and February and dropped to zero from June to August. Based on meteorologic data representing the time of the study, the average normal total sunlight hours from December to February (increased incidence of disease) were 195. For the period of May through August (lowest incidence of disease), the total sunlight hours were 721. While diet and other factors were not controlled, the authors postulated that inadequate UVB light exposure during the short winter days was the greatest risk factor for these birds developing rickets.[16] As a note, collared doves breed year round, are invasive to the United Kingdom and originate (at the end of the 19th century) from warm temperate and subtropical Asia from Turkey east to southern China and south through India to Sri Lanka.[17]

Additional potential therapies

It has been known that bone growth can be stimulated by heat. Additionally, heat treatment increases bone mineralization and heat shock protein expression in human cell cultures. 18 While a similar pathway has been suggested in avian species, heat therapy has not been proven to have similar effects in birds; however, considering this concept, low-level laser therapy has been used to improve the biomechanical properties and increase cortical bone volume in select osteoporotic bones of rats.[19,20] The author has been using low-level laser therapy in select cases of μCT confirmed osteoporosis in captive birds. The author only brings this up as a possible treatment option to explore and further study.

References

1. Tatara MR, Krupski W, Majcher P, Studzinski T. Long-term denervation of the humerus in turkeys as an experimental model for osteopenia. *Poult Sci.* 2005;84:718-722.

2. Dinev I. Clinical and morphological investigations on the incidence of forms of rickets and their association with other pathological states in broiler chickens. *Res Vet Sci.* 2012;92:273-277.

3. Dirrigl FJ, Dalsky GP, Warner SE. Dual-energy x-ray absorptiometry of birds: an examination of excised skeletal specimens. *J Vet Med. Assoc.* 2004;51:313-319.

4. Whitehead CC. Overview of bone biology in the egg laying hen. *Poult Sci.* 2004;83:193-199.

5. Riddell C, Helmboldt CF, Singsen EP, Matterson LD. Bone pathology of birds affected with cage layer fatigue. *Avian Dis.* 1968;12:285-297.

6. Gregory NG, Wilkins LJ. Broken bones in domestic fowl: handling and processing damage of end-of-lay battery hens. *Br Poult Sci.* 1989;30:555-562.

7. Fleming RH, McCormack HA, McTeir L, Whitehead CC. Relationships between genetic, environmental and nutritional factors influencing osteoporosis in laying hens. *Br Poult Sci.* 2006;47:742-755.

8. Beck MM, Hansen KK. Role of estrogen in avian osteoporosis. *Poult Sci.* 2004;83:200-206.

9. Tarlton JF, Wilkins LJ, Toscano MJ, et al. Reduced bone breakage and increased bone strength in free range laying hens fed omega-3 polyunsaturated fatty acid supplemented diets. *Bone.* 2013;52:578-586.

10. Zulauf-Fischer I, Liesegang A, Haessig M, et al. Differences in bone mineral content and density between male and female budgerigars (*Melopsittacus undulatus*) during the non-reproductive season. *J Vet Med Assoc.* 2006;53:456-457.

11. Martinez-Cummer MA, Hurtig M, Leeson S. Use of apparent transverse quantitative ultrasonography to assess skeletal integrity in layers. *Poult Sci.* 2006;85:1648-1651.

12. Bar A. Calcium transport in strongly calcifying laying birds: Mechanisms and regulation. *Comp Biochem Physiol A.* 2009;152:447-469.

13. Liu D, Veit HP, Wilson JH, Denbow DM. Long-term supplementation of various dietary lipids alters bone mineral content, mechanical properties and histologic characteristics of Japanese quail. *Poult Sci.* 2003;82:831-839.

14. Baird HT, Eggett DL, Fullmer S. Varying ratios of omega-6:omega-3 fatty acids on the pre- and postmortem bone mineral density, bone ash, and bone breaking strength of laying chickens. *Poult Sci.* 2008;87:323-328.

15. Kubiak M, Forbes N. Effects of diet on total calcium, vitamin D and parathyroid hormone in falcons. *Vet Rec.* 2012;171:504.

16. Cousquer GO, Dankoski EJ, Patterson-Kane JC. Metabolic bone disease in wild collared doves (*Streptopelia decaocto*). *Vet Rec.* 2007;160:78-84.

17. Eurasian collared dove. Available at: Wikipedia. Accessed June 2, 2016. https://en.wikipedia.org/wiki/Eurasian_collared_dove.

18. Johnsson M, Jonsson KB, Andersson L, et al. Genetic regulation of bone metabolism in the chicken: similarities and differences to mammalian systems. *PLOS Genetics*. 2015;11:e1005250:1-19.

19. Mohsenifar Z, Fridoni M, Ghatrehsamani M, et al. An evaluation of the effect of pulsed wave low-level laser therapy on the biomechanical properties of the vertebral body in two experimental osteoporosis rat models. *Lasers Med Sci*. 2016;31:305-314.

20. Mohsenifar Z, Fridoni M, Ghatrehsamani M, et al. Evaluation of the effects of pulsed wave LLLT on tibial diaphysis in two rat models of experimental osteoporosis, as examined by stereological and real-time PCR gene expression analyses. *Lasers Med Sci*. 2016;31:721-732.

Facilitating Falconers: Treating Performance Raptors and Accommodating a Unique Clientele

Marcie Logsdon, DVM

Session #2049

Affiliation: From Washington State University Veterinary Teaching Hospital, Exotics and Wildlife Department, 205 Ott Rd, Pullman, WA 99164, USA.

Abstract: The use of raptors for falconry and pest bird abatement is growing. Although the medical treatment of these patients may be familiar to many avian veterinarians, the specific needs of this clientele may not be. The following is an overview of falconry and abatement specific terms and concerns to help facilitate communication between the veterinary team and client. Establishing a good working relationship with these clients is vital to ensure your ability to provide timely and appropriate medical care to these patients. All regulations cited pertain to those found in the United States.

Part I. The Clients

Not everyone who works with raptors is the same. Most will fall into one of the categories below. Although there may be some overlap, note that the following groups of people do not necessarily get along, and have often had negative experiences with members of a different group. Possession of raptors is very heavily regulated and all individuals will have one or more government permits allowing them to keep their birds.

Falconers

Falconers *hunt* with their birds. The definition of falconry is the taking of wild game with a raptor. In 2010 UNESCO named falconry an Intangible Cultural Heritage.

Abatement operators

Abatement operators use falconry methods to chase pest birds away from crops (usually high-value fruit crops), airports (to decrease plane-bird collisions), or landfills. In this situation, the birds are a valuable business asset.

Breeders

Breeders breed birds for use in abatement, falconry, or educational facilities. Again, although there may still be a deep emotional attachment these birds a business asset and death of an individual can be a financial hardship as well.

Rehabilitators

Rehabilitators focus on rehabilitating injured raptors and returning them to the wild. In most states they are required to have a working relationship with a licensed veterinarian.

Educational Facilities

Raptors are used here for environmental education. These birds may be glove-trained or used for display only. Many facilities utilize raptors with permanent injuries that render them non-releasable; others may prioritize imprinted or fully flighted birds for use in flight demonstrations.

Part II. The Patients

Keep in mind, raptors are not pets. These birds have jobs. Continued ability to perform these jobs is an important consideration when discussing medical treatment.

Wild caught

A raptor that is taken from the wild is considered wild caught. Falconers and raptor breeders are allowed to take juvenile raptors from the wild, but this practice is heavily regulated. In this circumstance although possessed by an individual, the birds are still considered the property of the state. Each state regulates trapping permits and harvest numbers individually. Abatement operators are not allowed to use wild-caught raptors for their work as it is considered a commercial enterprise. Wild-caught raptors may be returned to the wild at any time. In fact, many falconers will trap a bird for the fall/winter hunting season and then release it in the spring.

Captive bred

A raptor that was bred in captivity is considered captive bred. A captive bred bird may have spent its entire growth period with its parents (chamber-raised) or have been taken away earlier for hand raising and human imprinting. Some breeders choose to remove eggs and incubate them artificially.

Native

A raptor species that breeds in North America is considered. For regulatory purposes, even if the breeding stock for a captive bred bird originated outside of North America the offspring will still be considered native.

Exotic

A raptor species not naturally found in North America is considered exotic. These birds often fall under different regulatory laws than native species.

Human imprint

A bird that was raised by humans and displays conspecific behaviors towards them is considered human imprinted. In practice, imprinting is more of a spectrum than a black and white—imprinted or not—situation. Imprinted birds may court or mount people instead of their own species. This behavior is sometimes desirable for performing artificial insemination or semen collection. Artificial insemination is often required for the production of hybrids.

Species

Table 1 provides a list of species commonly encountered and an appropriate spelling of their name. Once a front desk personnel kindly put "redneck falcon" down instead of red-naped Shaheen. Luckily that particular falconer had a good sense of humor. Table 1 also provides the native or exotic designation for each species. In many situations native and exotic species of raptor are regulated differently. The birds species listed are not all inclusive.

Table 1. Select list of raptors commonly encountered.[a]

Bird	*Genus species*	Native (N) or Exotic (E)
Red-tailed hawk	*Buteo jamaicensis*	N
Sharp-shinned hawk	*Accipiter striatus*	N
Cooper's hawk	*Accipiter cooperii*	N
Goshawk (Northern or Finnish),	*Accipiter gentilis*	N
Harris's hawk	*Parabuteo unicinctus*	N
American kestrel	*Falco sparvarius*	N
Merlin	*Falco columbarius*	N
Aplomado falcon	*Falco fenoralis*	N
Prairie falcon	*Falco mexicanus*	N
Peregrine falcon	*Falco peregrinus*	N
Gyrfalcon	*Falco rusticolus*	N
Barbary falcon	*Falco pelegrinoides*	E
Red-naped Shaheen	*Falco pelegrinoides babylonicus*	E
Saker falcon	*Falco cherrug*	E
Lanner falcon	*Falco biarmicus*	E

[a]Not an exhaustive list.

Part III. Glossary of Useful Terms

A wide range of terms is used by people who work with raptors. A glossary of terms is provided in Table 2.

Table 2. Glossary of Falconry and abatement terms.

Abatement (raptor-based or pest-bird)	Use of raptors to chase pest birds away from a high-value area such as ripened crops, airports, or landfills
Block	A perch with a flat top usually used for falcons
Broadwing	Hawk that falls in the genus *Buteo* or *Parabuteo*
Cast (verb)	To restrain a raptor, to regurgitate a cast
Cast (noun)	The pellet produced in the ventriculus that consists of indigestible portions of the meal (fur, feathers, +/- bone) that is regurgitated by raptors, a pair of two birds flown together
Cere	Fleshy area surrounding the nares
Coping	Trimming the beak or talons
Crop (verb)	To feed a raptor as much as it will consume in a single meal, to "crop out"
Crop (noun)	Widening of the esophagus used to store food, visible above the keel and below the head when full
Eyas	A young raptor taken from the nest before fledging
Foot (verb)	When a raptor grabs something (usually aggressively) with its foot, "footing" at things, being "footy"
Free-loft	To manage a bird untethered in a mew or enclosure.
Frounce/Canker	Trichomoniasis (Be careful, many raptor handlers will assume that any oral plaque seen in the mouth is caused by trichomonas when candida or capillaria may be just as likely.)
Gear, Equipment, or Furniture	Anklets: a strap (usually leather) that is placed around a raptor's legs Jess: Leather strap (or braided cord) that passes through the anklet and is used to hold the bird Hood (noun): a formed piece of leather placed on a bird's head over its eyes to reduce external stimulus
Haggard	A wild raptor in mature plumage (≥1yr old)
Hallux	The large back talon, digit I
Hood (verb)	To place a hood on a bird
Hybrid	Crossbreed between two different species of raptors, most commonly bred hybrids are falcons
Imping	Grafting new feathers onto the shaft of broken feathers in order to repair them
Keel	The sternum, to feel the keel means to assess the body condition of the raptor by palpating the large pectoral muscles and any overlying fat
Killer	Talon, digit II
Ledge	A flat shelf used for perching or nesting by falcons, a natural rock formation or an artificial platform in a chamber
Longwing	Large falcon
Longwinger	Someone who flies falcons
Manning	The process of acclimating a raptor to the presence of people, and other things, periods of time spent on the glove
Mews/chamber	Building dedicated to housing raptors
Microhawk	Merlin, sharp-shinned hawk, or kestrel
Microhawking	Flying a microhawk (generally any falconry that focuses on catching small passerines)
Molt (noun)	The time of year that birds are molting, as in "my bird is up for the molt," raptors are traditionally not hunted and are kept at higher weights during this time of the year (usually summer months)
Molting (verb)	Dropping and regrowing feathers
Mutes	Droppings, includes all three portions of the droppings: urates, urine, and feces
Passage	A wild raptor in its juvenile plumage, named for time when birds were commonly trapped during their first migration or passage through a location
Perch out (verb)	To tether a raptor to a perch, usually in an area that allows them access to sun
Scrape	A nesting depression made by a falcon in gravel or other substrate
Screamer	An imprinted raptor that vocalizes inappropriately and excessively
Sharp (adjective)	When used to refer to the keel or physical condition of a bird it means a skinny bird, when used to refer to the mental state of a bird it means that they are keen to hunt
Shortwing	Hawk that falls in the genus accipiter (eg, goshawk, sharp-shinned hawk, Cooper's hawk, etc.).
Spartrix	An over-the-counter trichomoniasis treatment (Carnidazolum) marketed for pigeon fanciers but used frequently by raptor owners
Tiercel	Male falcon
Turn over	To move food from the crop farther down the GI tract, as in, "my bird has yet to turn over his crop"
Weathering yard	A protected area where birds are perched out
Yarak	Physical and mental state at which a raptor is keen to hunt and chase, a state that takes into account many different factors including body weight, time of day, and whether or not the bird has cast yet

Part IV. Important History Questions

A thorough history is vital when dealing with raptors. In addition to the normal history questions Table 3 provides some that also should be asked. In addition, these are questions that you might be asked if you consult on a case with a raptor specialist.

Table 3. Questions to ask when obtaining a raptor history.

Question	Comments
What is the bird's age?	
Where and when was the bird acquired?	
What is the food source and how is it prepared?	Coturnix quail, day-old chick, and rodents are commercially available. Some falconers and other raptor facilities supplement with roadkill, which could be a potential source of lead exposure if the animal was shot at any point during its life. Others will feed wild-trapped house sparrows (*Passer domesticus*), rock pigeons (*Columba livia*) or European starlings (*Sturnus vulgaris*) which, although nutritious, may be a source of disease. Also ask if they remove part, or all, of the gastrointestinal tract prior to feeding. Many falconers feeding pigeon and dove will remove the upper GI tract (crop and above) to reduce the chance of *trichomoniasis* transmission. Others prefer to remove the entirety of the GI tract of all food items prior to feeding.
Are vitamin supplements used?	If so, brand, frequency and amount.
What is the bird's current purpose?	Is it used for hunting, breeding, abatement?
What species of birds are kept on the property?	Young Aplomado falcons are very susceptible to an adenovirus carried by peregrines.
What is the bird's current state?	Is it up for the molt, currently hunting, in a chamber with another bird for breeding?
Is the bird actively hunting?	If so, what game does it take? Does the owner allow it to feed on its kill?
What is the bird's weight?	Obtaining an accurate weight of a raptor is an important part of the physical exam but putting that weight in context with that individual's normal weight is even more so. Most handlers should be able to tell you the raptor's hunting weight, current weight, trap weight (if applicable) and molting weight.
How is the bird housed?	How a bird is housed will have implications for both possible disease exposure and treatment options. Ask if it is free-lofted or tethered.

Part V. Special Considerations on the Physical Exam

Equipment

Anklets and jesses should be examined closely as improperly placed or fitted anklets can cause abrasions or trauma to the legs or the tops of the feet. Occasionally the jesses of tethered birds may rub against the medial aspect of the foot. TrackPack® (Marshall Telemetry, North Salt Lake, UT) is a brand of transmitter harness that fastens around the body of a raptor and allows telemetry transmitters to be placed on the back of the bird. When fitted properly they are a very safe method for transmitter attachment but can cause discomfort and abrasions if placed too tightly.

Hooding

Properly fitting hoods are an invaluable tool for training raptors and for performing a physical exam in a low-stress manner. The use of hoods leads to several considerations.

Improperly fitted hoods can cause abrasions on the cere or eyes. Hoods should be properly fitted.

Ideally, birds are trained to take the hood calmly while standing on the glove. Depending on the age, species, and level of training the bird has, putting a hood on a bird improperly can do serious damage to the hood training process. Often the bird will come in hooded and this facilitates most of the exam; however, the exam should not be considered complete until the eyes are examined as well. Be clear in your intentions and ask if the falconer would prefer to hood and unhood the bird himself.

Hoods also can be used to keep a bird calm for extended periods of time. Again, if used properly they can be helpful during periods of stress (such as hospitalization) or recuperation.

Restraint

Most falconers have experience restraining their birds for placement of anklets or beak trims; however, they may not have experience holding a raptor to facilitate the procedures you need to perform. If you have trained personnel, it is best to have them do the restraint. If you do not have personnel trained in raptor restraint you may want to consider having the falconer restrain the bird–although this can turn into a fiasco. Use your judgment and the skills you have at your disposal.

Feathers

Feathers are as important to a working bird as hooves are to a racehorse. Nothing will tarnish a veterinarian's reputation amongst falconers faster than returning a bird with broken or bent feathers. Use care when handling to avoid damaging them and consider a tail guard for birds that have to spend any time in hospital (again, communicate with your clients, they may have a pre-fitted tail guard already for their bird, or it may be a species or individual that won't tolerate one).

Part VI. Common Reasons for Presentation

New Bird Exam

Many newly trapped birds will be presented for deworming. Juvenile American Kestrels in particular seem to be very susceptible to coccidiosis during periods of stress so many falconers will elect to treat preemptively.

Trauma

Every falconer's worst nightmare is a bird that gets injured on a hunt. Fractures, electrical burns from power lines or transformers, and cat bite wounds are some of the most commonly encountered injuries. Establishing a good working doctor-client relationship before an emergency presents will help set everyone at ease during a high-stress period.

Health certificate

The sale of a bird across state (or international) lines or travel to another state for a field meet often requires a health certificate. Many states also require import permit numbers so be sure to speak with the receiving state's department of agriculture, state veterinarian, and wildlife department prior to issuing a certificate of veterinary inspection. Some states (such as Oregon) have different requirements for the import of raptors depending on whether they are a native or exotic species, so be sure to be specific when asking for importation requirements.

West Nile virus vaccination

West Nile Virus began affecting birds in the eastern United States in the mid-1990's and has spread steadily west since. It is now present in Washington State. Many falconers and rehabilitation centers are now vaccinating birds annually. The vaccine currently used most frequently is the recombinant vaccine Recombitek® (Merial, Inc. Duluth, GA). This vaccine is designed for horses, but rarely stocked in most equine clinics (due to cost and the wide variety of combination vaccines available) so often veterinarians or falconers will have to purchase a flat (10 1-ml vials) specifically for use in avian patients.

Anesthesia

Some falconers will request anesthesia for placement of equipment (specifically TrackPacks®) or coping. The dispensing of injectable anesthetics such as ketamine/xylazine for in-home use is not recommended as most falconers will not have the appropriate training for anesthetic administration and monitoring (believe it or not, some will ask for it). Dispensing sedatives or tranquilizers may be appropriate at times but it is usually best to only dispense medications that you have extensive experience with so that the client can be forewarned as to what side effects to expect. For example, intranasal administration of midazolam can be a useful sedative, however some falcons will regurgitate after administration. Offering anesthesia in the clinic for procedures such as these will depend on your comfort level.

Sudden illness

As with any species, raptors (and their owners) may come to you for the first time because of a sudden illness. Keep in mind that birds are adept at hiding illness, and diseases can progress very quickly as compared to many companion animals so by the time they present to your clinic they are often very sick.

Conclusions

The ability to effectively communicate with falconers and instill confidence is vital to retaining this unique and engaging clientele. The above information provides some basic vocabulary and a snapshot into the complex world of performance raptors and the people that keep them, however nothing on paper can even approach the experience of seeing a hunting bird in action. Most states have a falconry association managed by the most outgoing of this reclusive group. Annual field meets hosted by these associations provide the opportunity to see these amazing birds in action. Nothing can better prepare you for dealing with these clients than seeing them with their hunting team in the field!

Recommended Reading

1. Arent L. *Raptors in Captivity: Guidelines for Care and Management.* Blaine, WA: Hancock House Publishers; 2007.

2. Macdonald H. *H is for Hawk.* New York, NY: Grove Press;2014.

3. Samour J. Management of Raptors. In: Harrison GJ, Lightfoot TL, eds. Clinical Avian Medicine. Vol 1. Palm Beach, FL: Spix Publishing;2006:915-934.

4. Scott D. *Raptor Medicine, Surgery and Rehabilitation.* 2nd ed. Boston, MA:CABI;2016.

5. Sibley D. *The Sibley Guide to Birds.* New York, NY: Random House;2000.

Advanced Imaging Techniques

Yvonne van Zeeland, DVM, MVR, PhD, DipECZM (Avian),
DipECZM (Small Mammal), CPBC, and
Nico Schoemaker, DVM, PhD, DipECZM (Small Mammal), DipECZM (Avian),
DABVP (Avian Practice)

Session #2055

Affiliation: From the Division of Zoological Medicine, Department of Clinical Sciences of Companion Animals, Faculty of Veterinary Medicine, Utrecht University, Utrecht, 3584 CM, The Netherlands.

Manuscript adapted from: van Zeeland YRA, Schoemaker NJ, Hsu E. Advances in diagnostic imaging. In: Speer BL, ed. *Current Therapy in Avian Medicine and Surgery.* St. Louis, MO: Elsevier Saunders;2016:531-549 (with permission). Parts of this Masterclass were presented at European Association of Avian Veterinarians 2011 and ICARE 2017.

Abstract: This Masterclass aims to provide an overview of the various advanced imaging techniques that are available, including CT, MRI and a variety of nuclear imaging techniques (scintigraphy, SPECT and PET). Information regarding the various procedures, including their practical applications and limitations, and examples of their use in avian medicine, will be discussed.

Introduction

Diagnostic imaging has revolutionized the practice of medicine by enabling non-invasive visualization of the body's anatomy and physiology. In avian medicine, a similar development has taken place, with the use of advanced imaging techniques gaining increasing interest over the past years. Of the various techniques that are available, conventional or projection radiography is still one of the most popular and commonly used techniques. However, in recent years, ultrasonography has also become an important diagnostic aid. In addition, other, more advanced diagnostic imaging techniques are also used with increasing frequency as they can help to achieve even more detailed and precise information compared to the aforementioned techniques. In particular, computed tomography (CT) is a technique that has proven its value, with many potential applications in the field of avian medicine. Magnetic resonance imaging (MRI) and nuclear imaging (eg, scintigraphy or positron emission tomography [PET]), on the other hand, are used less frequently. However, there is great potential for the use of these techniques in avian medicine, once their resolution increases and the cost of an examination decreases. Below, an overview of the various advanced imaging techniques, including their practical applications and limitations in avian medicine, will be discussed.

Computed Tomography

Computed tomography (CT) is a diagnostic imaging technique that uses X-rays to obtain cross-sectional images of the body. These images can subsequently be digitally processed with a computer to obtain 3D view and/or reconstruct the data into alternate planes. In both human and veterinary medicine, CT has become an invaluable tool for both diagnosing disease and planning, guiding and/or monitoring treatment of a patient. In avian medicine, CT is also increasingly used in both practices and university hospitals.

Compared to conventional radiography, the main advantage of CT imaging is that it allows for non-invasive imaging without interference of superimposition of the various tissues and structures. Moreover, compared to other advanced imaging techniques (eg, MRI) the time needed to perform a whole body scan is relatively short. Especially with modern scanners, investigation times may be as short as 1 or 2 minutes and can even be limited further if the bird is placed transverse to the gantry opening of the CT-scanner.

In veterinary medicine, both single-slice and multi-slice (helical) CT scanners are available for use. Whereas single-slice scanners obtain only one single tomographic slice or image in a single rotation of the X-ray beam, multi-slice (or multi-detector array) CT scanners are capable of acquiring several tomographic slices in each rotation. As a result, multi-slice scanners permit thinner slice-width image acquisition (ie, 0.5 mm versus 1 mm) and quicker scanning times. As a result, multi-slice scanners provide additional benefits, particularly for the small-sized avian patients.

Procedure

Despite the short duration of the procedure, general anaesthesia is typically recommended. Not only does this help to achieve proper positioning and prevent movement artifacts, but also to minimize stress of the patient. However, dependent on the goal of the procedure, health status and stress sensitivity of the patient, the choice can be made to sedate rather than anesthetize the patient, or even perform the CT awake. For example, in birds with respiratory disease, CT examination can be performed without anaesthesia, by placing the bird in an acrylic chamber. To optimize image quality and limit scanning time, the use of a multi-slice helical CT scanner is highly recommended.

During CT imaging, birds are usually placed in dorsal or ventral recumbency, the latter allowing for the bird to breathe more easily during the procedure. Larger sized birds are usually scanned using a transverse plane. Smaller sized birds can also be scanned using a sagittal plane, with the bird positioned with its long axis perpendicular to the gantry table. For reliable interpretation, the patient needs to be positioned as symmetric and straight as possible within the scanner. Various materials can be used to aid in the positioning (eg, towels, foam pads or IV fluid bags). Tape or a special avian restraint board may be used to prevent the bird from moving during the exam.

Contrast media can be administered to increase the radiodensity of tissues, organs and/or vascular systems, thereby helping to depict lesions or abnormalities (eg, neoplasia, inflammation). Both iodinated and non-iodinated contrast media can be used (eg, Telebrex®, 2 ml/kg IV). To allow the administration of contrast media, IV access should be obtained and maintained throughout the procedure without having to move the patient between the pre- and post-contrast scan. Due to the bird's fast circulation and thereby quick elimination of contrast, intervals between administration of the contrast agent and scanning need to be as short as possible (~1-2 minutes) in order to enable visualization of contrast uptake.

CT technology furthermore allows for the accurate placement of needles or biopsy instruments (eg, to collect tissue samples or fluids for diagnostic tests). After selecting the slice of interest, the position of this slice can be indicated by a thin light or laser beam following which a needle or biopsy instrument can be inserted. Before collecting the material, verification of placement can take place by scanning the region of interest.

After scanning the patient, the obtained CT images can be evaluated using advanced software programs. With these programs multiplanar reconstructions (MPRs) and (3D) image reconstructions and renderings are possible. With MPR the organs and structures can be reviewed in three different planes (eg, sagittal, longitudinal, transverse). Although all images are 2D, the combination of the three planes allows for a 3D reconstruction to be made. Most software programs also allow different types of measurements (eg, size, volume, density). "Windowing" and "levelling" may furthermore be used to adjust the contrast and brightness levels of the image, allowing for optimal display of the tissue of interest.

Applications

CT can be used to evaluate a wide variety of different organ systems. In birds, CT imaging is mainly used for visualization of abnormalities in the skeletal structures and respiratory tract.

For many years, CT has been considered as the most reliable and sensitive method for diagnosing lower respiratory tract disease in birds. Although larger abnormalities in the respiratory tract may be identified by conventional radiographs, many early and smaller lesions may be missed or misinterpreted. CT images allow for cross-sectional evaluation of the homogenous 'honey-comb like' lungs, thereby providing detailed information about alterations of the respiratory tract in birds, such as increased (multi)focal radiopacity (eg, in case of mycotic or mycobacterial pneumonia) or decreased volume of the lung parenchyma (typical for lung fibrosis). In addition to imaging abnormalities in the lung parenchyma, CT also enables visualization of abnormalities of the trachea (eg, intra- or extraluminal masses, strictures, stenosis or displacement), syrinx, bronchi and air sacs (eg, overinflation, air sacculitis, empyema, and compression by ascites, egg or intracoelomic masses).

Especially in areas with significant bone superimposition CT imaging has been found useful. In particular for the avian head, CT has a large range of relevant and practical applications. CT has been demonstrated as superior to radiographs for evaluating the nasal cavity, conchae and sinuses, allowing for detailed studying of the various anatomic structures. It furthermore allows for calculation of the volume of pneumatized areas as well as establishing the location and extent of pathologic changes in birds with (chronic) sinusitis or rhinitis. CT may also provide information about the shape and size of the avian eye, lens and scleral rings, although it is not able to differentiate intraocular structures such as the pecten oculi. In birds with neurologic disease due to hydrocephalus, other congenital malformations (eg, meningoencephaloceles in crested ducks), trauma, large intracranial masses or hemorrhage, CT examination may be used to get a quick diagnosis; for other brain diseases additional techniques, such as MRI, may be necessary.

Conventional radiographs are usually adequate for diagnosing the majority of cases that affect the skeletal system (eg, fractures). An important exception are spinal fractures, which are possibly missed in more than half of the cases. For these type of fractures, CT offers increased sensitivity for visualization and identification. Other examples of areas where CT shows superior imaging quality over conventional radiographs are the shoulder joint, pelvis (where much superimposition of overlying soft tissue structures and wings or legs is present), and cranium (in particular the hyoid bone and beak apparatus). Not only fractures or fissures, luxations and/or focal bony changes (eg, infections, lysis) may be visualized and assessed, but the bones may also be judged with regard to their radiodensity and thickness of the cortex.

Enlargement of various intracoelomic organs, such as liver, kidney and spleen, may easily be detected with CT imaging, especially when contrast medium is used. The gastrointestinal tract also can be visualized, especially the gas-filled proventriculus and grit-containing ventriculus. Recognizing pathology in the intestinal tract is often more difficult without the use of an intraluminal contrast medium such as barium sulphate. This especially allows visualisation of the thickness of the wall, aiding in distinguishing proventricular dilatation disease from proventricular dilatation syndrome. CT may furthermore be useful in determining the size and extent of neoplasia in various locations, thereby also helping to determine and plan the possibilities for therapeutic intervention and the associated prognosis.

Computed tomography angiography (CTA), in which contrast media are administered intravenously, allows for the visualization and assessment of the larger veins and arteries in larger psittacines and raptorial birds and may thus provide new insights and methods for diagnosing cardiovascular diseases, such as cardiomyopathy, atherosclerosis, aneurisms or congenital vascular anomalies. Because of the rapid circulation in birds, these CTA scans are preferably performed immediately after administration of the contrast.

Micro-CT imaging has recently become popular among anatomists because this technique provides high-quality visualization of the musculature without the need for time-consuming and difficult dissections. Staining of the anatomic specimen with an iodine solution greatly enhances differentiation between the different muscles. The technique typically involves specialized hardware, and requires longer scanning times than conventional CT, thereby making it less applicable to clinical avian cases in current times.

Limitations

The limitations to the use of CT for imaging structures and organs are mainly dependent on the spatial resolution of the CT scanner and size of the patient. Particularly in smaller sized patients (eg, budgerigars), resolution of regular CT scanners may be poor, limiting its diagnostic value for evaluating the presence of abnormalities. However, for these patients, the more modern (helical) multislice CT-scanners may produce images of high enough resolution to depict the different structures with sufficient detail and allow for identification of abnormalities. Similarly, micro-CTs may provide a good alternative for use in small-sized birds once they become available for use in patients in future.

Magnetic Resonance Imaging

Magnetic resonance imaging (MRI) is one of the latest cross-sectional imaging modalities to gain wide acceptance, particularly in the area of soft-tissue and neurologic imaging. Unlike CT scans or traditional radiographs, which use X-rays, MRI uses a powerful magnetic field in combination with high frequency radiowaves to align protons (nuclei of hydrogen atoms) in the body and obtain signals resulting from the change in movement of these atoms. As most of the hydrogen atoms are trapped in water molecules, the MRI image that is produced is mainly a representation of the different types, distributions and volumes of water in the various tissues. By changing the parameters of the MRI scanner (eg, pulse repetition time, echo time) different tissue weightings are obtained, which are subsequently used to analyze the tissue composition. The three types of weighting that are commonly used are:

T1-weighted images, also known as longitudinal relaxation or simply T1, relate to the rate of recovery of the proton spins after each perturbation (called excitation) required to elicit observable signal. Therefore, tissues with long T1 (eg, water) appear hypo-intense

T2-weighted images, also known as transverse relaxation or simply T2, relate to the rate of decay of the observable signal. Therefore, in contrast to T1, tissues with long T2 appear hyper-intense.

Proton–density-weighted or spin–density-weighted images (PDW): Contrast between tissues is primarily dependent on the amount of available spins (protons) (ie, brightness is highest in tissues with a high-protein density); these images minimize the T1 and T2 effects, therefore resulting in less contrast than T1 or T2 images; although these images may be useful for depicting normal anatomy, they are used less often.

As each type of tissue has its own characteristics for T1, T2 and PDW the images obtained from these three weighting types can be combined to identify the origin of the tissue involved.

Procedure

Magnetic imaging requires the patient to be anesthetized. This in turn warrants the use of special anaesthetic machines and monitoring equipment as many metals cannot be taken into the scanner room and radiowaves emitted by electric devices may interfere with the procedure, resulting in non-diagnostic images. In order to facilitate breathing, the patient should be placed in a semi-upright position with the use of MRI compatible materials (eg, pillows, sand bags). Prior to starting the procedure, it is also important to check for the presence of metal implants or objects that are present in the body, which may result in large-scale signal extinction and/or movement of the metal within the body, potentially causing damage to internal organs. Some experts therefore recommend screening birds for presence of metal in their gastrointestinal tract with a radiograph prior to performing the MRI. Most micro-transponders that are used to identify birds contain metal and result in signal extinction in MR scanners of 0.2 T, which can be a limiting factor. Magnets of 1.5 T, in contrast, are not affected by the metal from micro-transponders. Similarly, leg bands, which are usually made of aluminium, will not result in artifacts during imaging.

Gadolinium, a paramagnetic intravenous contrast agent, is one of the agents that can be used as a contrast medium for MRI studies (dose 0.25 mmol/kg IV). After administration of the compound via an IV catheter the gadolinium is distributed throughout the body, and changes the local magnetic field in tissues in which it is present in high concentrations (ie, highly vascularised tissues), allowing delineation from the surrounding structures. The contrast-enhanced images which are subsequently obtained may be particularly useful in detecting whether an active inflammation or neoplasm is present. Unfortunately, scientific studies on gadolinium use in birds are lacking.

Applications

MRI has been used in different bird species (eg, pigeons, poultry, pet birds) for anatomic studies of various organ systems, and is used in various scientific studies (eg, to analyse the topographic relationship between brain structures and singing ability in Passeriformes).

MRI has also been used with some success in avian patients. Because of its characteristics, it is primarily advantageous in diagnosing soft-tissue changes. Thus, the indications for which it has been applied in birds are mainly associated with neurologic disease (eg, brain abnormalities) and/or internal disease (eg, sinusitis). MRI has been found very useful for visualizing and evaluating various parts of the CNS, including the cerebral hemispheres, cerebellum, optic chiasm, brain stem and spinal cord. Its application in the diagnosis of a variety of different CNS diseases has also been demonstrated by various researchers. Among the diseases that have been diagnosed on MRI examination are (viral) encephalitis, lead poisoning, hydrocephalus and spinal cord trauma. MRI is also an excellent diagnostic tool for evaluating the eye, orbit and sinuses; identifying, localizing and characterizing lesions in these tissues; and planning the proper therapeutic approach.

MRI furthermore has potential value in visualizing the gastrointestinal tract, liver, spleen and urogenital tract, whereas it is found to be less useful for imaging the heart and lungs of birds, as cardiac motion creates artefacts obscuring the heart and adjacent lungs. The avian pancreas can be seen in some MRI scans. Adrenal glands are generally not identified in clinical MRI scans.

Along with unmatched versatility in differentiating soft tissues for anatomic visualization, MRI can use mechanisms to generate signal contrast beyond the standard T1, T2, and proton density for capturing the physiology of the body, including the circulatory system. Although CT provides faster scan speed and better spatial resolution, MR angiography does not require the use of exogenous agents to generate the desirable vessel-to-tissue contrast. This can be highly desirable because circulation in avian species can make contrast agent administration and subsequent imaging rather challenging. Examples of MRI techniques particularly suited and being applied to avian angiographic imaging include phase-contrast (PC) and time-of-flight (TOF) imaging.

To study brain activity during different tasks, functional MRI (fMRI) has become an important neuroimaging technique in cognitive neuroscience. This technique relies on blood oxygen level-dependent contrast, which reflects a local change in the ratio between oxygenated and deoxygenated hemoglobin in the blood induced by a local increase in oxygen consumption during neuronal activation. Only one research center is currently using this technique in birds (ie, zebra finches and pigeons) with the aid of a 7 T scanner.

Limitations

Although MRI has many potential uses in avian medicine, it is not frequently used due to the prolonged time that is needed to complete the scan. In combination with the need for multiple sequences, this results in prolonged anaesthesia times (ie, a full investigation may take up to 45 minutes), which may be detrimental to the patient, especially since the access to the patient for the anaesthesiologist is limited during the procedure. Additionally, the relative small size of most birds in combination with the relative low spatial resolution of most systems (< 0.5 T) limit the diagnostic value that MRI has in avian patients. Imaging may further be hindered by the high respiratory and heart rates, or by the presence of artifacts caused by the presence of air sacs and/or a micro-transponder. Costs may also be an issue as MRI is usually much more expensive than CT. With the increased availability of higher resolution systems, the use and practical implication of MRI in avian medicine is likely to become more prevalent.

Nuclear Imaging Techniques

Nuclear imaging techniques use radiopharmaceuticals, which are administered intravenously or per os. The radiation, emitted following the decay of these compounds, is captured by gamma cameras and used to form images. There are several techniques available within the field of nuclear imaging (ie, scintigraphy, single photon emission computed tomography [SPECT] and positron emission tomography [PET]).

In contrast to MRI and CT, which both focus on a particular section of the body (eg, head, chest, extremity), nuclear imaging techniques generally target a specific organ or tissue (eg, lungs, heart, bone, brain). Additionally, these techniques are primarily intended to show physiologic function of the investigated organ system rather than displaying an anatomic image. A change in function resulting from a pathologic process will be visible through a concurrent change in uptake of the radionuclide tracer in the organ system of interest. Diseases associated with an increase in physiologic function often leads to increased concentrations of the radionuclide tracer in the tissue, resulting in the appearance of a hot spot (ie, focal increase) or generalized accumulation throughout the tissue. Other disease processes may be resulting from a decrease in physiologic function, which results in the appearance of a cold spot (ie, focal decrease) because of exclusion or decreased uptake of the tracer.

A variety of different radionuclides or radioisotopes are available for use in nuclear imaging studies. The most commonly used radionuclides are technetium (^{99m}Tc), iodine (^{123}I and ^{131}I), thallium (^{201}Tl), gallium (^{67}Ga), fluorine-18 fluorodeoxyglucose (^{18}F FDG), xenon (^{133}Xe) and krypton (^{81m}Kr) which may be administered intravenously (eg, iodine) or by inhalation (eg, xenon). The choice for a specific radionuclide largely depends on the type of tissue that is investigated. For example, iodine is used for imaging of the thyroid gland, whereas technetium can be used for bone scans.

During nuclear imaging, a dataset is obtained which comprises of one or more images. Multi-image datasets may be dynamic (ie, represent a time sequence) or spatial (ie, represent a sequence where the gamma-camera has been moved relative to the patient).

Scintigraphy

Scintigraphy is a nuclear imaging technique that uses internal radionuclides to create 2-dimensional images, in contrast to PET and SPECT, which form 3-dimensional images. Scintigraphy has many applications, both in human and veterinary medicine. Main indications include lung scintigraphy or ventilation-perfusion scans to diagnose pulmonary embolism; hepatobiliary scintigraphy to diagnose liver shunts or bile duct obstruction; renal scintigraphy to assess kidney function (glomerular filtration rate); or thyroid scintigraphy to diagnose hyper- or hypothyroidism or thyroid neoplasia. Bone scintigraphy is furthermore used in orthopaedic patients to diagnose presence of neoplasia, including metastasis (eg, osteosarcoma) or identify the cause for lameness in cases where other imaging techniques have failed.

In avian medicine scintigraphy has mainly been used for research purposes. This includes the study of normal function (eg, GFR or hepatic clearance) as well as organ failure (ie, liver, kidney) after exposure to various toxic substances (eg, ethylene glycol, aflatoxin, gentamycin). Scintigraphy has also been performed in normal and radiothyroidectomized cockatiels, in which it was found helpful to establish hypofunction of the thyroid gland. For this purpose, the authors have also used the technique in a grey parrot suspected of hypothyroidism. Scintigraphy has furthermore demonstrated its potential use in avian patients with a variety of orthopaedic problems, as well as its usefulness in the therapeutic monitoring of patients with various types of neoplasia, including diagnosing the presence of metastasis.

Scintigraphy is generally performed under general anaesthesia with the animal placed in dorsal recumbency. As the radioisotopes usually need to be administered intravenously, an IV catheter should be placed. The dose that needs to be administered depends largely on the type of radioisotope that is used for the study and size of the animal. Scans are usually obtained at regular time intervals, which are mostly determined by the type of study and tissue examined, and may take up to one hour. Although some research protocols are available, no systematic studies have been performed on the use of scintigraphy in avian patients. This may in part be due to the low availability of the technique.

Single photon emission computed tomography

Single photon emission computed tomography (SPECT) is a technique that is very similar to conventional nuclear imaging techniques such as scintigraphy; however, it uses a rotating gamma camera which enables the object to be examined from a variety of different positions. The obtained images form a stack of slices of the body that can be reconstructed into a three-dimensional image, representing the distribution of the radionuclide within the patient. The information is typically presented as cross-sectional slices, but reconstructions can be made in different planes by manipulating the data. In human medicine, SPECT can be useful for functional imaging of heart and brain as well as imaging bone, thyroid tissue, neoplasia and/or infection. Although its use has also been described in veterinary medicine, no applications have been reported in (clinical) avian medicine.

Positron emission tomography

Positron emission tomography (PET) is similar to SPECT and scintigraphy in its use of radioactive tracer materials and detection of gamma rays. In contrast to SPECT, in which the emitted gamma radiation is measured directly, PET tracers emit positrons which annihilate with adjacent electrons, causing two gamma rays to be emitted in opposite directions. A PET scanner subsequently detects the emissions of these coincident pairs, thus providing a more accurate information about the localization of the radiation event.

A sugar, ^{18}F-fluorodeoxyglucose (FDG), is the most commonly used radiopharmaceutical agent in PET imaging. This tracer is mainly distributed to organs or tissues with a high blood flow or metabolic rate (eg, brain, liver, heart, kidney, neoplasia, inflamed tissue), and may be used as an indicator for altered tissue function. The characteristics of this specific tracer make the technique particularly important in the diagnosis of brain disease and monitoring of cancer treatments. Dependent on the definition of the problem and diagnostic/research question (ie, blood flow, metabolism, enzyme or receptor function), other radioisotopes may be used. PET typically involves the use of short-lived isotopes (eg, carbon [^{11}C], nitrogen [^{13}N], oxygen [^{15}O], and fluorine [^{18}F]), which may be incorporated into various compounds that are used in the body (eg, glucose, water, ammonia) or into molecules that bind to receptors or other sites of drug action. Because of the short longevity, the radioisotopes should be prepared in close proximity to the PET imaging facility, thereby warranting on-site cyclotrons (ie the apparatus that produces the radio-isotopes).

Over the last few years, there have been some studies performed with PET imaging in birds. These include studies into the functioning of the parrot brain, in particular with regard to its function in the perception of pain and use of analgesics as well as studies into the biodistribution of ^{18}FDG in the coelom of parrots and birds of prey. Results of the latter studies showed high variability in the intestinal radioactivity, which could be largely attributed to reflux of the compound into the cloaca. This finding should be taken into account when performing PET imaging of birds, particularly if the GI tract is the main object of interest, and may warrant the performance of "dynamic" rather than "static" studies. In addition to the aforementioned research studies, a number of applications in clinical cases have been described, which include the use of PET for planning and monitoring cancer therapy, determining the activity of lesions/nodules, but also the diagnosis of fractures of the skull and/or luxations of facial bones.

Despite the growing number of studies conducted, PET imaging is currently not expected to gain wide-spread use in the field of veterinary medicine. The main factors that are responsible for its limited use in clinical cases are the need for special equipment and associated higher expenses. It is expected that for the time being PET imaging will remain a technique that is mostly accessible for universities and in research settings.

Future Developments

In human medicine, hybrid technology, which uses a combination of techniques (eg, PET/CT or SPECT/CT) is becoming increasingly popular. With specific software or hybrid cameras nuclear scans can be superimposed on images obtained from modalities such as CT and MRI to highlight the part of the body in which the radiopharmaceutical is concentrated. This new fusion technique is a unique tool for obtaining information about the anatomy and function of specific tissues or organs, which would otherwise not be available and/or only obtained with more invasive procedures or surgery. Hybrid techniques are currently also making their entrance in veterinary medicine (including avian medicine) thereby providing many new opportunities and challenges for the future and creating a whole new field in diagnostic imaging to be explored further.

Conclusions

Diagnostic imaging should be considered an integral part of the medical and surgical management of avian patients. With the emergence of new imaging modalities the possibilities for diagnosing disease have increased drastically. Particularly in the last decade, advanced imaging modalities are used with increasing frequency in avian medicine, both for research purposes and clinical evaluation of patients. Each technique has its own advantages and limitations, thereby largely determining its potential applications in the diagnostic and therapeutic plan. Dependent on the goal of the procedure, availability and type of equipment, associated costs and experience of the examiner with performing the technique and interpreting the images, a well-thought-out choice can be made between the various available modalities.

References for Further Reading

1. Beaufrère H, Pariaut R, Rodriguez D, et al. Avian vascular imaging: a review. *J Avian Med Surg*. 2010;24: 174-184.

2. Beaufrère H, Rodriguez D, Pariaut R, et al. Estimating intrathoracic arterial diameter using CT angiography in Hispaniolan Amazon parrots (*Amazona ventralis*). *Am J Vet Res*. 2011;72:210-218.

3. Converse A, McCutcheon R, Sladky K, et al. 2-deoxy-2-[18F]fluoro-d-glucose positron emission tomography imaging of parrot brain. *Mol Imaging Biol*. 2005;7:1536-1632.

4. De Groof G, Verhoye M, Van Meir V, et al. In vivo diffusion tensor imaging (DTI) of brain subdivisions and vocal pathways in songbirds. *Neuroimage*. 2006;29:754-763.

5. de Groof G, Jonckers E, Güntürkün O, et al. Functional MRI and functional connectivity of the visual system of awake pigeons. *Behav Brain Res*. 2013;239:43-50.

6. Gumpenberger M, and Henninger W. The use of computed tomography in avian and reptile medicine. *Semin Avian Exot Pet Med*. 2001;10:174-180.

7. Hadley T, Daniel G, Rotstein D, et al. Evaluation of hepatobiliary scintigraphy as an indicator of hepatic function in domestic pigeons (*Columba livia domestica*) before and after exposure to ethylene glycol. *Vet Radiol Ultrasound*. 2007;48:155-162.

8. Harms A, Hoskinson J, Bruyette D, et al. Technetium-99M and Iodine-131 thyroid scintigraphy in normal and radiothyroidectomized cockatiels (*Nymphicus hollandicus*). *Vet Radiol Ultrasound*. 1994;35:473-478.

9. Jones MP, Morandi F, Wall JS, et al. Distribution of 2-deoxy-2-fluoro-d-glucose in the coelom of healthy bald eagles (*Haliaeetus leucocephalus*). *Am J Vet Res*. 2013;74:426-432.

10. Krautwald-Junghanns M, Schuhmacher F, Tellehlm B, et al. Evaluation of the lower respiratory tract in psittacines using radiology and computed tomography. *Vet Radiol Ultrasound*. 1993;34:382-390.

11. Krautwald-Junghanns M, Kostka V, Dörsch B, et al. Comparative studies on the diagnostic value of conventional radiography and computed tomography in evaluating the heads of psittacine and raptorial birds. *J Avian Med Surg*. 1998;12:149-157.

12. Krautwald-Junghanns M, Pees M, Reese S, Tully T. *Diagnostic Imaging of Exotic Pets*. Hannover, Germany: Schlütersche;2010.

13. Kusmierczyk J, Wall CR, Hoppes S, Budke CM, Spaulding KA. Comparison of computed tomographic images of birds obtained with sedation vs general anesthesia. *J Exot Pet Med*. 2013;22:251-257.

14. Mackey E, Hernandez-Divers S, Holland M, et al. Clinical technique: application of Computed Tomography in zoological medicine. *J Exot Pet Med.* 2008;17:198-209.

15. Romagnano A, Shiroma J, Heard D, et al. Magnetic Resonance Imaging of the brain and coelomic cavity of the domestic pigeon (*Columba livia domestica*). *Vet Radiol Ultrasound.* 1996;37:431-440.

16. Souza M, Greenacre C, Jones M, et al. Clinical use of microPET and CT imaging in birds. *Proc Annu Conf Assoc Avian Vet.* 2006;13-16.

17. Souza MJ, Wall JS, Stuckey A, et al. Static and dynamic ^{18}FDG - PET in normal Hispaniolan Amazon parrots (*Amazona ventralis*). Vet Radiol Ultrasound. 2011;52:340-344.

18. Stauber E, Holmes S, Deghetto D, et al. Magnetic resonance imaging is superior to radiography in evaluating spinal cord trauma in three bald eagles (*Haliaeetus leucocephalus*). *J Avian Med Surg.* 2007;21:196-200.

19. van Ruijssevelt L, van der Kant A, de Groof G, et al. Current state-of-the-art of auditory functional MRI (fMRI) on zebra finches: technique and scientific achievements. *J Physiol Paris.* 2013;107:156-169.

20. Wallack S. Basic magnetic resonance imaging principles used for evaluating animal patients with neurologic disease. *Vet Clin North Am Exot Anim Pract.* 2007;10:909-925.

Avian Endocrinology

Yvonne van Zeeland, DVM, MVR, PhD, DipECZM (Avian),
DipECZM (Small Mammal), CPBC, and
Nico Schoemaker, DVM, PhD, DipECZM (Small Mammal), DipECZM (Avian),
DABVP (Avian Practice)

Session #2057

Affiliation: From the Division of Zoological Medicine, Department of Clinical Sciences of Companion Animals, Faculty of Veterinary Medicine, Utrecht University, Utrecht, 3584 CM, The Netherlands.

Manuscript originally published: van Zeeland YRA. Endocrine diseases in parrots. *Br Small Anim Vet Assoc Congr Times*. 2016. Reprinted with permission.

Abstract: Compared to mammals, birds seem to seldom suffer from endocrine disorders. Hypo- and hyperadrenocorticism, as well as hyperthyroidism and thyroid neoplasia, have rarely been recognized in live birds. Hypothyroidism is suggested to occur more often, but in many cases lack of appropriate testing prevents a definite diagnosis from being made, thereby keeping the number of confirmed cases low. Secondary nutritional hyperparathyroidism, on the other hand, is among the most common endocrine diseases seen in birds which are fed diets low in calcium and/or vitamin D. In addition, a poor calcium:phosphorus ratio may result in osteodystrophy or hypocalcaemia. Imbalanced diets may furthermore result in iodine deficiency and goiter, most commonly seen in budgerigars. Pituitary neoplasia and diabetes mellitus are also recognized in this species, but do also occur in other bird species. In many cases, ante mortem diagnosis and treatment of endocrine disease remains challenging because of the limited pathophysiologic knowledge, unavailability of validated tests and lack of information on appropriate dosing regimens in birds. However, by remaining alert to the possibility of a potential underlying endocrine disorder and using the knowledge obtained from other companion animals, endocrine disease in birds may be more readily recognized, diagnosed and treated.

Introduction

In birds, the endocrine system consists of the hypothalamic-pituitary complex and pineal gland, carotid bodies, the thyroid, parathyroid, ultimobranchial, and adrenal glands, the pancreas and endocrine cells of the gastrointestinal tract, and the gonads.[1,2] All of these organs secrete hormones into the bloodstream that help to maintain homeostasis and regulate internal processes such as growth and development, metabolism and reproductive activity. Disease usually occurs as a result of a hypofunction or hyperfunction of the gland. Although endocrine diseases do occur in birds, their incidence appears to be relatively low, with most of the available information derived from single case reports or small case series. However, the low incidence may (in part) be explained by the lack of knowledge and appropriate tests to diagnose endocrine abnormalities. As a result, endocrine disease may occur more often and should certainly not be ruled out when compiling a differential diagnosis.

Of the various endocrine diseases that have been reported in birds, secondary nutritional hyperparathyroidism, resulting in hypocalcaemia and/or metabolic bone disease appears most common, especially in birds that are fed an incomplete diet (eg, seed diet in psittacine birds).[3] Similarly, goiter may be seen, most frequently in budgerigars, as a result of iodine deficiency.[4] Other endocrine diseases appear less common, with reports of their incidence being limited to a single or small number of birds.

Diseases of the Pituitary Gland

The pituitary gland, also referred to as the hypophysis, is a small gland that is located at the base of the brain, just below the hypothalamus. It consists of an anterior part (adenohypophysis) and posterior part (neurohypophysis) that both produce and secrete hormones (ie, luteinizing hormone [LH], follicle stimulating hormone [FSH], thyroid stimulating hormone [TSH], adrenocorticotropic hormone [ACTH], growth hormone [GH], prolactin, arginine vasotocin [AVT] and mesotocin, respectively).[1,2] Disease may occur as a result of various processes, most commonly as a result of a (benign) neoplastic growth. Such pituitary tumours have most commonly been described in budgerigars,[5-7] but may also occur in other species.[8-12] Generally, clinical signs will develop as a result of the mass pressing against the surrounding tissues, thereby causing neurologic signs, exophthalmos, and/or blindness.[13] Similarly, pressure against the normal pituitary cells may result in undersecretion of one or more of the pituitary hormones. Occasionally, oversecretion of a specific hormone may be noted, leading to clinical signs of, for example, Cushing's disease (eg, polyphagia, weight gain, polyuria/polydipsia, feather loss, liver lipidosis) as a result of overproduction of ACTH,[12] or diabetes insipidus (ie, polyuria/polydipsia) as a result of overproduction of AVT.[14] Although a presumptive diagnosis can be made based on the clinical signs and ruling out of other diseases, definite diagnosis requires visualization of the pituitary mass using advanced imaging techniques such as CT or MRI. In addition, hormonal hypersecretion may be detected using hormone-specific assays or function tests (eg, water deprivation test);[14-16] however, size of the patient and tumor may prevent ante-mortem diagnosis of the tumor, thereby rendering post-mortem examination as the only option to obtain a final diagnosis. Thus far, no treatment has been reported for these type of tumors, although one report has demonstrated long-term management of central diabetes insipidus in a grey parrot using desmopressin.[14]

Diseases of the Thyroid Gland

Disease of the thyroid gland may result from hyperplasia, neoplasia, infection, inflammation or atrophy, with clinical signs occurring as a result of enlargement, hypo- or hyperfunction of the gland.[13] The number of documented cases of thyroid tumors and hyperthyroidism in birds is scarce, with most cases involving non-functional thyroid adenomas or adenocarcinomas.[10,17,18] Clinical signs will usually result from the compression of the surrounding tissues by the tumor and may include dyspnea, regurgitation, and dysphagia.[10,17,18] In budgerigars fed an all-seed diet, these clinical signs were commonly seen as a result thyroid hyperplasia (goiter) which develops in an attempt of the body to compensate for the decreased production of thyroid hormone that results from due to iodine deficiency;[4] however, nowadays, the disease is considered more rare due to improvements in the diet. This disease is usually diagnosed based on the history, clinical signs and response to dietary iodine suppletion,[10] although the enlarged thyroid can sometimes be visualized on a radiograph or CT.

Hypothyroidism has often been suspected as the cause of feather loss (due to a lack of feather growth) and obesity in birds.[19,20] In many cases, diagnosis was based on the finding of a low plasma T4 and/or a positive response to thyroxine suppletion;[10,13,21] however, definite diagnosis requires the use of a TSH stimulation test, with hypothyroid birds showing less than two-fold increases in plasma T4 levels compared to baseline.[10,13,22-24] In addition, scintigraphy may be helpful to diagnose hypothyroidism, but is not widely available.[25] Following confirmation of the diagnosis, treatment may be initiated with oral thyroxine suppletion.[13,20]

Diseases of the Parathyroid Glands

Thus far, disease of the parathyroid glands in birds has primarily been linked to hyperplasia resulting from dietary calcium, phosphorus and/or vitamin D deficiencies or imbalances.[3] These dietary imbalances will subsequently result in a disturbance of the mineral homeostasis, which stimulates the parathyroid glands to produce more PTH to correct these imbalances (ie, secondary nutritional hyperparathyroidism).[26] In particular grey parrots appear prone to developing disease, potentially resulting from an increased dependency on ultraviolet light for their calcium metabolism.[26,27] Clinical signs often include weakness, ataxia, tetany and/or seizures. In young, growing birds, bone deformities and pathologic fractures of the long bones and spinal column are frequently seen, whereas female birds may present with egg binding or production of poorly calcified or deformed eggs. Diagnosis is usually based on low (ionized) calcium levels in the blood,[28] whereas radiographs may aid in the diagnosis of bone abnormalities.[13,26] Treatment consists of oral and parenteral calcium and/or vitamin D_3 supplementation, dietary corrections, and exposure to UV-B radiation through natural sunlight or artificial lighting.[3,26]

Aside from hypocalcaemia, hypercalcemia has also been reported to occur in birds. Most often, this is part of a physiologic phenomenon in egg laying females, with increased levels of estrogen resulting in increased calcium absorption in the gastrointestinal tract, increased production of calcium-binding proteins (eg, albumin), and mobilization of calcium from the medullary bone.[29] As a result of these physiologic changes, calcium concentrations can increase two- to even three-fold, with values up to 10 mmol/L (40 mg/dl) reported.[30] Pathologic hypercalcemia also occurs, most common in response to long-term provision of a diet containing excessive amounts of calcium and/or vitamin D_3.[29-31] The disease is commonly seen in young, growing birds (particularly macaws) that are fed excessive amounts of vitamin and mineral supplements. [29-31] Clinical signs include poor weight gain in growing animals, depression, anorexia, nausea, polyuria/polydipsia, ataxia, muscle weakness and joint pain. [29-31] Diagnosis is usually based on the clinical signs combined with high (ionized) blood calcium levels. In addition, uric acid and/or vitamin D_3 metabolites may be measured. [29-31] Radiography is often useful to assess bone density, renal size and detect tissue mineralization. [29-31] In case of advanced cases (ie, presence of tissue calcifications and/or visceral or articular gout as a result of uric acid deposition), prognosis is generally poor. In mild cases, prognosis may be more favorable following timely treatment with forced diuresis and dietary corrections. [29-31]

Diseases of the Adrenal Glands

The avian adrenal glands differ from those in mammals with regard to the lack of a distinct cortex and medulla and the production of corticosterone rather than cortisol as the predominant glucocorticoid hormone.[21,32] Similar to other animals, disease may result from hyper- or hypofunction of the adrenal glands, resulting in excess or lack of one of the adrenal hormones (glucocorticoids, mineralocorticoids, [nor]epinephrine). Thus far, the number of birds diagnosed with spontaneous hyperadrenocorticism has been scarce,[12,32,33] with only one case with an ante mortem diagnosis of Cushing's syndrome as a result of a primary adrenal gland tumor reported in a bird.[34] In this bird, a Senegal parrot, clinical signs included polyphagia, polydipsia, mild alopecia, obesity and hepatic lipidosis. Following diagnosis, which was achieved using a modified dexamethasone suppression test, the bird was successfully treated with trilostane for over one year.[34] Aside from primary adrenal hyperfunction, hyperadrenocorticism may also result from overstimulation by a pituitary tumor (ie, Cushing's disease) or exogenous corticosteroid administration, resulting in similar signs as those described above.[12,18] In addition, exogenous corticosteroids exert profound effects on the immune system and may suppress both humoral and cell-mediated immunity, resulting in increased susceptibility to infections, in particular aspergillosis.[32,35,36]

Thus far, there have been no reports documenting the presence of spontaneous hypoadrenocorticism in pet birds. It is, however, likely that adrenal hypofunction occurs in these species. Similar to other animals, adrenal cell atrophy or necrosis may be induced by neoplasia, bacterial or viral infections, immune-mediated disease, amyloidosis, trauma, or drugs (eg, mitotane).[18] Diagnosis may be achieved using an ACTH stimulation test.[32,37,38] However, since the condition of animals with hypoadrenocorticism may quickly deteriorate as a result of hyponatremia-associated dehydration and shock, and hyperkalemia-associated cardiac disturbances, it is highly likely that fatal complications develop too rapidly in birds, without the possibility of obtaining an ante mortem diagnosis or initiating appropriate treatment.

Diseases of the Pancreas

In birds, glucagon, secreted by the α cells of the pancreatic islands, is considered the major glucose-regulating hormone, with insulin playing a less well-defined role in the glucose metabolism.[39] As such, glucagon excess or inadequate insulin:glucagon ratios rather than hypoinsulinemia has been proposed as the cause of avian diabetes.[40] Histologically, destruction of the β cells (consistent with type 1 diabetes in humans) has been documented in birds, with underlying causes varying from paramyxovirus or herpesvirus infections to hemochromatosis (iron storage disease).[41,42] Similarly, causes for peripheral insulin resistance (consistent with type 2 diabetes), such as obesity, high levels of endogenous or exogenous corticosteroids, and increased levels of other hormones (ie, glucagon, growth hormone, epinephrine), have also been identified.[41-43] Neoplasia associated with diabetes mellitus has been diagnosed in one bird.[44]

In parrots, diabetes mellitus is most commonly reported in budgerigars and cockatiels, although it may occur in larger species as well.[18,41-43,45,46] A presumptive diagnosis can usually be made based on the clinical signs (polyuria, polydipsia, weight loss despite good appetite) in combination with the finding of a persistent hyperglycemia and glycosuria.[13,42,45] The normal range for plasma glucose is significantly higher than that of mammals and may range from 10.0 to 27.9 mmol/L (180-500 mg/dl), dependent on the species, with values > 44.4 mmol/L (> 800 mg/dl) considered diagnostic for diabetes mellitus.[45] Elevated fructosamine levels (reference: 113 to 238 mmol/L)[47] and/or ketonuria may also be found, supporting the diagnosis.[13] Furthermore, determination of plasma glucagon and insulin has been recommended, but remains challenging due to the lack of appropriately validated tests.[13,48] Treatment of birds with diabetes mellitus can be challenging, especially given the small size of many patients and the lack of clear-cut therapeutic protocols.[13] Prior to initiating treatment with glucose-regulating drugs, attempts should be made to rule out and treat any underlying or concurrent conditions. Similarly, birds may require supportive care in the form of fluids or force feeding. In birds with persistent hyperglycemia, treatment may be attempted with insulin and/or oral hypoglycemic agents (eg, glipizide), although the success of this treatment may vary from case to case.[13,41,45,48,49] Hospitalization of the bird is often recommended to enable gradual titration and regular monitoring of the patient (eg, serial blood glucose monitoring or urinalysis to check for glycosuria) until the bird is stable, following which treatment can be continued at home.[13,45]

References

1. Orosz SE. Anatomy and physiology of the endocrine system—protein hormones. In: Speer BL, ed. *Current Therapy in Avian Medicine and Surgery.* St. Louis, MO: Elsevier Saunders;2016:378-385.

2. Ritchie M, Pilny AA. The anatomy and physiology of the avian endocrine system. *Vet Clin North Am Exot Anim Pract.* 2008;11:1-14.

3. Stanford M, Keen J, Campbell-Ward ML. Hypocalcemia in other species. In: Rand J, ed. *Clinical Endocrinology of Companion Animals.* Ames, IA: Wiley-Blackwell;2013:335-355.

4. Schmidt RE. Avian thyroid metabolism and diseases. *Semin Avian Exot Pet Med.* 2002;11:80-83.

5. Bauck L. Pituitary neoplastic disease in nine budgies. *Proc Annu Conf Assoc Avian Vet.* 1988;87-89.

6. Langohr IM, Garner MM, Kiupel M. Somatotroph pituitary tumors in budgerigars (*Melopsittacus undulatus*). *Vet Pathol Online.* 2012;49:503-507.

7. Schlumberger HG. Neoplasia in the parakeet I. Spontaneous chromophobe pituitary tumors. *Cancer Res.* 1954;14:237-245.

8. Curtis-Velasco M. Pituitary adenoma in a cockatiel (*Nymphicus hollandicus*). *J Assoc Avian Vet.* 1992;6: 21-22.

9. Latimer KS. Oncology. In: Ritchie BW, Harrison GJ, Harrison LR, eds. *Avian Medicine: Principles and Applications.* Lake Worth, FL: Wingers Publishing Inc;1994:640-672.

10. Lumeij JT. Endocrinology. In: Ritchie BW, Harrison GJ, Harrison LR, eds. *Avian Medicine: Principles and Applications.* Lake Worth, FL: Wingers Publishing Inc;1994:582-606.

11. Romagnano A, Mashima TY, Barnes HJ, et al. Pituitary adenoma in an Amazon parrot. *J Avian Med Surg.* 1995;9:263-270.

12. Starkey SR, Morrisey JK, Stewart JE, Buckles EL. Pituitary-dependent hyperadrenocorticism in a cockatoo. *J Am Vet Med Assoc.* 2008;232:394-398.

13. De Matos R, Monks D. Diseases of the endocrine system—protein hormones. In: Speer BL, ed. *Current Therapy in Avian Medicine and Surgery.* St. Louis, MO: Elsevier Saunders;2016:385-399.

14. Starkey SR, Wood C, de Matos R, et al. Central diabetes insipidus in an African grey parrot. *J Am Vet Med Assoc.* 2010;237:415-419.

15. Alberts H, Halsema WB, De Bruijne JJ, Lumeij JT. A water deprivation test for the differentiation of polyuric disorders in birds. *Avian Pathol.* 1988;17:385-389.

16. Lumeij JT, Westerhof I. The use of the water deprivation test for the diagnosis of apparent psychogenic polydipsia in a socially deprived African grey parrot (*Psittacus erithacus erithacus*). *Avian Pathol.* 1988;17: 875-878.

17. Doukaki C, Kik MJL, Ducatelle R, et al. Thyroid adenocarcinoma in two parrots. *Vet Rec.* 2014;2:e000101.

18. Rae M. Endocrine disease in pet birds. *Semin Avian Exot Pet Med.* 1995;4:32-38.

19. Kritchevsky JS. Hypothyroidism in other species. In: Rand J, ed. *Clinical Endocrinology of Companion Animals.* Ames, IA: Wiley-Blackwell;2013:278-290.

20. Oglesbee BL. Hypothyroidism in a scarlet macaw. *J Am Vet Med Assoc.* 1992;201:1599-1601.

21. Oglesbee B. Diseases of the endocrine system. In: Altman RB, Clubb SL, Dorrestein GM, Quesenberry K. *Avian Medicine and Surgery.* Philadelphia, PA: Saunders;1997:482-488.

22. Greenacre CB. TSH testing in birds using canine, bovine, or human TSH. *Proc Annu Conf Assoc Avian Vet.* 2011;49.

23. Lothrop CD, Loomis MR, Olsen JH. Thyrotropin stimulation test for evaluation of thyroid function in psittacine birds. *J Am Vet Med Assoc.* 1985;186:47-48.

24. Zenoble RD, Kemppainen RJ, Young DW, Clubb SL. Endocrine response of healthy parrots to ACTH and thyroid stimulating hormone. *J Am Vet Med Assoc.* 1985;187:1116-1118.

25. Harms CA, Hoskinson JJ, Bruyette DS, et al. Technetium-99M and iodine-131 thyroid scintigraphy in normal and radiothyroidectomized cockatiels (*Nymphicus hollandicus*). *Vet Radiol Ultrasound.* 1994;35:473-478.

26. Stanford M. The effect of UV-B lighting supplementation in African grey parrots. *Exot DVM* 2004;6:29-32.

27. Stanford M. Calcium metabolism. In: Harrison GJ, Lightfoot TL, eds. *Clinical Avian Medicine.* Vol 1. Palm Beach, FL: Spix Publishing;2006:141-151.

28. Stanford M. Clinical pathology of hypocalcaemia in adult grey parrots (*Psittacus e erithacus*). *Vet Rec.* 2007;161:456.

29. Campbell-Ward ML. Hypercalcemia in other species. In: Rand J, ed. *Clinical Endocrinology of Companion Animals.* Ames, IA: Wiley-Blackwell;2013:385-395.

30. De Matos R. Calcium metabolism in birds. *Vet Clin North Am Exot Anim Pract.* 2008;11:59-82.

31. Schoemaker NJ, Lumeij JT, Beynen AC. Polyuria and polydipsia due to vitamin and mineral oversupplementation of the diet of a salmon crested cockatoo (*Cacatua moluccensis*) and a blue and gold macaw (*Ara ararauna*). *Avian Pathol.* 1997;26:201-209.

32. De Matos R. Adrenal steroid metabolism in birds: anatomy, physiology, and clinical considerations. *Vet Clin North Am Exot Anim Pract.* 2008;11:35-57.

33. Cornelissen H, Verhofstad A. Adrenal neoplasia in a scarlet macaw (*Ara macao*) with clinical signs of hyperadrenocorticism. *J Avian Med Surg.* 1999;13:92-97.

34. Van Zeeland YRA, Bastiaansen P, Kooistra H, Schoemaker NJ. Diagnosis and treatment of Cushing's syndrome in a Senegal parrot. ExoticsCon 2015;119.

35. Verstappen FALM, Dorrestein GM. Aspergillosis in Amazon parrots after corticosteroid therapy for smoke-inhalation injury. *J Avian Med Surg.* 2005;19:138-141.

36. Westerhof I. Pituitary-adrenocortical function and glucocorticoid administration in pigeons. *Vet Q.* 1997;19:60-61.

37. Lothrop CD, Olsen JH, Loomis MR, et al. Evaluation of adrenal function in psittacine birds, using the ACTH stimulation test. *J Am Vet Med Assoc.* 1985;187:1113-1115.

38. Walsh MT, Beldegreen RA, Clubb SL, Chen CL. Effect of exogenous ACTH on serum corticosterone and cortisol concentrations in the Moluccan cockatoo (*Cacatua moluccensis*). *Am Vet Res.* 1985;46:1584-1588.

39. Hazelwood RL. Pancreatic hormones, insulin/glucagon molar ratios, and somatostatin as determinants of avian carbohydrate metabolism. *J Exp Zool.* 1984;232:647-652.

40. Sitbon G, Laurent F, Mialhe A, et al. Diabetes in birds. *Horm Metab Res.* 1980;12:1-9.

41. Gancz AY, Wellehan JF, Boutette J, et al. Diabetes mellitus concurrent with hepatic haemosiderosis in two macaws (*Ara severa, Ara militaris*). *Avian Pathol.* 2007;36:331-336.

42. Pilny AA. The avian pancreas in health and disease. *Vet Clin North Am Exot Anim Pract.* 2008;11:25-34.

43. Candeletta SC, Homer BL, Garner MM, Isaza R. Diabetes mellitus associated with chronic lymphocytic pancreatitis in an African grey parrot (*Psittacus erithacus erithacus*). *J Assoc Avian Vet.* 1993;7:39-43.

44. Ryan CP, Walder EJ, Howard EB. Diabetes mellitus and islet cell carcinoma in a parakeet. *J Am Anim Hosp Assoc.* 1982;18:139-142.

45. Campbell-Ward ML, Rand J. Diabetes mellitus in other species. In: Rand J, ed. *Clinical Endocrinology of Companion Animals.* Ames, IA: Wiley-Blackwell;2013:191-200.

46. Desmarchelier M, Langlois, I. Diabetes mellitus in a nanday conure (*Nandayus nenday*). *J Avian Med Surg.* 2008;22:246-254.

47. Gancz AY, Wellehan JF, Boutette J, et al. Diabetes mellitus concurrent with hepatic haemosiderosis in two macaws (*Ara severa, Ara militaris*). *Avian Pathol.* 2007;36:331-336.

48. Bonda M. Plasma glucagon, serum insulin, and serum amylase levels in normal and hyperglycemic macaw. *Proc Annu Conf Assoc Avian Vet.* 1996;88.

49. Pilny AA, Luong R. Diabetes mellitus in a chestnut-fronted macaw (*Ara severa*). *J Avian Med Surg.* 2005;19:297-302.

Applying Critical Thinking and Clinical Reasoning in Avian Medicine

Yvonne van Zeeland, DVM, MVR, PhD, DipECZM (Avian), DipECZM (Small Mammal), CPBC and

Nico Schoemaker, DVM, PhD, DipECZM (Small Mammal), DipECZM (Avian), DABVP (Avian Practice)

Session #2060

Affiliation: From the Division of Zoological Medicine, Department of Clinical Sciences of Companion Animals, Faculty of Veterinary Medicine, Utrecht University, Utrecht, 3584 CM, The Netherlands.

Manuscript adapted from: van Zeeland YRA, Schoemaker NJ. Basic diagnostic and therapeutic decision making: a guide to apply evidence-based veterinary medicine in your practice. *ExoticsCon.* 2015 and van Zeeland YRA, Schoemaker NJ. Applying critical thinking and clinical reasoning in exotic animal medicine. *ExoticsCon.* 2016;521-532. Parts of this Masterclass were presented at *ICARE* 2015.

Abstract: In practice, veterinarians are commonly faced with many diagnostic and therapeutic challenges in treating a great variety of avian species. In particular the lack of available scientific information and inability to perform certain diagnostic tests or treatment modalities may refrain veterinarians from practicing evidence-based veterinary medicine (EBVM). However, veterinarians are encouraged to think out of the box and apply knowledge obtained from other species. During this process it will become clear that, by following the basic tenets of problem-oriented medicine, critically appraising the current literature as well as applying sound clinical reasoning, a high quality of EBVM can be delivered. When the principles of evidence-based veterinary medicine are applied, whereby the best available research evidence, the perceived value of the bird and expertise of the clinician are integrated, our patients will be provided with the best possible care which is tailored to the individual bird and is satisfactory for the clinician, the owner, as well as the patient. In addition, this process will stimulate practitioners to critically evaluate the choices and decisions that they make on a day-to-day basis, thereby enabling them to work according to the highest standards of care and advance the field of avian medicine.

Introduction

Birds comprise a reasonable portion of the consultations seen in companion animal practice. Patients can be presented for a variety of different reasons, ranging from a health check, grooming or relative straightforward disease process requiring standardized tests and routine medical care to more complex problems necessitating a more extensive work-up and/or novel treatment options which have previously not been attempted in the species involved. Regardless of the complexity of the problem we are faced with, critical appraisal of the current literature as well as sound clinical reasoning are essential to deliver high-quality, evidence-based veterinary medicine (EBVM), allowing us to treat our patients with the highest standards of care. This in particular can be challenging due to the great variety of species that need to be dealt with, including their relative small size, as well as the limited availability of data-based, empirically-validated therapies and documented best clinical practice guidelines in this field, thereby emphasizing the importance for veterinarians to be able to think outside of the box and apply knowledge obtained from other species.

Many clinicians get nervous when confronted with the term "evidence-based medicine" as they perceive that in the field of avian and exotic animal medicine too little scientific data is available thereby preventing them to work according to the rules of EBVM. It is important to realize; however, that working according to the EBVM principle merely means that the clinician uses the best-available medicine when implementing a diagnostic and therapeutic plan.

To help clinicians understand the process of evidence-based (veterinary) medicine, the process that can be used is described, including the use of so-called patient, intervention, comparison, and outcome (PICO) questions. In addition the diagnostic process, including the use of a problem-oriented veterinary medical record (POVMR) system, will be explained. The latter will greatly help clinicians to work systematically and help them make sound decisions. Each of these techniques is described further below.

During the interactive Masterclass the participants will be presented with a selection of clinical cases involving a variety of bird species in which the aforementioned tools and techniques are to be applied. The selected cases range from common and routine disorders, to patients with more complex and unusual diseases. Participants will be asked and encouraged to critically think about the work-up of the presented case, whereby they will be asked to consider all relevant aspects and motivate each of the steps they take in the process towards establishing a definite or presumptive diagnosis and treatment plan. With the varying backgrounds of the participants (ie, number of years in practice, level of experience with exotics or number and type of continuing education hours followed, but also where, in which country, and when training was received), fun and interesting discussions are likely to arise, whereby challenging questions may be raised that require sound clinical reasoning and critical thinking in order for an informed decision to be made concerning the diagnosis and/or treatment of the avian patient.

Evidence-based (Veterinary) Medicine

Evidence-based medicine (EBM) is the "conscientious, explicit and judicious use of current best evidence in making decisions about individual patients."[1] Many factors play a role in making these decisions, including the best available (scientific or clinical research) evidence, patient's needs, client's wishes and the individual clinician's clinical expertise and judgment.[2,3] Similar to EBM, EBVM involves a clinical decision-making process that allows veterinarians to find, appraise, and integrate the same aforementioned factors (Fig 1).[2]

As circumstances and conditions will vary for each individual case whereby different weights and values are attributed to each of these factors, a cookbook approach is therefore neither applicable nor justifiable when attempting to provide optimal patient care to the individual, regardless of the patient being a human, dog, cat, small mammal or bird.

Although our desire is to be able to provide a definitive diagnosis to our patients and clients and inform them about the best possible therapy for their animal, all of us are aware that opinions and views can differ greatly among experts, especially in the field of non-exact sciences such as (veterinary) medicine. As such, every incidence of a disease should be regarded as a unique entity, which needs a diagnostic and therapeutic protocol specifically tailored to the individual patient. As academic teachers, we try to educate our students that veterinary medicine is a dynamic profession in which previously held beliefs about a treatment or a diagnosis will change or need adjustment over time.[4] A dean of Harvard Medical School (Sydney Burwell) once said during a graduation ceremony "Half of what you are taught as medical students will in 10 years have been shown to be wrong. And the trouble is, none of your teachers knows which half."[5]

Each veterinarian is therefore advised to look at every patient as an individual and approach every problem as a unique case whereby it needs to be determined whether the individual patient will benefit from the available evidence, or that a different approach should be used to solve the problem that this animal is encountering. Throughout this process, the veterinarian is strongly encouraged to make decisions based on sound background

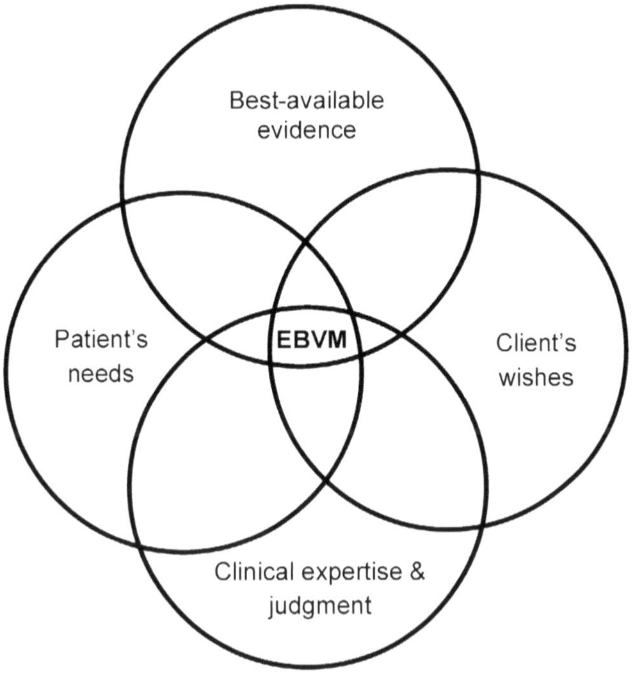

and pathophysiologic information.[3]

Figure 1. Evidence-based veterinary medical decisions integrate the best-available clinical research evidence; client preferences, values, or beliefs; individual patient circumstances and needs; and the clinical judgment and expertise of the attending veterinarian. Taking all of the aforementioned factors in consideration allows us to make the best clinical decision for the individual patient (Gibbons PM, Mayer J. Evidence in exotic animal practice: a "how-to guide." *J Exot Pet Med.* 2009;18:174-180 and Schmidt PL. Evidence-based veterinary medicine: evolution, revolution, or repackaging of veterinary practice? *Vet Clin North Am Small Anim Pract.* 2007;37:409-417).

The process of making an EBVM decision

In order to make EBVM decisions, veterinarians are encouraged to follow the following five steps:

1. Ask an answerable clinical question.

   Asking the right, answerable question is essential when practicing EBVM. Although it sounds easy, correctly formulating a question so that it can be scientifically answered is not always that simple. However, this step is essential to start the search for the most up-to-date, relevant and robust evidence that is available in the literature to successfully help solve a clinical problem or answer a clinical question that is raised during the evaluation or management of an individual case. The standard method that can be used to formulate a searchable and answerable clinical question is using a model known by the acronym PICO:

   Patient, problem or population of interest, including its most important and relevant characteristics (eg, blue and gold macaw diagnosed with proventricular dilatation disease);

Intervention or indicator that is being considered; this could be an exposure or etiologic factor, diagnostic test, treatment, or prognostic factor (eg, treatment with a selective COX-2 inhibitor such as meloxicam or celecoxib);

Comparison, control or comparator, comprising the main alternative to which the intervention or indicator is compared; this can either be a decision between two tests or treatments, but also a comparison to a control or placebo in order to determine whether the alternative is better or worse than the intervention that is being considered (eg, placebo); and

Outcome of interest to the patient or problem (ie, what needs to be accomplished, measured, improved or affected such as clinical improvement and/or longer survival time)?

Learning to ask these questions in the PICO structured format helps to ensure clear and precise formulation of a question. Although this can be difficult at first, incorporating this technique into everyday practice will allow veterinarians to master this technique in a relatively short period of time thereby facilitating the process for future patients.

It is important to be as specific as possible during the formulation of your question. Keep in mind, however, that being too specific in narrowing your question to a specific problem or population may complicate the search for evidence to answer this question. This is especially true when dealing with exotic animals as many diseases, including their diagnosis and treatment, have not yet been described in these species. This does not mean, however, that these diseases do not occur or cannot be diagnosed or treated. Searches may therefore be extended to other species, thereby expanding the possibilities to come across results that may potentially be extrapolated to the species of interest and can help to design a novel diagnostic test or treatment protocol for the 'new' disease you may be dealing with.

2. Find the best available evidence by searching the (veterinary) medical literature.

   To find the best available evidence it is essential to be aware of the best ways to search for high quality literature as well as have knowledge of the best and most useful sources of information that are out there.

   Internet-based search engines are highly effective for searching literature as long as the correct terminology is being used during a search. Commonly used search engines are PubMed (http://www.ncbi.nlm.nih.gov/pubmed), Scopus (http://www.scopus.com) [although this search engine from Elsevier requires an institutional login], Google Scholar (http://scholar.google.com), Web of Knowledge (http://www.isiwebofknowledge.com), BioONE (http://www.bioone.org), Veterinary Information Network (http://www.vin.com), and the International Veterinary Information Service (http://www.ivis.org). Moreover, lots of journals can nowadays be accessed online (frequently for free in open access journals), thereby rendering it easier for the private practitioner to find and access specific literature compared to 15 years ago.[2]

   When searching for literature, it is generally recommended to use more than one search engine to find appropriate references, especially since each search engine has its own methodology incorporated to search and sort the found results (eg, Google Scholar searches within the full text and sorts results partly by the number of citations rather than limits the search to the title, abstract and keywords and sorting results by date or relevance such as Scopus does). Similarly, multiple search strategies may be employed, including the systematic search (using original and relevant search terms); snowball method (starting with one or multiple known publications based on which a search is conducted to find older literature using characteristics of the known publications such as author, references, citations, keywords); citation searches (in which the snowball method is used to track citations to find more recent literature relevant to the subject); and catalogue method (during which searches are conducted within specific digital or printed collections like Worldcat, Google Books, LibraryThing). Dependent on the results obtained, searches can subsequently be narrowed or broadened by respectively simultaneously using multiple or more specific search terms, exact phrases and/or filters; versus less and broader search terms, alternative sources and/or alternative search terms (eg, synonyms, translations) during consecutive searches.

3. Critically appraise the found evidence.

   Once literature has been found that addresses the formulated question, the evidence provided within this literature needs to be appraised for its validity (closeness to the truth), relevance (appropriateness), impact (size of the effect), and application (usefulness in our clinical practice). Unfortunately, it is not uncommon that the literature contains sources of bias that may represent potential flaws in the study or limits the ability to extrapolate the findings to the question that was raised. The three key elements to critical thinking and critical appraisal of evidence are incorporated in the term RED:

   R = Recognize assumptions
   E = Evaluate arguments
   D = Draw conclusions

   To correctly appraise the value of the gathered information, the value of the source in which the infor-

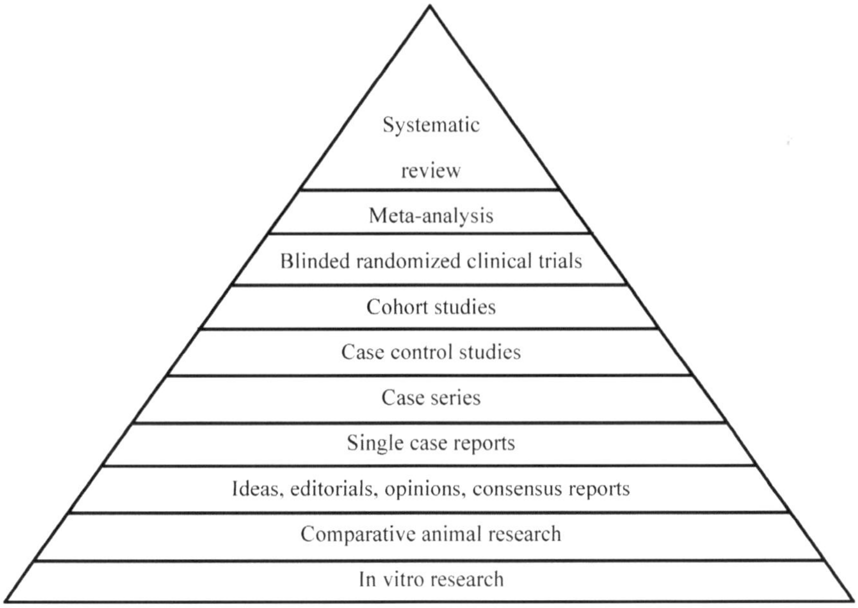

mation was found also needs to be appraised. For this purpose, the pyramid of evidence is considered extremely useful (Fig 2).

Figure 2. Pyramid of evidence. This pyramid is widely used to judge the quality of the evidence used to make clinical decisions. On the top of the pyramid the strongest sources of evidence are represented, while those at the bottom of the pyramid are considered the weakest sources of evidence (Holmes MA. Evaluation of the evidence. *Vet Clin North Am Small Anim Pract*. 2007;37:447-462).

When scanning the literature it is recommended to always first read the methods, as it is crucial to judge whether these are valid or not. If the latter is not the case, the article is generally not worth using for the evaluation of your case. Upon deciding that the methods are valid, the next section to be read and evaluated is the results, as this will allow the veterinarian to draw its own conclusions about the data provided rather than relying on the author's interpretation and conclusions, which may be biased by his/her beliefs or past experiences. The discussion is also considered valuable as in this section new questions may arise about the topic of investigation by comparing previous work with the newly gathered information from the study. Last but not least, the conclusion needs to be critically analyzed and appraised, as authors may be biased whereby they can sometimes draw conclusions that may not be supported by the actual results.[2]

4. Integrating the found evidence.

   Once evidence has been found in the literature, all the collected evidence should not only be mapped and balanced against each other, but also be critically appraised against your own clinical expertise as well as the patient's specific and unique biology and circumstances. Only by combining all factors should a further plan be developed.

5. Implement and evaluate the findings.

   As a final step, all findings should be implemented and evaluated in your patient or population, looking at outcomes that are important to you, the patient, and the client.

Diagnostic Process

Before being able to apply EBVM, veterinarians need to be able to diagnose what is going on with their patient. To be able to do so, many diagnostic decisions need to be made. These decisions, however, can only be made after properly identifying the clinical abnormalities and risk factors associated with disease. This is then followed by generating a tentative or definite diagnosis or list of differential diagnoses based on clinical reasoning. Just as with EBVM five separate steps can be followed during the clinical decision making process.

The first step (Step 1) is the collection of data by performing a physical examination, including a thorough history, and diagnostic tests (eg, imaging, laboratory tests). Upon collecting this data, it is essential to be able to distinguish normal from abnormal findings (eg, by identifying species-specific anatomic or physiologic features or comparing laboratory findings to established reference ranges (also taking into account the type of diagnostic test that is used).

Following the gathering of the data and generation of a problem list, all possible differential diagnoses may be recalled (Step 2). Generating an accurate and complete list of differential diagnoses is important, as your final diagnosis should ideally be included within this list. The DAMNIT mnemonic tool can greatly help during the generation of an adequate differential list. Within this mnemonic the D stands for degenerative or developmental; the A for anomalous, autoimmune, atmospheric or allergic; the M for metabolic, mechanical or mental; the N for neoplastic or nutritional; the I for infectious, ischemic, immune-mediated, inflammatory, inherited, idiopathic or iatrogenic; and the T for toxic or traumatic.

Following the composition of a list of differential diagnoses, the diagnoses on this list should then be ranked according to their likelihood (Step 3). This ranking, in which diseases may be excluded or considered more or less likely based on the data collected during Step 1, may take place using various methods, including pattern matching; probabilities and pathophysiologic reasoning. Generally this step requires much practice and experience, thereby making it one of the most difficult steps to master. As a result, many veterinary students can become frustrated upon repeatedly attempting to learn to master this part of the process of clinical reasoning.

Once a list of probable diagnoses has been made, further diagnostic investigations are subsequently needed to further differentiate between the potential diagnoses and exclude or confirm a suspected causation for the encountered clinical signs (Step 4). This may occasionally include the gathering of further clinical information by taking an additional history from the owner or performing organ-specific examinations.

Often additional diagnostic tests (eg, hematology, biochemistry, urinalysis, fecal examination, diagnostic imaging, endoscopy, culture and sensitivity, cytology and/or histopathology) are necessary to obtain a (tentative or definite) diagnosis. Owners (and some veterinarians) often believe that a (single) test result will guarantee whether a disease is present or not; however, each test has a risk of resulting in an erroneous test outcome, as reflected by its sensitivity, specificity, positive and negative predictive values (Fig 3). Unfortunately, the accuracy and predictive values of many of the tests that are used in (exotic animal) medicine are often unknown. In addition, errors in test results may also arise due to incorrect sampling or procedural mistakes (eg, erroneous handling, transport or processing of the sample). An example of such a mistake is omitting to spin down a blood sample directly following its collection, thereby resulting in unreliable plasma potassium concentrations in birds. Simi-

		Presence of condition (as determined by the "gold standard")		
		Condition present	Condition absent	
Test outcome	Test outcome positive	True positive	False positive (Type I error)	Positive predictive value = Σ True positive / Σ Test outcome positive
	Test outcome negative	False negative (Type II error)	True negative	Negative predictive value = Σ True negative / Σ Test outcome negative
		Sensitivity = Σ True positive / Σ Condition present	Specificity = Σ True negative / Σ Condition absent	

larly, collection of blood in tubes with red rubber stoppers for plasma zinc concentrations is deemed to result in unreliable results as these stoppers (used to) contain high concentrations of zinc which could then dissolve into the plasma during transport and result in abnormally high concentrations of zinc to be measured.

Figure 3. Overview of the relationship between accuracy and predictive values of a test outcome. Accuracy of a test, as determined by the test's sensitivity and specificity, is a fixed test characteristic that remains constant. The sensitivity of a test is defined as the proportion of animals affected by the condition which turn up positive in the used test, thereby indicating the probability of an animal with the condition correctly being identified as positive in the test. The specificity of the test, on the other hand, indicates the proportion of healthy animals which indeed turn up negative in the used test, thereby indicating the probability of the test being negative when the animal is not sick. The test may, however, also incorrectly identify an animal as being positive or negative, respectively resulting in a type I error (false positive result) or type II error (false negative result).

Positive and negative predictive values are other important characteristics of a test. The positive predictive value demonstrates the chance of the animal truly being positive when the test outcome is positive and is reflected by the percentage of true positive results from all the positive test results. Negative predictive value is defined as the percentage of true negative results from all the negative test results thus indicating the chances of an animal truly being healthy when the test outcome comes back negative. In contrast to the accuracy of a test, predictive values of a test are highly dependent on the prevalence of the condition within the population (O'Connor A, Evans RB. Critically appraising studies reporting assessing diagnostic tests. *Vet Clin North Am Small Anim Pract.* 2007;37:487-497).

Moreover, rather than merely comparing results to the reference ranges, test results need to be critically evaluated to determine whether the obtained results match the patient's clinical presentation. Considering the greater risk of false positive outcomes in a population with low prevalence of the disease, one must furthermore be critical towards the type and number of tests that are performed, in which animals to perform the test, and the meaning of a positive test outcome (eg, histopathology versus serology versus culture). In fear of being sued for missing a diagnosis, a tendency among (veterinary) medical experts exists to perform as many tests as possible. This not only results in a huge financial burden for the owner, but is also more likely to result in a false positive outcome that subsequently needs to be addressed. Thus, selecting the appropriate test(s) to confirm or exclude specific conditions is recommended over the use of diagnostic tests as a screening tool.

Once all the data have been collected and a definite or tentative diagnosis is reached, further decisions may then be made on how to manage the case, and determine the best therapeutic and/or preventive plan for the patient, also taking its prognosis into account (Step 5).[6]

Based on the knowledge and experience of the clinician, the clinical reasoning process may be worked through more quickly or slowly using either one of two reasoning processes as described by Croskerry.[7] During the so-called type 1 reasoning process, clinicians use pattern recognition to quickly obtain a diagnosis for a specific clinical presentation of the patient. Less-experienced clinicians (eg, students, interns or residents) often have

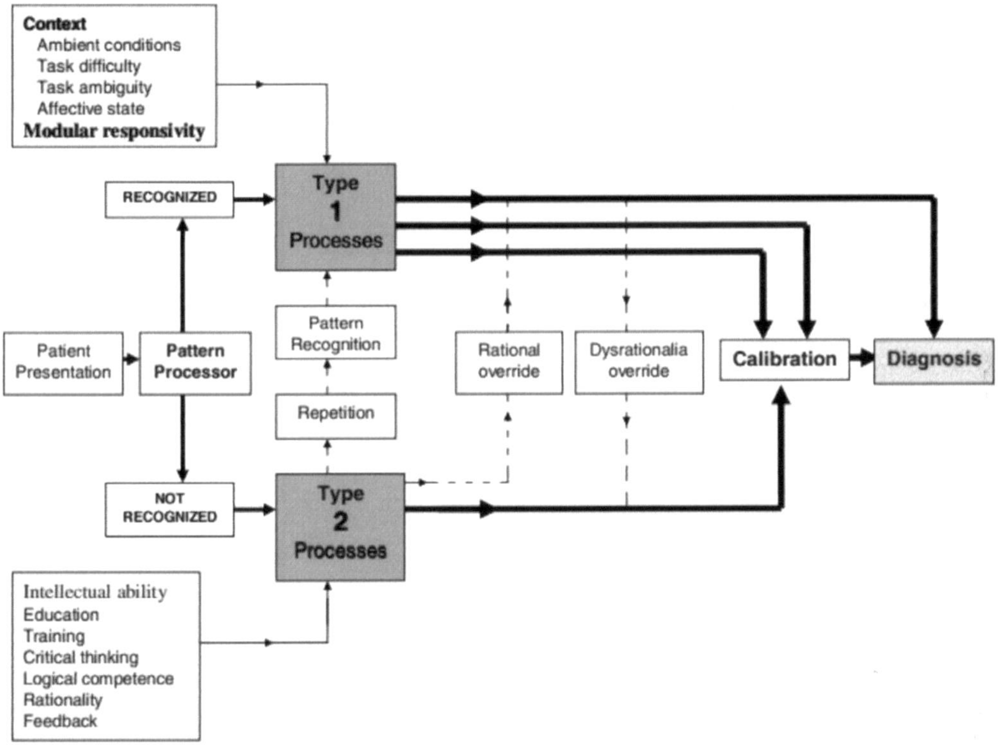

not established such pattern recognition skills. As a result, the more thorough and extensive type 2 reasoning process will be followed, during which findings are continuously (re)evaluated and more extensive diagnostic work-up is needed (Fig 4).[7]

Figure 4. Dual process model of Croskerry. This model proposes that problems are solved using so-called Type 1 and/or Type 2 processes. Type 1 processes involve intuition and are largely automatic, fast, and unconscious (but also more vulnerable to bias) whereas Type 2 processes are more deliberate, analytical, and systematic (thereby also requiring more cognitive effort). Veterinarians use both types of processes when making a diagnosis, but the relative emphasis varies with the type of disease, context and experience of the clinician. Experienced veterinarians may generally be able to take quick actions based on pattern recognition (ie, Type 1 processes), whereas students or newly graduated veterinarians often rely on Type 2 processes and need more time and tests to come to a final diagnosis. Upon repeated exposure to a similar situation, pattern recognition and switch from a Type 2 to a Type 1 process may occur. Similarly, under certain circumstances (eg, time pressure or individual clinical judgement) Type 1 processes may override Type 2 processes (ie. a dysrationale override). This increases the risk of diagnostic errors, as the clinician quickly jumps to the "most probable diagnosis" without proper evaluation and ruling out other conditions; however, following the recognition of incorrect assumptions, a so-called rationale override may occur, whereby the Type 1 decision-making process is overridden by a Type 2 process that involves a more detailed evaluation, using deductive, analytic reasoning (Croskerry P. A universal model of diagnostic reasoning. *Acad Med.* 2009;84:1022-1028).

Once a tentative or definite diagnosis has been reached, a treatment plan can be initiated. In an ideal world a specific treatment plan would be in place for each diagnosis that can be made. As we all know, this is unfortunately not the case in the real world. Decisions will therefore have to be made on the best treatment option for a certain diagnosis. This is where the use of EBVM comes into place again. Although the human brain is well equipped to rapidly process complex information, it has a limited capacity to interpret competing strategies objectively and with sufficient accuracy and reliability. It is therefore important that all the available options are first explored and considered prior to conducting a critical analysis and appraisal of the various treatment options that may be attempted.[8] For this purpose, a 7-step decision-making process can be followed:

Step 1: Identify the problem (Define the problem you wish to solve).
Step 2: Gather information (Explore the possible interventions for your problem).
Step 3: Identify alternatives (Decide what the potential interventions are).
Step 4: Predict immediate and long-term consequences (List the pros and cons of each potential intervention).
Step 5: Make the best choice (Decide which of the alternatives is the best option for this patient).
Step 6: Act (Implement the decision regarding the intervention to be followed).
Step 7: Evaluate (Reflect on the decision that has been made and decide *if* and *what* needs to be done differently next time).

Using the aforementioned decision-making process forces the clinician to consider all options and outcomes prior to making a decision regarding the treatment options that can best be pursued. Several examples of such analyses exist in human and veterinary medicine, which ultimately even led to changed thoughts regarding disease etiologies and/or changes in treatment protocols which for long had been considered the standard. In human medicine for example, a critical appraisal of the literature, including a decision analysis, was needed to convince skeptics that *Helicobacter pylori* indeed played an important role in the establishment of gastric ulcers. Similarly in veterinary medicine, a thorough analysis of the pros and cons of preventive orchiectomy in cryptorchid dogs led to the conclusion that there actually is no need to perform such a surgery, as the pros of tumor prevention did not outweigh the cons of surgical complications and cost of the preventive surgery.[9]

In order to draw valid conclusions, the latter type of analysis requires a thorough evaluation process to be conducted based on a sufficient amount of evidence that has been gathered on each of the therapeutic options. Unfortunately, the latter is frequently lacking in avian and exotic animal medicine. Would it not be nice to be able to provide owners with clear evidence whether ovariohysterectomy is preferred over ovariectomy in rabbits? Similarly, would it not be beneficial to use decision analyses to determine the optimal technique for wing trimming or the best way to treat cloacal prolapse in cockatoos? The sooner we start asking ourselves these questions, the sooner we can commence the collection of this much needed data so that in the near future similar analyses can be conducted to answer the aforementioned questions and advance the knowledge within the field of avian and exotic animal medicine.

Problem-oriented Veterinary Medicine

A final "tool" that is at the disposal of veterinarians that would like to work according to the rules of evidence-based medicine is the problem-oriented veterinary medical record (POVMR) method. Using this method, one does not necessarily need prior experience to solve new problems, because by integrating new knowledge into the basic POVMR method, clinical success can be achieved through identifying knowledge gaps, seeking new information, and assessing available evidence.[10]

By using the problem-oriented veterinary medicine (POVM) method a form of clinical reasoning (ie, hypothetico-deductive) is applied that can be recorded and easily understood by others. It is a stepwise process of gathering case information; defining problems; making plans for diagnosis, treatment, and client education; and evaluating patient progress over time.[6] Veterinarians frequently refer to POVM as a patient SOAP, which is an acronym for the process of subjectively (S) and objectively (O) recording information, followed defining and assessing (A) each problem. Finally, based on this information a diagnostic, therapeutic, and client education plan (P) is drawn so that each problem may be addressed. The system of POVM (or SOAP) is commonly used in academic institutions to train residents and students to systematically evaluate their patients. Similarly, this system would be beneficial in private practice, especially in larger ones which employ multiple veterinarians as this methodology allows each clinician to easily see what the reasoning behind a certain diagnostic procedure or treatment plan was, thereby allowing him or her to continue with that plan or make informed decisions about a potential adjustment of the diagnostic-therapeutic plan for the patient without compromising the patient's health and welfare or upsetting the client.

Conclusions

Different tools are at the disposal of clinicians to provide their patients with the best-available care. Once the clinician starts to use the POVMR method systematically during the work-up of patients, it then becomes easier to see what the potential differentials for the disease are and to establish a most likely diagnosis based on clinical reasoning using the data gathered from the history, physical examination and additional diagnostic tests. By using PICO questions, the clinician can then quickly scan and critically appraise the literature to find the best-available scientific evidence to treat that individual patient. This evidence may subsequently be combined with the veterinarian's judgment and clinical expertise; the client's wishes and beliefs; and the patient's circumstances and needs to allow an optimal treatment plan to be designed for the patient according to the rules of EBVM. In doing so, EBVM may become an integral part of the veterinary practice, whereby veterinarians are assured to deliver sound, evidence-based, cost effective and high quality care to their patients, even in those situations where scientific evidence is scarce.

References

1. Sackett DL, Straus SE, Richardson WS, et al. *Evidence-based Medicine: How to Practice and Teach EBM.* 2nd ed. Edinburg, TX: Churchill Livingston;2000:1-27.

2. Gibbons PM, Mayer J. Evidence in exotic animal practice: a "how-to guide." *J Exot Pet Med.* 2009;18: 174-180.

3. Robertson RS. Refining the clinical question: the first step in evidence-based veterinary medicine. *Vet Clin North Am Small Anim Pract.* 2007;37:419-431.

4. Holmes MA. Evaluation of the evidence. *Vet Clin North Am Small Anim Pract.* 2007;37:447-462.

5. Pickering GW. The purpose of medical education. *Br Med J.* 1956;2(4985):113-116.

6. Gibbons PM. Problem-oriented exotic companion animal practice. *J Exot Pet Med.* 2009;18:181-186.

7. Croskerry P. A universal model of diagnostic reasoning. *Acad Med.* 2009;84:1022-1028.

8. Cockcroft PD. Clinical reasoning and decision analysis. *Vet Clin North Am Small Anim Pract.* 2007;37: 499-520.

9. Peeters MAJ, van Sluis FJ. Decision analysis tree for deciding whether to remove an undescended testis from a young dog. *Vet Rec.* 2002;150:408-411.

10. Gibbons PM. Introduction: evidence-based, problem-oriented medicine. *J Exot Pet Med.* 2009;18:172-173.

11. Schmidt PL. Evidence-based veterinary medicine: evolution, revolution, or repackaging of veterinary practice? *Vet Clin North Am Small Anim Pract.* 2007;37:409-417.

12. O'Connor A, Evans RB. Critically appraising studies reporting assessing diagnostic tests. *Vet Clin North Am Small Anim Pract.* 2007;37:487-497.

Pelicanidae: Pelican Medicine

J. Jill Heatley, DVM, MS, DABVP (Avian Practice), DABVP (Reptile and Amphibian Practice), Dipl ACZM,
Jordan Gentry, DVM,
Brian Mihura, MS,
Sharon Schmalz, BS

Session #2062

Affiliation: From the Department of Small Animal Clinical Sciences (Heatley, Gentry), College of Veterinary Medicine and Biomedical Sciences, Texas A&M University, College Station, TX 77843, USA and from the Wildlife Center of Texas, Houston, TX 77024, USA (Mihura, Schmalz).

Abstract: Pelicans have long held cultural significance in mythology, and in Christian and heraldic iconography.[1-3] As a panglobal species group, Pelicans are affected by numerous infectious and many human related noninfectious diseases. Pelicans have often been persecuted by humans based on perceived competition with commercial and recreational fishing. Population declines are based on habitat destruction, disturbance, disease and environmental pollution. Despite being commonly encountered in rehabilitation, within the zoo, and aviculture community, this group has received relatively little specific attention to common injuries, maladies and appropriate veterinary care. This master class will review handling, anesthesia, diseases, treatment and care in captivity for these species, based on review of rehabilitation and zoo records for these species.

Introduction and Overview

Pelicans are a genus of large coastal birds characterized by a long, large bill and a large throat pouch (Table 1). Within the class Aves, Order Pelicaniformes, they are Family Pelecanidae. These species are most closely related to the shoebill, hamerkop, and other birds within the Order Pelicaniformes, including ibis, spoonbill, heron, and the desolate bittern. Fossil evidence for these species dates back at least 30 million years ago, when they evolved in the old world and later spread to the Americas. They have a patchy distribution ranging from tropical to temperate zones both inland and coastally.[4] They are near-surface piscivores, which may plunge dive or scoop their prey. As a gregarious species they nest colonially, and hunt cooperatively. Light- to white-colored species nest on the ground while brown and darker-color species nest in trees or rocks. Here we review the anatomy and physiology, appropriate handling, anesthesia, analgesia, infectious and noninfectious diseases, and general care and husbandry of these species.

Table 1. Pelican species of the world.[a]

Common name (*Genus species*)	IUCN[b] status	Size	Range
American white pelican (*Pelecanus erythrorhynchos*)	Least concern	Weight 5-9 kg (10-20 lb) Plumage white, except black primaries and secondaries	Inland North America, winters in Mexico
Brown pelican (*Pelecanus occidentalis*)	Five subspecies Least concern Endangered in Texas	Weight 3.6-4.5 kg (7.9-9.9 lb) Smallest pelican, brown plumage, plunge diver	Coastal distribution North America, Caribbean to northern South America and Galapagos
Peruvian pelican (*Pelecanus thagus*)	Near threatened	Weight 7 kg (15 lb) Dark, white stripe runs crown down the sides of neck, sea rock nests	Pacific Coast of South America, Ecuador to Peru to southern Chile
Great white pelican (*Pelecanus onocrotalus*)	Least concern	Weight 10-11 kg (22-24 lb) Plumage white, pink facial patch and legs	Patchy distribution East Mediterranean east to Indochina and Malay Peninsula, to South Africa
Australian pelican (*Pelecanus conspicillatus*)	Least concern	Weight 4-8.2 kg (8.8-18.1 lb) White, black along primaries, large pale pink bill	Australia and New Guinea, vagrant to New Zealand, Solomons, Bismarck Archipelago, Fiji and Wallacea.
Pink-backed pelican (*Pelecanus rufescens*)	Least concern	Weight 3.9-7 kg (8.6-15.4 lb) Grey and white plumage, occasionally pinkish on back, yellow upper mandible, grey pouch	Africa, Seychelles and southwestern Arabia, extinct in Madagascar
Dalmatian pelican (*Pelecanus crispus*)	Vulnerable	Weight 10-12 kg (22-26 lb) Largest pelican, differs from great white pelican based on curly nape feathers, grey legs, greyish-white plumage.	South-eastern Europe to India and China
Spot-billed pelican (*Pelecanus philippensis*)	Near threatened	Weight 5 kg (11 lb) Mainly grey-white, grey hind neck crest in breeding season, pinkish rump, spotted bill pouch	Southern Asia from southern Pakistan across India east to Indonesia, extinct in the Philippines and possibly eastern China

[a]Johnsgard PA. *Cormorants, Darters, and Pelicans of the World.* Washington, DC: Smithsonian Inst Press;1993.
[b]International Union for Conservation of Nature.

Anatomy and Physiology

Anatomy and physiology have been well described in pelican species.[5-11] The Australian pelican has the longest bill of any bird, ranging from 15-20 inches in length. The crop is absent and replaced with the gular pouch. Pouch anatomy is unique to pelicans and the vasculature has been described.[8] Muscular anatomy of the pelican has been documented at the gross anatomic as well as the histologic level.[7] Histologically, the superficial muscle belly of the pectoral muscle is type A white or fast muscle and the smaller deep pectoral muscle is type B slow or dark muscle.[12] Therefore the superficial pectoral is likely used for flapping flight while the deep pectoral is used for soaring and gliding flight.[12] Gross anatomic studies on the sternum and clavicle of spot-billed pelican *(Pelecanus philippensis)* revealed fusion of the hypocledium to the ventral part of the cranial border of the sternal crest and clavicle fusion to the cranial border of the sternal crest. The most characteristic feature of the sternum of the pelican was the hypocledium of the clavicle fused to the ventral part of the cranial border of the sternal crest. This fusion makes the pectoral girdle more rigid to facilitate soaring with relatively small breast muscles. The greatly inflated clavicles and sternum may regulate excess heat generated by the flight muscles by heat radiation into the hollow core of the bones and then passed out of the body with the exhaled air.[13] As in most other avian species the humerus is pneumatic and connected directly with the *sacci claviculari* through a *foramen pneumaticum* arising in the *fossa pneumotricipitalis*.[9]

Extensive studies of the embryonic and young adult pelican lung reveal them to begin relatively smaller than many other volant species; however, as the pelican matures the lung becomes one of the largest and most capable of air exchange among the avian species.[14] The macroanatomy of the heart valves of 5 white pelicans recently was reported with only minor anatomic changes from other avian species observed.[15]

In comparison to other birds, adrenal of the brown pelican, *Pelecanus occidentalis*, has intermingled masses of interrenal and chromaffin cells like other birds, but resembles the mammalian adrenal in certain respects and also has pelican-specific features.[16] The pelican adrenal gland differs from most avian adrenals as it has less chromaffin tissue and more abundant interrenal tissue. The interrenal component has 3 zones similar in position and general appearance, but not in relative proportions, to the zona glomerulosa, zona fasciculata and zona reticularis of the mammalian adrenal; however, the zona reticularis forms the central mass of the gland and is broader and the fasciculata narrower than in mammals. Another feature of the pelican adrenal is the organization of interrenal cells into solid cord-like or cylindrical formations, composed of a single layer of elongated radially arranged cells whose distal ends form the surface facing the blood channels and opposite (proximal or central) ends converge to form the axial cores. This regular disposition of the cells in the cords establishes a distinctive structural and functional polarity peculiar to the pelican. Peculiar to pelicans, interrenal masses are organized into cords of a single row of tall columnar cells radially arranged like those of a gland tubule, with broad ends forming the surface facing the blood channels and their opposite, narrower, nucleated ends converging to form the solid central cores of the cords creating a polarity to cortical cells.

Handling

Pelican defenses include snapping with the bill which may be accompanied by neck extension, as well as beating the wings.[17] Most pelicans have a hook on the end of the bill (Fig 1). In addition, they beat their wings, attempting to engulf the would-be holder's hands or head, as well as regurgitate pouch or stomach contents, and expel feces. Generally, 2 people are recommended to be present for all but the most docile of pelicans. One person has the carrier ready (generally a large dog crate or large cardboard box) and serves as safety buddy in case the primary pelican wrangler gets into trouble. Eye protection is recommended as well wearing pants. Gloves are generally not necessary. A sheet also is useful to protect the handler's body and to wrap the pelican to avoid soiling of clothing. The bill is secured first with a finger between the rhamphotheca and gnathotheca as these species often lack or have only vestigial nares. In some cases, the animal may be now be blindfolded and walked around lead by the bill; however in many cases the body should now be covered by the sheet and the wings gathered against the body as the pelican is picked up and held under the arm and elbow against the holder's body. Beware that pelicans, like cranes, are very prone to capture myopathy and capture time and holding should be minimal. Planned capture events should ideally be performed in cooler temperatures. Preemptive or protective treatment with drugs to avoid capture myopathy is recommended for consideration (see Treatments Section). When alarmed, pelicans may inflate their subcutaneous air sacs, which may be palpable as subcutaneous crepitus. This normal finding does not indicate trauma or severe gas-forming bacterial infection as in companion mammals.[5] Large open-weave nets also may be used for capturing pelicans from water or enclosures. Net guns and cast nets also may be used to capture pelicans but should restricted to use by experienced personnel as they can further injure or drown the animal if used incorrectly.

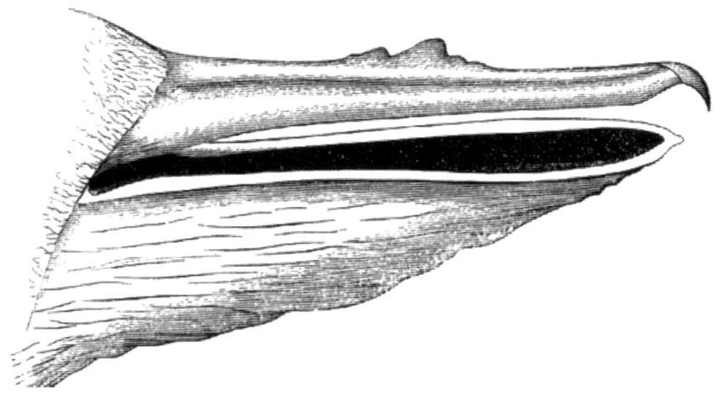

Figure 1. Pelican bill.

Anesthesia and Analgesia

Anesthesia and analgesia often is indicated for pelicans based on the need for traumatic injury repair, or fishhook removal. For sampling and physical examination I rarely find anesthesia necessary, although sedation can be helpful.[17] The anatomy of the bill pouch and trachea make induction of anesthesia cumbersome for the beginner. I do not recommend face-mask-only induction because of the large pouch surface area and long bill, which results in extensive environmental gas contamination. A disposable livestock rectal glove also may be used for a makeshift elongated induction mask.[5] Nitrous oxide use is avoided because of subsequent inflation of the subcutaneous air sacs of pelicans.[18] Pelecaniformes also tolerate direct tracheal intubation for induction via uncuffed tube. At Texas A&M University Veterinary Medical Teaching Hospital, a group of pelicans recently presented for fishhook removal via endoscopy (manual removal had been unsuccessful). The analgesia, sedative, anesthetic protocol used for these pelicans was: midazolam 0.5 mg/kg IM, buprenorphine 0.05 mg/kg IM, intubate and maintain on isoflurane 5-L bag, 2 L/min flow rate oxygen. Other authors have advocated ketamine 3 mg/kg IM and butorphanol 1 mg/kg as preanesthetic drugs for these species.[5] The steroid anesthetic saffan (alfaxalone/ alfadolone) also has been used successfully for field anesthesia in a pelican.[19]

Non-infectious Diseases

Trauma

In my experience traumatic injury, followed by toxin exposure, are the 2 most common causes for presentations of pelicans for health care and rehabilitation. Humeral fractures (presumed secondary to plunge diving in brown pelicans, but also reported in whites as early as 1937),[20] as well as pouch laceration, and fishhook trauma occur most commonly. Constrictive (fishing line) or traumatic lesions occur to the extremities. Bumblefoot is a common disease of captivity in pelicans associated with repeated trauma to the foot. Hailstorms, powerline strikes (electrocution), freezing in lakes and gunshot have been reported as causes of mortality in free-living species.[21,22]

Toxicity

Ingestion of other fishing-related objects (sinkers, lead, zinc) and toxicity associated with these objects may occur. Toxicity caused by petrochemicals or red tide (domoic acid, brevetoxin) have been reported.[24] Clinical signs of pelicans affected by domoic acid poisoning include the following: slow side to side head motion, fine motor tremors, scratching at the pouch, and vomiting.[5] Blood urea nitrogen (BUN) and plasma uric acid (UA) may have increased concentrations be in affected birds, with males more frequently affected than females.[24] At necropsy, abnormalities may be limited to hemorrhage and necrosis of muscle. Pelicans are commonly affected by oil and have been documented to dive through it to feed; however, pelicans are fairly sturdy and adaptable patients in rehabilitation and can be expected to have a reasonable rehabilitation rate.

Nutritional diseases

Diseases of captive pelicans and are often associated with feeding frozen fish or vitamin supplementation. Hypovitaminosis E has been associated with steatitis in the American white pelican and myopathy in brown pelicans.[25] Death may occur suddenly after capture events or several days after the last handling event based on the authors experience. Post mortem may reveal pale skeletal muscles with linear streaking, while histopathologic findings may include degeneration and mineralization of skeletal muscles and myocardium as well as granulomatous steatitis. Hypervitaminosis E was associated with coagulopathy in pink backed pelicans based on the interference with vitamin K function.[26] Clotting abnormalities responded to vitamin K injections and correction of vitamin E concentrations in the diet.

Neoplasia

Neoplastic diseases are apparently rare in pelicans but skin melanoma of the throat pouch, squamous cell carcinoma of the foot, multicentric T-cell lymphoma, osteosarcoma with metastases, and xanthoma have been reported in pelicans.[27-32]

Infectious Diseases

Viruses

Viral infectious diseases that infect or affect pelicans include avian influenza, West Nile virus, Newcastle disease virus and pox virus. West Nile virus exposure has been detected serologically and via PCR from stable flies parasitizing American white pelicans (*Pelecanus erythrorhynchos*).[33,34] West Nile virus resulted in significant population decline in 2002-2003 of American white pelicans within the major northern plains breeding colonies for this species based on an increase in chick mortality from 4-40%.[35,36] In pelicans, clinical signs of West Nile virus infection include neurologic abnormalities and general lethargy caused by myocarditis and encephalomyelitis. Multiple influenza viruses have been isolated from pelican species from 2007-2009. Avian influenza virus H13N9 was isolated from hatchling American white pelican breeding colonies in Minnesota, USA in 2007-2008 and H3N6 has been isolated from a free-living white pelican in Zambia in 2009.[37,38] Vaccination with a commercially available poultry inactivated vaccine AIV H5N2 elicited no immune response in pelicans.[39] Pox viral-like inclusion particles have been found in a squamous cell carcinoma on the foot of a pink backed pelican.[30]

Bacterial and Fungal Diseases

Bacterial diseases of pelicans are multiple and often associated with food or environmental exposures. Ten of 50 American white pelicans sampled harbored *Vibrio* species of 5 different serovars.[40] While carriage rate did not appear greater than many other similar aquatic birds, pelicans may be capable of vectoring *Vibrio cholerae*. The sea bird tick *Cario* (*Ornithodoros*) *capensis* infests the nest of brown pelicans and is associated with nest abandonment. These ticks harbor *Coxiella* sp. *Borrelia lonestari*, as well as *Rickettsia felis* and 2 undescribed *Rickettsia* spp. which may be of zoonotic risk for humans that visit rookeries.[41] Type C botulism has caused over 44 die-offs and an estimated mortality of over 11,588 pelicans by 2003. Invasion of *Clostridium*-tolerant tilapia may actually reduce sustainability of pelican populations in the Salton Sea.[42] Fatal *Clostridium perfringens* infection via necrotic enteritis, presumed contracted from contaminated fish, occurred in a brown pelican.[43] Necrotic enteritis is caused by this gram-positive bacterium. Bacterial enteritis of pelicans has been associated with *E. coli*, *Proteus* spp., *Salmonella* spp., and *Campylobacter* spp.[5] *Erysipelothrix rhusiopathiae*,a g+ facultative intracellular bacterium commonly found in fish slime, is an uncommon cause of sea bird dies offs but has caused dies offs in brown pelicans.[44] Chlamydiosis has been reported as a cause of death in few pelicans.[22] Avian cholera, caused by *Pasteurella multocida*, has caused death in pelicans.[45] Aspergillosis uncommonly affects pelicans but has caused mortality in dalmation, and brown and white pelicans.[21,22]

Parasites

Pelicans are known to suffer a number of parasites and based on the author's experience and other reports, parasitic loads may be quite heavy; *Contracecum* spp. are some of the most commonly observed.[46-49] Pelican foraging style may affect parasite load. Mean intensities for *Contracaecum* spp. for free-living white pelicans were approximately 4 times higher than those reported for brown pelicans in Florida. As diving birds, brown pelicans prey on larger fish; whereas, white pelicans are bill dippers that feed on smaller fish as they swim.[47] General types of parasites recovered from pelicans are provided in Table 2, along with expected clinical signs, the organ(s) parasitized and recommended diagnostic methods. A checklist of pelican endoparasites is available.[50] Hepatitis, splenitis, and mortality based on systemic infection tetratrichomoniasis in an American white pelican with exposure to anseriform and galliform birds at the zoological institution may have led to cross-species infection and severe disease in this novel host. [51] The flagellate *Tetratrichomonas* is a common inhabitant of the intestinal tract of anserformes and galliformes that is rarely associated with disease in these species. The cosmopolitan host generalist sea bird tick *Ornithodoros capensis* carries five different viruses, including the zoonoses West Nile virus and soldado virus.[53-55] Rookery nest treatment with permethrin and piperonyl butoxide significantly reduced nest infestation.[56]

Diagnostics

Clinical pathology

Hematology and biochemistry values of multiple Pelican species have been reported. Pelicans are a heterophilic species, and as with other shore birds tend toward relatively high concentrations of eosinophils, possibly based on their high parasitic load in the free-living state.[57-61] Urine concentrations have been determined for the white pelican *P. erythrorhynchos*.[62] Measurements were made on the urine of normal birds with water and food available, followed by similar measurements on the same birds during water deprivation. Urine osmolality via freezing point depression, and sodium and potassium concentrations were obtained with a flame photometer (Table 3). The authors concluded that water reabsorption from the cloaca may occur based on the osmotic and electrolyte concentrations observed. Serology is also a useful clinical tool in pelicans for determination of avian influenza and West Nile virus exposure.[33,37]

Table 2. Select common parasites of pelicans.[a,b,c]

Class	Species	Organs affected	Clinical signs	Diagnostic(s)
Ascarid	*Contracaecum* spp.	Upper GI	Gastric ulcers/ Necrosis emaciation	Fecal flotation, Scrape of lesion
Trematode	*Phagicola longa, Mesostephanus appendiculatoides*	Small intestine	Ill thrift	Fecal Sedimentation
Trematode	*Riberoia ondatrae*	Proventriculus	??	Fecal sediment
Nematode	*Microtetrameres pelecani*	Proventriculus	Ill thrift	Fecal float, radiograph, endoscopy
Lice	*Piagetiella* spp.	Pouch	Hemorrhagic stomatitis	Visual
Tick	*Carios capensis* *Ornithodorus capensis*	Skin	Nest abandonment	Visual
Mite	*Phalacrodectes spp*	SC/mesenteric	??	??
Flukes	*Renicola* spp. *Schistosoma* spp.	Renal, mesenteric vasculature	Anorexia/ Emaciation	Fecal direct or sediment Swimmer's itch
Nematode	*Cyathostoma phenisci* *Syngamus* spp.	Lung	Cough/ Gurgling while breathing, increased resp rate	Fecal flotation or check of sputum
Schistosome	*Austrobilharzia variglandis*	Blood	Blood smear	Swimmers itch
Nematode	*Capillaria* spp.	GI	Ill-thrift, none	Fecal flotation
Nematode	*Eustrongylides* spp.	Proventriculus /coelom	ADR/ Coelomitis/ Sepsis	Fecal flotation, Necropsy

[a]Dyer WG, Jr EH, Mignucci-Giannoni AA, Jiménez-Marrero NM, et al. Helminth and arthropod parasites of the brown pelican, *Pelecanus occidentalis*, in Puerto Rico, with a compilation of all metazoan parasites reported from this host in the Western Hemisphere. *Avian Pathol.* 2002;31(5):441-448.
[b]Kinsella JM, Spalding MG, Forrester DJ. Parasitic helminths of the American white pelican, *Pelecanus erythrorhynchos*, from Florida, USA. *Comp Parasitol.* 2004;71(1):29-36.
[c]Weber M. Pelecaniformes In: Fowler ME, Miller RE, eds. *Zoo and Wild Animal Medicine.* 5th ed. Philadelphia, PA: Saunders;1999:118-121.

Table 3. White pelican urinalysis values (n = 9), mean ± SD (range).[a]

Condition	n	Total mOsm	Na, mEq/L	K, mEq/L
Normal	9	213 ± 52 (52-434)	35 ± 10 (7-78)	16 ± 4 (4-43)
Food only 34-38h	4	518 ± 67 (359-646)	47 ± 22 (10-100)	39 ± 4 (32-51)
NPO 23-24h	4	580 ± 98 (292-734)	18 ± 6 (8-33)	114 ± 26 (64-187)

[a]Calder WA, Bentley PJ. Urine concentrations of two carnivorous birds, the white pelican and the roadrunner. *Comp Biochem Physiol.* 1967;22(2):607-609.

Imaging

As relatively large birds, good positioning and technique for radiography provides excellent images for diagnosis in these species; however, computed tomography has been described as a useful adjunct for diagnosis of chronic shoulder arthritis in a white pelican.[63] Transesophageal echocardiography has also been shown feasible in brown pelicans.[64]

Surgery

Short-term success of fresh homologous penetrating keratoplasty (PK) of the cornea of a young adult female California brown pelican (*Pelecanus occidentalis*) for the treatment of vision-threatening corneal scarring and granulation tissue has been reported. The animal died based on subacute necrotizing myositis at 6 weeks postoperatively.[65] Surgical repair of the gular pouch has also been reported for the brown pelican. This surgery requires a meticulous technique and approach based on the tendinoelastic pouch tissues and tenuous vascularization of the pouch.[66] After establishing that both subramal and ventral gular blood supplies are intact, the pouch is closed in separate layers with a simple interrupted pattern to assure that vasculature remains uncompromised.[66]

Orthopedic injuries are common in the pelican species. Of 233 pelicans in taken at the Wildlife Center of Texas from 2012-2017, 85 were wing injuries, including only 4 patagial injuries, and only 8 were leg and foot injuries (author's unpublished data). Surgical repair of humeral fracture and radius ulna fractures via amputation have long ago been reported. The tolerance of pelicans to minimal care, major injury, and surgical techniques far from modified for avian species remains remarkable.[19,20] A specific technique for pinioning of pelicans has not been reported. Surgical approaches to the limbs of the brown pelican are provided in Table 4.[10]

Table 4. Surgical approaches for orthopedic surgery of the appendicular skeleton in the brown pelican.[a]

Bone	Recommended Approach	Comments
Humeral head or scapula	Dorsal	Avoid division of the pectoral mm necessary for ventral approach to the shoulder joint.
Humerus, distal	Dorsal or ventral	Pneumatized but easily exposed based on foreshortened muscle bellies of the upper wing muscles, = absence of muscle bellies around the distal two-thirds of the humerus.
Radius/ulna	Dorsal or ventral	Avoid structures in interosseous space.
Femur	Anterolateral, separate muscularis iliofemoralis externus and iliotibialis lateralis	Heavy musculature obstruction. Nerve and vascular supply courses medially.
Tibiotarsus, distal	Medial: elevate, retract muscularis gastrocnemius pars medialis posteriorly	
Tibiotarsus, proximal	Medial: separate muscularis gastrocnemius pars medialis and pars intermedia, retract pars medialis anteriorly and pars intermedia and muscularis flexor perforans et preforatus digiti III caudally	
Tarsometatarsus	Lateral	Avoid medial vasculature and abductor digitorum II and muscularis extensor hallucis longus.

[a]McSweeney T, Stoskopf M. Selected anatomical features of the neck and gular sac of the brown pelican (*Pelecanus occidentalis*). *J Zoo Anim Med.* 1988;19(3):116-121.

Treatments

To the author's knowledge, no pharmacokinetic data or drug approvals exist specific to pelican species, make the following recommendations and most information available anecdotal and/or off-label.

Non-steroidal anti-inflammatories

Meloxicam, and carprofen have been used in pelicans. Necropsy of birds that received nonsteroidal anti-inflammatory medication showed no organ lesions associated with toxicity.[27,32]

Anthelmintics

Anthelminitics are used commonly in pelicans based on high parasitic burdens; however, fatal enteritis and bone marrow damage in pink-backed pelicans (*Pelecanus rufescens*) after the administration of fenbendazole has been reported. Affected birds were found dead or exhibited respiratory distress and died shortly after presentation.[5] Presumptive fenbendazole (50 mg/kg q24h for 5 days) toxicity in an American white pelican has been reported. The pelican became lethargic and inappetent on day 3 of fenbendazole treatment and was found dead on day 7. Necropsy confirmed intestinal crypt cell necrosis, stomatitis, and splenic lymphoid depletion consistent with fenbendazole toxicity.[68] Brown pelicans were successfully treated for 3 parasites (*Contracaecum multipapillatum*, *Mesostephanus appendiculatoides*, *Phagicola longus*) with the relatively low fenbendazole dose of 22 mg/kg.[69] The author prefers the use of ivermectin, pyrantel pamoate and praziquantel in these species based on the high likelihood of flukes and lung worms. Multiple doses are often required during rehabilitation. Supportive care and nutrition as well as consideration of vitamin supplementation and anti-inflammatories are also recommended in cases of high parasitic load.

Antioxidants

Use of antioxidants and kidney sparing anti-inflammatories deserve consideration in these species which are prone to capture myopathy and toxin (particularly oil) exposure.[70,71] The author prefers the use of BO-SE (Intervet/Merck Animal Health, Merck Animal Health, Madison, NJ, 07940). Each milliliter contains 2.19 mg sodium selenite (equivalent to 1 mg selenium), 50 mg (68 USP units) vitamin E (as d-alpha tocopheryl acetate), at a dose rate of 1 ml/pelican as a prophylaxis. Vitamin EAD Intervet/Merck Animal Health injectable A+D tocopherol (vitamins A, D3, and E) contains 100,000 IU vitamin A, 10,000 IU vitamin D3 and 300 IU vitamin E (as d-alpha-tocopherol, a natural source of vitamin E) per milliliter; however, this formulation contains a relatively high amount of vitamin D, which most pelicans do not need and a relatively low amount of vitamin E. Supplementation of vitamin E orally is continued after the injection; however beware of over supplementation of vitamin E (see Nutritional Diseases).

Antibiotics

Many antibiotics have been used on these species without known untoward effects. For short-term broad spectrum perioperative coverage the author prefers ceftiofur while for slightly longer-term broad spectrum coverage the author has used ceftiofur crystalline free acetate formulations. Antibiotic perfusion therapy as an adjunct treatment for digital lesions in pelicans has been described.[72] Many other antibiotics have been used in these species without known untoward effects.

Antifungals

As fungal diseases are not particularly common in these species specific doses for antifungals are not available. The use of intraconazole, enilconazole or terbinafine would be reasonable options.

References

1. De Grazia M, Quilligan M, Stallybrass P. *Subject and Object in Renaissance Culture.* New York, NY: Cambridge University Press;1996.

2. Graham VE. The pelican as image and symbol. *Revue de Littérature Comparée.* 1962;36(2):235.

3. Evans EP. *Animal Symbolism in Ecclesiastical Architecture.* London, UK: Heinemann; 1896.

4. Johnsgard PA. *Cormorants, Darters, and Pelicans of the World.* Washington, DC: Smithsonian Inst Press;1993.

5. Redrobe S. Pelecaniformes (pelicans, tropicbirds, cormorants, frigatebirds, anhingas, gannets). In: Miller E, Fowler ME, eds. *Fowler's Zoo and Wild Animal Medicine.* Vol 8. St. Louis, MO: Elsevier Saunders;2015:96-99.

6. Owen R. Notes on the anatomy of the red-backed pelican of Dr. Lantham, *Pelecanus rufescens. Proc Zool Soc Lond I.* 1835;9-12.

7. Kuroda N. Fragmental notes on avian morpho-anatomy: Myological sketches of a brown pelican *Pelecanus occidentalis. J Yamashina Inst Ornithol.* 2001;32:80-85.

8. McSweeney T, Stoskopf M. Selected anatomical features of the neck and gular sac of the brown pelican (*Pelecanus occidentalis*). *J Zoo Anim Med.* 1988;19(3):116-121.

9. McSweeney T, Stoskopf, M. Selected features of the abdominal and thoracic anatomy of the brown pelican (*Pelecanus occidentalis*). *J Zoo Wildl Med.* 1989;20(2):184-190.

10. McSweeney T, Stoskopf M. Aspects of the limb anatomy of the brown pelican (*Pelecanus occidentalis*). *J Zoo Wildl Med.* 1989;20(2):191-198.

11. McSweeney T. The topographical anatomy of the brown pelican, *Pelecanus occidentalis*. Dissertation;1984.

12. Rosser BW, Waldbillig DM, Wick M, Bandman E. Muscle fiber types in the pectoralis of the white pelican, a soaring bird. *Acta Zoologica.* 1994;75(4):329-36.

13. Sathyamoorthy OR, Thirumurugan R, Kumar KS, Jayathangaraj MG. Gross anatomical studies on the sternum and clavicle of spot-billed pelican (*Pelecanus philippensis*). *Tamilnadu J Vet Anim Sci.* 2012;8(3):166-70.

14. Runciman S, Seymour RS, Baudinette RV, Pearson JT. An allometric study of lung morphology during development in the Australian pelican, *Pelicanus conspicillatus*, from embryo to adult. *J Anat.* 2005;207(4): 365-80.

15. Yoldaş A, Dayan MO. A macro-anatomic study of white pelican (*Pelecanus onocrotalus*) hearts. *Eur J Vet Sci.* 2012;28(3):177-81.

16. Knouff RA, Hartman FA. A microscopic study of the adrenal of the brown pelican. The *Anat Rec.* 1951;109(2):161-87.

17. Heard D. Birds: Miscellaneous. In: West G, Heard D, Caulkett N, eds. *Zoo and Wildlife Immobilization and Anesthesia.* 2nd ed. Ames, IA: Wiley Blackwell;2014:507-513.

18. Reynolds WT. Unusual anaesthetic complication in a pelican. Vet Rec. 1983;113(9):204.

19. Haigh J. The use of Ct1341 for surgical anesthesia in a pelican. *J Zoo Anim Med.* 1976;7(3):39-39.

20. Oehlenschlaeger EA. Orthopedic surgery on a white pelican. *Auk.* 1937;520-523.

21. Rocke T, Converse K, Meteyer C, McLean B. The impact of disease in the American white pelican in North America. *Waterbirds.* 2005;28:87-94.

22. Pyrovetsi M, Papazahariadou M. Mortality factors of Dalmatian pelicans (*Pelecanus crispus*) wintering in Macedonia, Greece. *Environ Conserv.* 1995;22(04):345-51.

23. Crivelli AJ, Jerrentrup H, Mitchev T. Electric power lines: a cause of mortality in *Pelecanus crispus* Bruch, a world endangered bird species, in Porto-Lago, Greece. *Col Waterbirds.* 1988;11(2):301-305.

24. Work TM, Barr B, Beale AM,et al. Epidemiology of domoic acid poisoning in brown pelicans (*Pelecanus occidentalis*) and Brandt's cormorants (*Phalacrocorax penicillatus*) in California. *J Zoo Wildl Med.* 1993;1:54-62.

25. Campbell G, Montali RJ. Myodegeneration in captive brown pelicans attributed to vitamin E deficiency. *J Zoo Anim Med.* 1980;11(2):35-40.

26. Nichols DK, Wolff MJ, Phillips Jr LG, Montali RJ. Coagulopathy in pink-backed pelicans (*Pelecanus rufescens*) associated with hypervitaminosis E. *J Zoo Wildl Med.* 1989;57-61.

27. Di Girolamo N, Lane EP, Reyers F, Gardner BR. Subcutaneous xanthomatosis in a great white pelican (*Pelecanus onocrotalus*). *J Zoo Wildl Med.* 2014;45(1):153-156.

28. Molenaar FM, Strike T, Dodds J, et al. Metastasizing osteosarcoma in an eastern white pelican (*Pelecanus onocrotalus*). *J Zoo Aquar Res.* 2015;3(3):77-79.

29. Pauly A, Szentiks CA, Wibbelt G, Speck S. Epithelial neoplasm at the upper bill—a rare malign tumor in a pelican. *Zool Garten.* 2009;78:141-148.

30. Pesaro S, Biancani B, Fabbrizi G, Rossi G. Squamous cell carcinoma with presence of poxvirus-like inclusions in the foot of a pink-backed pelican (*Pelecanus rufescens*). Avian Pathol. 2009;38(3):229-231.

31. Redtenbacher I, Hebel C, Schmidt V, et al. Multicentric malignant lymphoma of idiopathic etiology in a captive pink-backed pelican (*Pelecanus rufescens*). Available at: http://awwp.alwabra.com/wp-content/uploads/2014/04/113-Multicentric-malignant-lymphoma-of-idiopathic-etiology-in-a-captive-Pink-back-Pelican-Pelecanus-rufes.pdf Accessed May 30, 2017.

32. Schmidt V, Philipp HC, Thielebein J, et al. Malignant lymphoma of T-cell origin in a Humboldt penguin (*Spheniscus humboldti*) and a pink-backed pelican (*Pelecanus rufescens*). *J Avian Med Surg.* 2012;26(2): 101-106.

33. Bureau E, Murri S, Kane Y, et al. Serological evidence of West Nile virus in the great white pelican (*Pelecanus onocrotalus*). Vet Rec. 2008;163(2):52-53.

34. Johnson GP. Surveillance for West Nile virus in American white pelicans, Montana, USA. *Emerg Infect Dis.* 2010;16(3):406-411.

35. Sovada MA, Pietz PJ, Hofmeister EK, Bartos AJ. West Nile virus in American white pelican chicks: transmission, immunity, and survival. *Am J Trop Med Hyg.* 2013;88(6):1152-1158.

36. Sovada MA, Pietz PJ, Converse KA, et al. Impact of West Nile virus and other mortality factors on American white pelicans at breeding colonies in the northern plains of North America. *Biol Conserv.* 2008;141(4): 1021-1031.

37. Lebarbenchon C, Sreevatsan S, Ramakrishnan MA, et al. Influenza A viruses in American white pelican (*Pelecanus erythrorhynchos*). *J Wildl Dis.* 2010;46(4):1284-1289.

38. Simulundu E, Mweene AS, Tomabechi D,et al. Characterization of H3N6 avian influenza virus isolated from a wild white pelican in Zambia. *Arch Virol.* 2009;154(9):1517-1522.

39. Oh S, Martelli P, Hock OS, et al. Field study on the use of inactivated H5N2 vaccine in avian species. *Vet Rec.* 2005;157(10):299-300.

40. Ogg JE, Ryder RA, Smith HL. Isolation of *Vibrio cholerae* from aquatic birds in Colorado and Utah. *Appl Environ Microbiol.* 1989;55(1):95-99.

41. Reeves WK, Loftis AD, Sanders F,et al;. *Borrelia*, *Coxiella*, and *Rickettsia* in *Carios capensis* (Acari: Argasidae) from a brown pelican (*Pelecanus occidentalis*) rookery in South Carolina, USA. *Exp Appl Acarol.* 2006;39(3-4):321-329.

42. Chattopadhyay J, Bairagi N. Pelicans at risk in Salton sea—an eco-epidemiological model. *Ecol Model.* 2001;136(2):103-112.

43. Ankerberg Jr CW. Pelican deaths in the vicinity of a sewage lift station: a bacteriological investigation. *Microbios Lett.* 1984;26(101):33-42.

44. Franson JC, Galbreath EJ, Wiemeyer SN, Abell JM. *Erysipelothrix rhusiopathiae* infection in a captive bald eagle (*Haliaeetus leucocephalus*). *J Zoo Wildl Med.* 1994;446-448.

45. Crawford RJ, Cooper J, Dyer BM. Conservation of an increasing population of great white pelican, *Pelecanus onocrotalus*, in South Africa's Western Cape. *South Afr J Mar Sci.* 1995;15(1):33-42.

46. Dyer WG, Jr EH, Mignucci-Giannoni AA, Jiménez-Marrero NM, et al. Helminth and arthropod parasites of the brown pelican, *Pelecanus occidentalis*, in Puerto Rico, with a compilation of all metazoan parasites reported from this host in the Western Hemisphere. *Avian Pathol.* 2002;31(5):441-448.

47. Kinsella JM, Spalding MG, Forrester DJ. Parasitic helminths of the American white pelican, *Pelecanus erythrorhynchos*, from Florida, USA. *Comp Parasitol.* 2004;71(1):29-36.

48. Liu SK, Edward AG. Gastric ulcers associated with *Contracaecum* spp. (Nematoda: Ascaroidea) in a steller sea lion and a white pelican. *J Wildl Dis.* 1971;7(4):266-271.

49. Sreedevi C, Prasuna K, Lavanya K, Latha KK. *Contracaecum rudolphii* Hartwich, 1964 (Nematoda: Anisakidae) in a wild spot-billed pelican (*Pelecanus philippensis*): a case report. *J Parasit Dis.* 2017:1-4.

50. Weber M. Pelecaniformes In: Fowler ME, Miller RE, eds. *Zoo and Wild Animal Medicine.* 5th ed. Philadelphia, PA: Saunders;1999:118-121.

51. Dronen NO, Blend CK, Anderson CK. Endohelminths from the brown pelican, *Pelecanus occidentalis*, and the American white pelican, *Pelecanus erythrorhynchus*, from Galveston Bay, Texas, USA, and checklist of pelican parasites. *Comp Parasitol.* 2003;70(2):140-154.

52. Burns RE, Braun J, Armién AG, et al. Hepatitis and splenitis due to systemic tetratrichomoniasis in an American white pelican (*Pelecanus erythrorhynchos*). *J Vet Diagn Invest.* 2013;25(4):511-514.

53. Clifford CM. Tick-borne viruses of seabirds. In: Kurstak E, ed. *Arctic and Tropical Arboviruses.* New York, NY: Academic Press;1979:83-100.

54. Hutcheson HJ, Gorham CH, Machain-Williams C, et al. Experimental transmission of West Nile virus (Flaviviridae: Flavivirus) by *Carios capensis* ticks from North America. *Vector Borne Zoonotic Dis.* 2005;5(3):293-295.

55. de la Fuente J, Estrada-Pena A, Venzal JM, et al. Overview: Ticks as vectors of pathogens that cause disease in humans and animals. *Front Biosci.* 2008;13(13):6938-6946.

56. Norcross NL, Bolen EG. Effectiveness of nest treatments on tick infestations in the Eastern brown pelican. *Wilson Bull.* 2002;114(1):73-78.

57. Puerta ML, Del Campo AG, Huecas V, Abelenda M. Hematology and blood chemistry of the white pelican (*Pelecanus onocrotalus*). *Comp Biochem Physiol A.* 1991;98(3):393-394.

58. Ferguson LM, Norton TM, Cray C, et al. Health assessments of brown pelican (*Pelecanus occidentalis*) nestlings from colonies in South Carolina and Georgia, USA. *J Zoo Wildl Med.* 2014;45(4):802-812.

59. Zaias J, Fox WP, Cray C, Altman NH. Hematologic, plasma protein, and biochemical profiles of brown pelicans (*Pelecanus occidentalis*). *Am J Vet Res.* 2000;61(7):771-774.

60. Wolf SH, Schreiber RW, Kahana L, Torres JJ. 1985. Seasonal, sexual and age-related variation in the blood composition of the brown pelican (*Pelecanus occidentalis*). *Comp Biochem Physiol A.* 1985(82);837-846.

61. Shmueli M, Izhaki I, Zinder O, Arad Z. The physiological state of captive and migrating great white pelicans (*Pelecanus onocrotalus*) revealed by their blood chemistry. *Comp Biochem Physiol A.* 2000;125(1):25-32.

62. Calder WA, Bentley PJ. Urine concentrations of two carnivorous birds, the white pelican and the roadrunner. *Comp Biochem Physiol.* 1967;22(2):607-609.

63. Whitehead MC, Parker DL. The use of computed tomography to diagnose chronic shoulder arthritis in an American white pelican (*Pelecanus erythrorhynchos*). *Can Vet J.* 2015;56(3):261.

64. Beaufrere H, Pariaut R, Nevarez JG, Tully TN. Feasibility of transesophageal echocardiography in birds without cardiac disease. *J Am Vet Med Assoc.* 2010;236(5):540-547.

65. Lynch GL, Scagliotti RH, Hoffman A, Dubielzig RR. Penetrating keratoplasty in a California brown pelican. *Vet Ophthalmol.* 2007;10(4):254-261.

66. Williams TD, Gawlowski PQ, Strickland DM,et al. Surgical repair of the gular sac of the brown pelican (*Pelecanus occidentalis*). *J Zoo Anim Med.* 1988;122-125.

67. Anderson K, Garner MM, Reed HH, et al. Hemorrhagic diathesis in avian species following intramuscular administration of polysulfated glycosaminoglycan. *J Zoo Wildl Med.* 2013;44(1):93-99.

68. Lindemann DM, Eshar D, Nietfeld JC, Kim IJ. Suspected fenbendazole toxicity in an american white pelican (*Pelecanus erythrorhynchos*). *J Zoo Wildl Med.* 2016;47(2):681-685.

69. Grimes J, Suto B, Greve JH, Albers HF. Effect of selected anthelmintics on three common helminths in the brown pelican (*Pelecanus occidentalis*). *J Wildl Dis.* 1989;25(1):139-142.

70. Giri DK, Miller DL, Thompson LJ, et al. Superoxide dismutase expression and oxidative damage in a case of myopathy in brown pelicans (*Pelecanus occidentalis*). *J Vet Diagn Invest.* 2007;19(3):301-304.

71. O'Brien MF, Lee R, Cromie R, Brown MJ. Assessment of the rates of injury and mortality in waterfowl captured with five methods of capture and techniques for minimizing risks. *J Wildl Dis.* 2016;52(2s):S86-95.

72. Fiorello CV. Intravenous regional antibiotic perfusion therapy as an adjunctive treatment for digital lesions in seabirds. *J Zoo Wildl Med.* 2017;48(1):189-195.

Avian Acute Phase Response Diagnostics: Past, Present, Future

Carolyn Cray, PhD

Session #2064

Affiliation: From the Division of Comparative Pathology, Department of Pathology & Laboratory Medicine, University of Miami Miller School of Medicine, Miami, FL 33136, USA.

Abstract: Viewing and tracking the acute phase response via protein electrophoresis (EPH) has been established to have sensitivity in health assessments and prognostic value in avian patients with diverse clinical conditions. EPH provides the only valid quantitation of albumin, an important negative acute phase protein, while also acting as a snapshot of changes in globulin fractions which are reflective of ongoing inflammatory and humoral immune responses. In recent years, a growing body of literature has demonstrated the application of assays for specific acute phase proteins in companion and large animals as well as non-domesticated mammals. As the sensitivity of these measures often rival EPH, the potential applications of these type of assays as well as other biomarker testing is of growing interest in avian species.

Introduction

The acute phase response (APR) is a systemic process that is an essential part of the innate immune response.[1-3] Incited by stimuli including cytokines produced locally in response to trauma or infection, the liver produces perhaps well over 200 acute phase proteins (APP).[1] Each has a known biological function including enhancing opsonization or chemotaxis and binding complement or hemoglobin.[2] Recent studies have even shown interactions between APP and the cell mediated immune response. The importance of the role of these proteins is reflected in the high conservation of expression from species ranging from mollusks to reptiles to humans.[2]

Applications of Acute Phase Protein Measurement

APP are categorized broadly as negative or positive.[2-4] The primary negative APP is albumin; transferrin is also considered a negative APP in mammals. Positive APP are subcategorized by their level of increased expression: major (10- to 1000-fold), moderate (2- to 10-fold), and minor (< 2-fold). Major APP are commonly the focus of studies and clinical applications as they are present in negligible levels in normal animals. The increase is rapid within 16-24 hours of stimulation. The proteins have a short half-life and, with suitable negative feedback seen with a positive response to treatment, there is a rapid decrease. Mild and moderate APP are present in detectable levels in normal animals as is seen with the APP fibrinogen. Increases are often delayed with peak levels noted 4-6 days after insult; elevated levels persist after resolution.

Considerable studies have been conducted in mammals comparing and contrasting traditional measures of inflammation including complete blood count, fibrinogen, and A/G ratio. When inflammatory agents like turpentine oil or complete Freund's adjuvant are injected into laboratory animals, there is often a mild but transient increase in total WBC count but the significant sustained increase is not observed until day 4 to 7. Albumin will decrease but the lowest level is not observable until day 5. Major APP, however, peak by 48 hours. As these and other measures have different timelines of expression versus APP, it is not unexpected that the correlation between these types of results are poor. For example, the correlation between WBC count and SAA was weak (r=0.11) in horses with various clinical presentations.[5] In this same study, SAA had the best diagnostic accuracy (vs. total WBC, A/G ratio, and fibrinogen) at 75%.

Elevated APP levels are not diagnostic of any one disease but are reflective of a systemic inflammatory process. Much as EPH has been embraced for use especially in routine health assessments of avian, exotic, and wildlife species, APP measurements can provide the same or better value. In addition, the rapid change of the major APP provides improved accuracy for the detection of inflammation while also proving to have key prognostic value given their potential for a rapid decrease with a positive response to treatment.

Methods of Acute Phase Protein Measurement

The measurement of globulins can be done accurately for many mammalian species via chemistry analysis where total protein and albumin are quantitated via separate methods and the globulins (and A/G ratio) are calculated; however, such measurements are not valid in avian and reptile species as many globulins cross react with the bromcresol green reagent which is used to determine albumin concentration.[6,7] Automated methods for fibrinogen quantitation also require validation with avian species as many are antibody based and have poor cross reactivity. The heat precipitation method should be a primary tool consider for routine use.[8]

Protein electrophoresis

The use of EPH in avian medicine has been readily applied for more than 20 years.[9] It has been demonstrated to provide reproducible results and have application as an adjunct tool in the diagnosis of infectious disease.[10-12] There is variability by diagnostic laboratory which is a function of equipment variation and the technician's application of protein fraction delimits.[13]

In EPH, total protein is determined by refractometry or biuret methods. The sample is applied to a thin layer agarose gel and the proteins are separated by charge. After staining, the fractioned sample can be evaluated by densitometry or scanning where fraction delimits can be placed to quantitate each fraction. The percent of each fraction is then multiplied by the total protein to obtain the absolute value. The A/G ratio is calculated by dividing the sum of prealbumin and albumin by the sum of the globulin fractions.[10] Hemolysis and lipemia can both cause artifactual changes in results.[10]

Immuno- and colorimetric assays

Several reagents for specific APP have been validated for use in mammals and some avian and reptile species.[2,14] These are antibody based methods including immunoturbidimetric assays and ELISA. The former type of assay can be run on automated machinery which enhances accuracy and decreases turnaround time. For haptoglobin quantitation, a colorimetric assay is available which can be used on any species.

Acute Phase Proteins in Avian Species

EPH provides a good sensitivity for the assessment of ongoing APR in avian species. While providing key information on globulins, it also provides a valid quantitation of albumin.[6] In a retrospective review of CBC versus EPH data from African grey parrot samples submitted to the Avian & Wildlife Laboratory, a marginal correlation was observed between elevated total WBC and decreased A/G ratio (Cray, et al, unpublished data, May 2016).

To this point, in 49% of cases with a decreased A/G ratio, an elevated WBC count was found; however, of note, 43% of cases had a WBC count within normal limits when a decreased A/G ratio was present. As with APP information in mammals, this data also illustrates the different sensitivity for inflammatory processes for these two methods. It also supports the use of EPH in avian medicine.

For the most part, APP studies in avian species have been conducted primarily in chickens. In various experimental models, serum amyloid A (SAA), transferrin (TN), PIT-54 (the avian analog of haptoglobin), and alpha -1 acid glycoprotein (AGP) have been documented to increase.[15,16] Stress related APP changes were also observed in chickens under high stocking density and it including changes in AGP, ceruloplasmin and TN.[17] These studies indicate TN is a positive APP in birds which contrasts that observed in mammals.

While our laboratory experience with currently available APP reagents has found limited cross reactivity with avian species, a commercial human reagent has been found to quantitate SAA in falcons. Significant increases have been observed in falcons with aspergillosis indicating that this is a major APP in this species.[18,19] In one study, significant changes in protein electrophoretic values were also concurrently present.[19]

Future of Acute Phase Protein Testing

Methods including mass spectrometry are allowing for the rapid expansion of the field of proteomics so more possible biomarkers including acute phase proteins will have the opportunity for study for applications to veterinary medicine. One such spectroscopy study of samples from falcons with aspergillosis revealed that 3-hydroxybutytrate and LDL are significantly altered versus normal falcons.[20] Recent preliminary work in our laboratory has successfully employed 3-hydroxybutyrate testing in a study of aspergillosis in African penguins (Cray, et al, unpublished data, April 2017).

In addition, point-of-care testing is becoming available for CRP and SAA in traditional species. As antibody differences are present vs. the validated tests at the reference laboratory level, each POC assay needs to be independently validated for each species. While laborious, these types of assays have the potential to greatly aid in the immediate assessment of clinically ill patients as well as allow APP testing to become part of routine blood work.

References

1. Kaneko JJ. Serum proteins and the dysproteinemias. In: Kaneko JJ, Harvey JW, Bruss ML, eds. *Clinical Biochemisty of Domestic Animals*. 5th ed. San Diego, CA: Academic Press. 1997;117-138.

2. Cray C. Acute phase proteins in animals. *Prog Mol Biol Transl Sci. 2011; 105:113-150.*

3. Eckersall PD. Proteins, proteomics, and the dysproteinemias. In: Kaneko JJ, Harvey JW, Bruss ML, eds. *Clinical Biochemistry of Domestic Animals*. 6th ed. San Diego, CA: Academic Press. 2008;117-155.

4. Cray C, Zaias J, Altman NH. Acute phase response in animals: a review. *Comp Med*. 2009;59:517-526.

5. Belgrave R, Dickey M, Arheart KL, Cray C. Assessment of serum amyloid A (SAA) testing and its clinical application in a specialized equine practice. *J Am Vet Med Assoc*. 2013;243(1):113-119.

6. Cray C, Wack A, Arheart KL. Invalidity of albumin measurement of specimens from clinically ill birds by bromcresol green methodology. *J Avian Med Surg.* 2011; 25:14-22.

7. Muller K, Brunnberg L. Determination of plasma albumin concentration in healthy and diseased turtles: a comparison of protein electrophoresis and the bromcresol green dye-binding method. *Vet Clin Pathol.* 2010;39(1):79-82.

8. Millar HR, Simpson SG, Stalker AL. An evaluation of the heat precipitation method for plasma fibrinogen estimation. *J Clin Pathol.* 1971;24:827-830.

9. Cray C, Tatum L. Application of protein electrophoresis in avian diagnostic testing. *J Av Med Surg.* 1998;12:4-10.

10. Cray C, Rodriguez M, Zaias J. Protein electrophoresis of psittacine plasma. *Vet Clin Pathol.* 2007;36(1): 67-72.

11. Cray C, Zieleienski-Roberts K, Bonda M, Stevenson R, Ness R, Clubb S, Marsh A. Antemortem diagnosis of sarcocystosis in psittacine birds. *J Av Med Surg.* 2005;19(3):208-215.

12. Cray C, Watson T, Rodriguez M, Arheart K. Application of galactomannan analysis and protein electrophoresis in the diagnosis of aspergillosis in avian species. *J Zoo Wildl Med.* 2009;40(1):64-70.

13. Cray C, Wack A, Arheart KL. Invalidity of albumin measurement of specimens from clinically ill birds by bromcresol green methodology. *J Avian Med Surg.* 2011;25(1):14-22.

14. Cray C. Biomarkers of inflammation in exotic pets. *J Exot Pet Med.* 2013;22:245-250.

15. O'Reilly EL, Eckersall PD. Acute phase proteins: a review of their function, behaviour, and measurements in chickens. *World Poult Sci J.* 2014;70:27-44.

16. Chamanza R, van Veen L, Tivapasi MT, et al. Acute phase proteins in the domestic fowl. *World Poult Sci J.* 1999;55:61-71.

17. Najafi P, Zulkifli I, Jajuli NA, et al.. Environmental temperature and stocking density effects on acute phase proteins, heat shock protein 70, circulating corticosterone and performance in broiler chickens. *Int J Biometereol.* 2015;59(11):1577-1583.

18. Caliendo V, McKinney P, Bailey T, et al. Serum amyloid A as an indicator of health status in falcons. *J Avian Med Surg.* 2013;27:83-89.

19. Fischer D, Van Waeyenberghe L, Cray C, et al. Comparison of diagnostic tools for the detection of aspergillosis in blood samples of experimentally infected falcons. *Avian Dis.* 2014;58:587-598.

20. Pappalardo L, Hoijemberg PA, Pelczer I, et al. NMR-Metabolomics study on falcons affected by aspergillosis. *Curr Metabolomics.* 2016;2:155-161.

Herbals and Nutraceuticals in Avian Practice: Liver, Cardiac, and Integument Applications

Cathy A. Johnson-Delaney, DVM, and
Susan E. Orosz, PhD, DVM, DABVP (Avian Practice), DipECZM (Avian)

Session #2070

Affilation: From Avian and Exotic Consulting, 13813 65th Ave W #7, Edmonds, WA 98026, USA (Johnson-Delaney) and Bird and Exotic Pet Wellness Center, 5166 Monroe St, Suite 305, Toledo, OH 43623, USA (Orosz).

Abstract: Use of supplements to the diet are well-accepted by most pet owners despite the lack of controlled studies of efficacy. Practitioners are being introduced to products developed specifically for dogs, cats, horses, and other companion animals, although few of these have had scientific studies proving efficacy. The term "supplement" includes nutraceuticals, botanicals and herbals. In the United States, the Food and Drug Administration does not regulate this dietary industry, so potency and label claims may not be accurate. Formulations for pet birds are being used by practitioners adapted from studies done in poultry or small animals, or from anecdotal effects on humans. Some of the most commonly used products for liver, cardiac and skin health include omega-3 and omega-6 fatty acids, L-carnitine, milk thistle, taurine, coenzyme Q, ginseng, quercetin, and dandelion. Formulations used by the authors have seemed to augment traditional therapies. These are well-accepted by both owners and the birds. While clinical improvement using these may be subjective, they do no harm and may be efficacious as therapy.

Introduction

The use of complementary, "alternative" or integrative medicine is becoming more widespread among both the veterinary and the human medical community.[1-3] Some of the information in this presentation has been collected and updated from previous papers on this subject.[2-4] While nutraceuticals, botanicals, or herbal products are essentially chemicals, they are not treated like pharmaceuticals by the Food and Drug Administration in the United States. Because of this, products may or may not live up to the guaranteed analysis on the label, may contain unwanted contaminants, and may make label claims that are not supported by science. Quality control for manufacturing is essentially managed by the manufacturer and it may result in inconsistent products. Many of these are all or partially manufactured overseas, and tracing country of origin of an ingredient may be impossible. Many practitioners and owners will use these products as they are seen to be "natural," "organic," and "not really medicine," thereby reinforcing the idea that they are safer than regular medications, will do no harm and *are "good for you." Ideally, information on efficacy, dosage and toxicity of a pertinent chemical would* be available from an actual study. Unfortunately, this is rarely the case. Extrapolation from the literature and other species is difficult in that the base chemicals may be listed or compounded for a different absorption or metabolism, plant Latin names may vary, and controlled studies have not been done. The practitioner should research labels and products available. Consumer Labs (www.consumerlabs.com) membership is extremely useful as their on-line information on nutraceuticals and herbals list brands that have been evaluated for potency and guaranteed analysis. The authors list only the chemical name in this presentation as all are available from numerous manufacturers and sources.

Classification of the different nutraceuticals and herbals make researching an exact clinical effect difficult. For purposes of this presentation, the listing of RIA International (www.riausa.com) is useful. This company is a supplier of chemicals including intermediate chemical products for manufacturing in addition to nutraceuticals and botanicals. While the term "botanical" is used by some for the flower formulas, in the broader sense "botanical" is interchangeable with "herbal." Chinese herbal medicine is not being included in this presentation.

Nutraceuticals

Omega fatty acids receive a lot of attention and investigation. Various ratios of omega-3 to omega-6 as well as omega-9 are found from natural sources such as flax seed, chia seed, pumpkin seed, coconut and red palm oil, and fish or krill oil. Companion birds appear to utilize plant-based sources, although studies have not been done to determine optimal utilization or ratio (3:6:9). The omega-3 family of fatty acids includes linolenic acid and its products. The omega-6 fatty acids include linoleic acid and its products. The balance between these two fatty acid families appears to be important for clinical effects. It appears that both types are required and in proper balance. Higher levels of omega-3 to omega-6 appear to have anti-inflammatory activity whereas if the ratio of omega-6 to omega-3 is higher, it appears to promote inflammation. The benefit of using a balanced product such as Optomega (USANA Health Sciences, Salt Lake City, UT, USA) is that it has a high proportion of omega-3 to balance out the high level of omega-6 typically found in high fat diets. They are considered necessary for cardiovascular function and in proper nerve transmission in addition to the immune-modulating effects.[3] Lipid metabolism in birds differs from that in mammals, atherosclerotic plaques resemble those of mammals. Feeding polyunsaturated fatty acids decreased the prevalence of atherosclerosis. It appeared that the balance of omega-3 to omega-6 was important as well in decreasing the amount of body fat, reducing some degree of insulin resistance and having some boosting of the immune system. Further studies may determine optimum levels in the different bird species.[5,6] A study has been published looking at the effects of dietary omega-3 fatty acids on red blood cell lipid composition and plasma metabolites in cockatiels. The study compared flax oil to fish oil and concluded that fish oil may be beneficial as it caused greater reductions in cholesterol and triacylglyderides.[7] Additional effects in skin health and other disease as are being studied in dogs have not been studied in pet bird species, although many studies have been conducted in chickens.[8]

Amino acids may also be considered nutraceuticals and many have been associated with clinical effects when dosed as supplementation. Although formulated diets may contain maintenance amounts of essential amino acids, species, life stages and disease may necessitate supplementation with specific amino acids. Essential amino acids are ones the bird either cannot make or cannot manufacturer fast enough during times of need. Nonessential amino acids and amino nitrogens are used to synthesize proteins and essential amino acids. Essential, nonessential and amino nitrogens make up the "protein" content of the diet. Nutritional imbalances can occur if some nutrients are supplied in excess by interfering with the normal metabolism of another nutrient. This may cause deficiencies in other nutrients, so the clinician needs to be aware of potential interferences and clinical signs associated with nutrient excess and deficiency.

Coenzyme Q_{10} (ubiquinone, ubidecarenone, coenzyme Q abbreviated CoQ_{10}) is a fat-soluble substance present in most cells, required for the generation of energy in the form of ATP. It is present in the highest concentration in the heart, liver, and kidney. It is being advocated as a supplement for many disease processes in human medicine, and therefore may be included in supplemental therapies for cardiac and liver disease in avian medicine.

Dimethylglycine (DMG) is an amino acid derivative that has been found to have multiple effects. It is considered an adaptogen that works with other cofactors in the body. By increasing oxygen utilization and serving as an antioxidant, it has a role in preventing free radical damage found in degenerative disorders. It has been linked with improving nerve transmission, boosting immune system and metabolism, as well as affecting proper cardiovascular function. It is considered to aid in hepatic detoxification, regulation of cholesterol metabolism and possibly increasing overall energy levels.[3] Its exact effect in birds has not been studied.

Amino Acids

Role in liver disease

Liver disease in birds is very common, and has numerous underlying etiologies. With any liver disease, an adequate amount of energy must be supplied to the avian patient. This is important to prevent catabolism of protein for use as an energy source. Branched-chain amino acids (isoleucine, leucine, and valine) supplementation decreases the work of the liver's transamination reactions and allows the needed metabolism to take place in peripheral tissues. Phenylalanine and tyrosine are poorly metabolized when birds have liver disease and dietary amounts should be minimized.[9]

In hepatic lipidosis or "fatty liver," supplementation with choline and methionine may be needed as they may become rate limiting during fasting or anorexia. Protein and amino acid deficiencies may result in lipid accumulation in an already lipid laden liver. This is due to the fact that protein is needed for lipoprotein synthesis which, in turn, is needed for normal lipid metabolism and transport.

High levels of taurine are found in cardiac muscle, white blood cells, skeletal muscle and the central nervous system. It is also a key component of bile. Bile emulsifies fats in the GI tract for the absorption of fat soluble vitamins, fatty acids and cholesterol. It has been shown to play a role in sparing the loss of potassium in heart muscle, thereby reducing the likelihood of potentially dangerous or fatal arrhythmias. Taurine is found in eggs, fish and meat but not in vegetable proteins. If the liver is functioning normally, it can be synthesized from cysteine and from methionine in other parts of the body as long as there are sufficient levels of vitamin B_6. While carnitine is not a true amino acid, it structurally is similar to the amino acids. Its main function is to help in the transport of long chain fatty acids which are used commonly as an energy source. Carnitine increases the use of fat and helps prevent fatty buildup in the heart, liver and skeletal muscles. It also has been shown that carnitine supplementation can reduce the levels of triglycerides in the blood. Carnitine has also been shown to enhance the effectiveness of the antioxidant vitamins C and E. Choline deficiency may play a role in liver disease. In humans with deficiency, the liver enzyme ALT is elevated. Isoleucine is an essential amino acid (not manufactured, and requires biotin (vitamin B_7) for metabolism) necessary for liver and fatty acid metabolism. Leucine is needed for liver metabolism as well as muscle protein synthesis. Demonstration of deficiencies in these in pet birds has not been documented.[4]

Role in cardiac disorders

Use of amino acids taurine and L-carnitine in avian cardiovascular disease is based on the use in mammals. L-carnitine is utilized by cardiac and skeletal muscle. Taurine deficiencies in cats with dilated cardiomyopathy (DCM) has been speculated as potentially contributing to the same disease in other animals, although specific requirements along with dosages to treat DCM in other species is still lacking. Choline deficiency may play a role in atherosclerosis.

Role in integument disorders

Lysine deficiency in chickens and Japanese quail is associated with achromatosis of the wing feathers. Cockatiels and pigeons fed lysine-deficient diets do not demonstrate achromatosis.[10] Instead, a choline deficiency in cockatiels is associated with achromatosis of both the wing and tail feathers.

Birds with herpes virus infections may benefit from increased levels of lysine with a decrease in arginine as the ratio of these two amino acids are important for reducing replication of herpes viruses. Almonds and peanuts have high levels of arginine and low levels of lysine and should not be fed to birds with cloacal papillomas and those that are herpes positive. Foods containing lysine include cheese, eggs, red meat, lima beans, soy products and yeast.[4]

Tryptophan is an essential amino acid necessary for the production of vitamin B_6 or niacin. It is used by the brain for the production of the neurotransmitter, serotonin. Serotonin is an important transmitter for sleep and to stabilize moods. In children, it helps control hyperactivity and alleviates stress responses. In rhesus monkeys, supplementation at 100 mg/kg q12h, in those animals with self-mutilation reduced that problem.11 Tryptophan has been used in cases to reduce feather damaging behavior. The author (J-D) has used the dosage of 100 mg/kg q8-12h dosage in birds with feather destructive disorder and found it of benefit in some. Dosage can be adjusted per the individual bird. High dosages may result in drowsiness.

Herbal Medicine

Herbal medicine represents the melding of botany, history, ethnomedicine, and pharmacology. It may augment conventional pharmacology and support additional methods of healing. Many herbal substances have been used for generations on humans, but controlled, scientific study is ongoing. Veterinarians may be extrapolating from use in any species for use in birds. Dosages and usage is empirical and anecdotal for the most part.

In the United States, the Dietary Supplement Health and Education Act of 1994 created formal recognition of a diverse array of supplements including herbs for human consumption. This allowed these products to be over-the-counter products and for non-food species (ie, the companion animals) this in general is included. However, the intent of the law was to exclude animals so that they would remain regulated by the veterinary community. The FDA Center for Veterinary Medicine allows that these products, for non-food producing animals, can be sold to the lay public over the counter, further eroding the clinical judgment of the veterinary community for animal health.

The European Parliament of the European Union developed and adopted the Directive on the Food Supplements in 2005. It contains a list of vitamins and mineral ingredients that may contain herbs that are allowed in food supplements. They have also adopted guidelines for human and veterinary medicinal product quality in supplements as well as for herbal products. The European Cooperative in Phytotherapy (ESCOP) has developed a coordinated framework for the scientific assessment herbal medicines for humans and for supporting phytotherapy research.

There have been studies looking at animals' use of medicinal herbs. One example in birds is the use by European starlings of a number of aromatic herbs when building their nests. In North America, it was found that starlings preferentially select for wild carrot (*Daucus carota*), yarrow (*Achillea millifolium*) agrimony (*Agrimonia parviflora*) elm-leaved and rough goldenrod (*Solidago* sp.) and fleabane (*Erigeron* sp.) even when they are not the most common plants available. These preferred plants contain monoterpenes and sesquiterpenes that are harmful to a number of bacteria including *Streptococcus* spp., *Staphylococcus* spp., *Pseudomonas* spp. and mites and lice. Chicks with nests where these herbs were used in the lining had a much higher rate of survival than nests without these herbal materials.[4,12]

Use in liver disease

Liver disease may include hepatitis from bacteria (including *Chlamydia psittaci)*, hepatic lipidosis, toxic hepatopathies, neoplasia, and extending into hepatic encephalopathy. Conventional therapy is limited to antimicrobials, possibly surgery, and largely supportive care. Herbs such as milk thistle *(Silybum marianum)* with the active components of 3 flavonolignans: silybin (silybinin), silychristin, and silydianin (these are called "silymarin") studied in both humans and animals. A concentrated extract is used because of the degradation of the herb in gastric juices (low pH). Only 20-50% is absorbed through the intestinal wall. It is considered to be hepatoprotective, a cholagogue, and an antioxidant. Work in rats has shown it to increase bile flow and bile salt secretion.[2] It increases secretion of bile into the duodenum and provides a gastroprotective effect in preventing ischemic mucosal injury. These properties may explain the hepatoprotective effect which may be due to its anti-fibrotic activity. In a rat model of complete bile duct occlusion, fibrosis was blocked by administration of silymarin.[2] In hepatocyte cell cultures, adding silymarin reduces the uptake of hepatotoxic agents such as amantadine and thioacetamide.[2] Silymarin is involved in stimulating nucleolar polymerase A, which results in an increase in ribosomal protein synthesis, which stimulates the regeneration of hepatocytes. In addition to liver activity, silymarin has been shown to have anticarcinogenic effects in human prostrate carcinoma DU145 cell cultures and in mouse skin tumors. It has anti-inflammatory and anti-arthritic action thought to be the result of the inhibition of 5-lipooxygenase activity.[13] Silybinin inhibits leukotriene formation by Kupffer cells which reduces proinflammatory mediators.

The plant is indigenous to the Mediterranean region but has spread to many areas of North and South America and across Europe. It is considered a pest species in Australia. It is a member of the Composite family and is an annual or biannual. The flowers are spiny and a deep violet color with rows of spiny bracts that mature to seeds which technically are fruits. The leaves are dark green, oblong, and have spiny margins. Most herbal preparations are derived from the seeds.[2]

Recently ConsumerLab.com released information of analysis of milk thistle. They found that products can be contaminated with yeast and molds which create toxins that are potentially harmful to the liver. This occurs if the product is not dried and stored properly after harvest. Yeasts and mold, and significant levels of toxins were found most commonly in products made of whole milk thistle seed and cut herbs. It was less likely to be found in other types of milk thistle supplements. However, there was an analysis of products in the United States that found 100% of seed powders (14 of 14 samples) and herb powders (7 or 7) were contaminated with yeast or mold. 88% of whole seed products (30 of 34 samples) and 71% of cut herb products (10 of 14) were contaminated. None of the supplements sold as tea bags, alcohol-based seed extracts, oil-based seed extracts, capsules or soft gels had yeast or mold.[14] Another study looked at mycotoxins in commercial products and found that 28% (9 of 32) had 30-75% of the daily tolerable intake amount (humans).[15] ConsumerLab.com concluded that it may be prudent for people with liver disease to avoid milk thistle sold as whole seed, cut herb, or powders and to use supplements with caution. They tested 10 commercial products available in the US: only 2 passed with the level of silymarin listed on the label.[16]

Overuse of milk thistle has caused elevations of liver enzymes and has the potential for slowing liver function.[3] Birds with liver disease placed on milk thistle should have repeated blood chemistries run at regular intervals. In a study looking at the effects of dietary milk thistle in white carneaux pigeons challenged with B_1 aflatoxin, the dosage of 10 mg/kg orally q 24h had some effects of decreasing liver enzyme parameters as compared to those that did not receive the milk thistle. The study did recommend further work as the authors could not conclude the degree of hepatoprotectant effects against aflatoxins.[17]

Dandelion (*Taraxaum officinale*) root is considered a strong liver tonic. It has been used in Native American cultures for heartburn, dyspepsia and stomach pain. It is licensed in Germany for use as a medicinal tea for biliary disorders, digestive and gastrointestinal problems as well as for diuresis. Its leaves are a potent diuretic. Interestingly, the high potassium content of the leaf seems to replace the potassium loss during diuresis.[2] The roots, leaves and flowers contain sesquiterpene lactones, triterpenes, carotenoids, fatty acids, flavonoids, minerals and vitamin A. It also has significant concentrations of vitamins D, B complex, and C, iron, magnesium, sodium, potassium, zinc, manganese, copper and phosphorus.[2] The primary therapeutic action is from taracin, the various terpenoids, inulin, and the high concentration of nutrients. Dandelion increases bile production, enhances the flow of bile (choleretic effect), and causes contraction and release of stored bile (cholagogue effect).[2] Dandelion and milk thistle together are complementary in supporting proper liver function and healing including promotion of normal bile flow.

Quercetin is an antioxidant and has been linking with reducing chromosomal aberrations associated with viral hepatitis. It may have some anti-inflammatory activity. It can be found in high concentrations in evening primrose, green tea, apples and cranberries and number of vegetables.

Use in cardiac disease

Hawthorn berry (*Crataegus* sp.) has strong cardiovascular effects through various modes of action. The active components are proanthocyanidin and flavonoids (quercetin, rutin, hyperoside; vitexin; vitexin-rhamnoside), catechin, and epicatechin. It may lower blood pressure, increase myocardial contractility and reduce peripheral vascular resistance through Na^+/K^+ ATPase and angiotensin-converting (ACE) inhibition.[3] It may have antilipidemic effects decrease cholesterol levels and reduce atherosclerosis. In humans it has been used along with cardiac glycosides to treat congestive heart failure.[18]

Use in integument disease

Skin diseases may reflect systemic illness, so any underlying conditions must be treated as well. Common skin disorders in birds including traumatic wounds, feather plucking/destructive disorder, PBFD, papillomatosis. Topical therapies cannot include oils, particularly volatile oils such as eucalyptus and melaleuca (tea tree oil is toxic) and goldenseal (*Hydrastis* sp.).

Aloe (*Aloe vera*, although there are also a number of *Aloe* sp. that may be used) is commonly known for its ability to heal burns when used topically, but it is also effective in treating wounds. Aloe squeezed directly from a freshly cut leaf or a gel extract is administered topically to burns, abrasions or lacerations.

There are a number of herbs that may be of benefit to treat skin conditions. *Calendula officinalis*, *Hypericum perforatum*, *Stellaria media* (chickweed) all have anti-pruritic properties. *Calendula officinalis*, *Matricaia recutita* (German chamomile), *Hypericum perforatum*, *Plantago* (plantain) all have anti-inflammatory properties. Comfrey (*Symphytum officinale*) has been used for wound healing.

Formulations

These formulations are suggestions and should be used with caution as they have not been proven. Dosages are empirical. They have been used by the authors based on individual case needs. It is by no means comprehensive. The authors recommend the practitioner study the herb or nutraceutical for its properties and potential side effects before using any of the below listed suggested treatments.

Cardiac disease

Although taurine and the effects of Coenzyme Q_{10} have not been directly linked with cardiovascular disease in birds, there is evidence in mammals of beneficial effects of supplementation.[19] Based on work done in mammals, the author (J-D) has developed several formulations that have clinically improved cardiac function and liver disease. Dosages are empirical. Blood levels of the active chemicals have not been determined. True pharmacokinetic, dynamic, and toxicokinetic studies have not been done. The formulas have not been chemically evaluated for potency over time, although formulas are given a 30-day refrigerated shelf-life.

Dr. CAJD's Avian Cardiac Formula (perpetually under evaluation and revision):

> 1000 mg L-carnitine
> 1000 mg taurine
> 1000 IU Vitamin E
> 400 mg CoQ_{10}

If lipemic, if there are high levels of cholesterol, and/or if atherosclerosis is confirmed on echocardiography, then 20-30 mg of policosanol are added, higher levels for more severe disease.

Optomega oil (30 ml; USANA Health Sciences, Salt Lake City, UT, USA) is a blend of organic flax seed, sunflower, pumpkin, olive to provide omega-6, and omega-3 fatty acids. If Optomega (USANA) is unavailable, then organic flax seed oil can be used; however, the mixture may become viscous and require warming prior to administration.

Stir well, keep refrigerated. Label with a 30-day expiration date. Dose at 0.1-0.5 ml/kg bird PO q24h.

Vitamin E, CoQ_{10} are gelcaps. Puncture with an 18-gauge needle and squeeze contents into the bottle. Light-resistant amber glass bottles are preferred for dispensing as it is unknown if light degradation would occur. Glass also will not leach chemicals into the oil mixture.

Liver disease

Lactulose is used as a base in this liver formulation. Lactulose is a synthetic non-digestible sugar used to treat hepatic encephalopathy. It also treats constipation as it draws water into the bowel. It treats hyperammonemia by drawing ammonia from the body. It is metabolized in the colon by bacterial flora to short chain fatty acids, producing lactic and acetic acid. It acidifies colonic contents, favoring the formation of nonabsorble NH_4^+ which traps NH_3 in the colon thereby reducing plasma NH_3 concentrations. It has been used in human for treatment of cirrhosis. It is not absorbed from the gut. It also limits the proliferation of ammonia produced by gut flora and increases the clearance of protein load from the gut.

Adverse reactions to milk thistle were observed in 1 parrotlet and 1 blue-fronted Amazon, which had a reaction with swelling of the rictus following oral administration. (SO)

Dr. CAJD's Avian Liver Formulation (perpetually under evaluation and revision):

> 750 mg silymarin (milk thistle active components are flavonoids: silybine, isosilybine, silycristine, silidianine, taxifoline and flavonolignans: dehydrosilybin, siliandrin,silyhermin)
> 10 ml lactulose

The dosage per bird is 75 mg/kg silymarin PO q12h. Each milliliter has 75 mg, making total volume acceptable to birds based on body weight.

Birds should have liver enzymes measured frequently during liver disease treatment. In birds with elevated serum cholesterol, lipidemia, policosanol (100 mg/10 ml lactulose) is added. Although Chinese red ginseng root powder has been shown to decrease cholesterol and triglycerides in chicken livers, the author has not yet been adding it as a supplement.[7] SAM-e is also being investigated for its hepatoprotective properties, however in mammals it must be given in a formulation not destroyed by stomach acid, and is usually in an enteric-coated pill. It is unknown how it can effectively be administered to birds.

Herbs including quercetin, dandelion, licorice, ginseng may be added to either the above formulation or the below formulation. Dosages may be left to the practitioner and particularly licorice may affect the palatability.

In addition to this, based on mammalian information about nutrients needed by the liver, the author (J-D) also administers:

Avian Liver Adjunct (perpetually under evaluation and revision):

250 mg L-carnitine
500 mg taurine
200 mg CoQ_{10}
100 mg Choline
100 mg Isoleucine
100 mg Leucine
100 mg Valine
100 mg Methionine

Optomega oil (30 ml; USANA Health Sciences, Salt Lake City, UT, USA) is a blend of organic flax seed, sunflower, pumpkin, olive to provide omega-6, and omega-3 fatty acids. If Optomega (USANA) is unavailable, then organic flax seed oil can be used; however, the mixture may become viscous and require warming prior to administration.

Stir well, keep refrigerated. Label with a 30-day expiration date. Dose at 0.1-0.5 ml/kg bird PO q24h.

Integument disorders

Nutraceuticals to be combined include vitamin E, vitamin A, beta carotenes using an omega 3-6-9 oil such as Optomega or a flax seed oil as the suspending liquid. Herbs that can be added to this include: Aloe, *Calendula officinalis, Hypericum perforatum, Stellaria media* (chickweed), *Matricaia recutita* (German chamomile), *Plantago* (plantain), *Althaea officinalis* (marshmallow), *Ulmus ruba* (slippery elm), *Achillea millifolium, Hamamelis virginiana* (witchhazel), *Symphytum officinale* (comfrey). A suspension would be administered orally, usually once daily.

With feather destructive disorder, in addition to promoting skin health, herbs and nutraceuticals added should have calming properties and nervous system effects. The omega 3-6-9 oil is used as the base solution. Nutraceuticals include the use of L-theanine, L-tryptophan, CoQ_{10}, Vitamin B_6, Vitamin B_{12}, Vitamin E. Herbs may include valerian, *Hypericum perforatum* (St. Johns wort) *Humulus* (hops), lavender, *Melissa officinalis (*lemon balm*)*, *Pulsatilla*, *Avena* (oatmeal), *Echinacea, Astragulus (*this is used for chronic immune support, itching and inflammation*), Eleutherococcus* (Siberian ginseng).

Note: *Melissa officinalis (*lemon balm*)* can be obtained as a powder or non-alcohol tincture. It has been used topically on papillomas in birds.

Conclusions

In conclusion, birds may well benefit from nutraceutical and herbal formulations, but at present, clinical research to evaluate efficacy, metabolism, and establish doses has not been published. Veterinarians should inform owners of the experimental nature of using these formulations and supplements. Veterinarians also need to ask owners what they are giving their birds as supplements, as many may have direct influence on disease and treatments. Formulations are empirical and may be adjusted by volume and palatability.

References

1. Burns K. Assessing pet supplements. Use widespread in dogs and cats, evidence and regulation lacking. *J Am Vet Med Assoc.* 2017;250(2):116-124.
2. Orosz SE. Common herbs and their use in avian practice. *Proc Annu Conf Assoc Avian Vet.* 2006;239-247.
3. Ness RD. Herbal and natural health supplements in avian medicine. ExoticsCon. 2016;113-117.
4. Orosz SE, Johnson-Delaney CA, Echols MS. Application of nutritional supplements and herbs in avian practice. *Proc Annu Conf Assoc Avian Vet.* 2007;175-189.
5. Petzinger C, Heatley JJ, Cornejo J, et al. Dietary modification of omega-3 fatty acids for birds with atherosclerosis. *J Am Vet Med Assoc.* 2010;236(5):523-527.
6. Petzinger C, Heatley JJ, Bailey CA, et al. Lipid metabolic dose response to dietary alpha-linolenic acid in Monk parrot (*Myiopsitta monachus*). *Lipids.* 2014;49(3):235-245.
7. Heinze CR, Hawkins MG, Gillies LA, et al. Effect of dietary omega-3 fatty acids on red blood cell lipid composition and plasma metabolites in the cockatiel, *Nymphicus hollandicus*. *J Anim Sci.* 2012;90: 3068-3079.
8. Khatibjoo A, Kermanshahi, H, Golian A, et al. Immunology, health, and disease: the effect of dietary n-6:n-3 ratio and sex on broiler breeder immunity. *Poult Sci.* 2011;90:2209-2216.
9. Smith JM, Roudybush TE. Nutritional disorders. In: Altman RB, Clubb SL, Dorrestein GM, Quesenberry K, eds. *Avian Medicine and Surgery.* Philadelphia, PA: Saunders;1997:501-516.
10. Roudybush TE. Nutrition. In: Altman RB, Clubb SL, Dorrestein GM, Quesenberry K, eds. *Avian Medicine and Surgery.* Philadelphia, PA: Saunders;1997:27-44.
11. Weld KP, Mench JA, Woodward RA, et al. Effect of tryptophan treatment on self-biting and central nervous system serotonin metabolism in rhesus monkeys (*Macaca mulatta*). *Neuropsychopharm.* 1998;19:314-321.
12. Clark L, Mason JR. Use of nest material as insecticidal and anti-pathogenic agents by the European startling. *Oecolgia.*1985;67:169-176.
13. Marazzani P, Bombardelli E. Silybum marianum (*Carduus marianus*). *Fitoterapia.* 1995;66:3-42.

14. Tournas VH, Rivera Calo J, Sapp C. Fungal profiles in various milk thistle botanicals from US retail. *Int J Food Microbiol*. 2013;164:87-91.

15. Veprikova Z, Zachariasova M, Dzuman Z, et al. Mycotoxins in plant-based dietary supplements: hidden health risks for consumers. *J Agric Food Chem*. 2015;53:6633-6643.

16. ConsumerLab.com. Milk Thistle. Available at: http://www.ConsumerLab.com. Accessed March 27, 2017.

17. Grizzle J, Hadley TL, Rotstein DS, et al. Effects of dietary milk thistle on blood parameters, liver pathology, and hepatobiliary scintigraphy in white carneaux pigeons (*Columba livia*) challenged with B1 alflatoxin. *J Avian Med Surg*. 2009;23(20):114-124.

18. Skidmore-Roth L. *Mosby's Handbook of Herbs & Natural Supplements*. St. Louis, MO: Mosby, Inc;2001.

19. Gompf RE. Nutritional and herbal therapies in the treatment of heart disease in cats and dogs. *J Am Anim Hosp Assoc*. 2005;41:355-67.

Gross Lesion Recognition for the Avian Practitioner

Michael M. Garner, DVM, DACVP

Session #2100

Affiliation: From Northwest ZooPath, 654 West Main, Monroe, WA 98272, USA.

Gross lesion recognition, like other imaging modalities, is a bit of an art form. Vast zones of grey may confront the investigator when considering what the lesion is, what it could be, and what to do with it. And nobody is good at all of it. A seasoned livestock pathologist could possibly clear an entire days' necropsies without the help of a single histology slide. A bee keeper would recognize immediately the maggot stages that parasitize his colony. A dog-cat pathologist in a busy private laboratory or a companion animal clinician may not have a clue about gross lesions in livestock or bees. Historically, our attempts at accurately diagnosing disease by gross lesion recognition have been a humbling experience, and that has been the impetus for all further diagnostic specialties. So now that we have put gross pathology in its proper place, why even bother with such an imperfect science? Obviously, gross lesions, or lack thereof, are the first visual indication of what may be wrong with the patient. These are the lumps, the effusions, the asymmetrical oddities, the discolorations, the odiferous clues to disease that (hopefully) stimulate a "scientific" thought process culminating in a list of differential diagnoses. The purpose of this workshop is to present images of common and not so common avian gross lesions in live and necropsy specimens, and in a participatory manner, establish a differential diagnosis and means for establishing a definitive diagnosis. Proper photographic documentation of gross lesions will also be addressed.

References

1. Campbell TW, Ellis C. *Avian and Exotic Animal Hematology and Cytology*. Ames, IA: Iowa State University Press;2007.

2. Garner MM. An overview of biopsy, cytology and necropsy techniques for clinical veterinarians. In: Aguilar RF, Hernandez SM, Divers SJ, Perpinan D. eds. *Atlas de Medicina de Animales Exoticos*. 2da. Buenos Aires, Argentina: Inter-Medica SAICI;2010:437-443.

3. Work TM. *Avian Necropsy Manual for Biologists in Remote Refuges*. U.S. Geological Survey, National Wildlife Health Center, Hawaii Field Station. 2000. Available at: https://www.nwhc.usgs.gov/hfs/Globals/Products/Sea%20bird%20manual%20english.pdf Accessed May 30, 2017.

4. Latimer KS, Rakich PM. Necropsy Examination. In: Ritchie BW, Harrison GJ, Harrison LR, eds. *Avian Medicine: Principles and Applications*. Lake Worth, FL: Wingers Publishing Inc;1994:356-376.

Satellite Telemetry of Water Birds

Scott Ford, DVM, DABVP (Avian Practice)

Session #2104

Affiliation: From Avian Specialty Veterinary Services, 519 S 68th St, Milwaukee, WI 53214, USA.

Abstract: Since 2009, the author has implanted and deployed over 273 intracoelomic transmitters in 21 species of water birds including loons, gannets, geese, and sea ducks in locations all over North America. This talk will summarize those experiences, look at an analysis of factors most influential on survival, and discuss the evolution of care protocols.

Introduction

Satellite transmitters, also known as platform transmitting terminals (PTTs), are sealed radio transmitters. In water birds, they are implanted within the right abdominal cavity. There is an external antenna that exits the skin dorsally near the tail. A synthetic collar is present at the base of the antenna to discourage ascending infection and slow the rate of ejection by the body. The body of the PTT is also enveloped in fine synthetic mesh to facilitate anchoring inside the body. A veterinarian is required to perform the surgery and it is performed under sterile conditions with the assistance of general anesthesia and local and general analgesics. There are minor variations in surgical techniques that vary by individual veterinarian or the particular circumstances of the project. This master class reviews the author's techniques and experiences with special attention to recent adaptations of techniques and trends for future research.

Overview

Rationale

Following water birds throughout their annual cycle has traditionally been very difficult. Their long migratory distances, often over inhospitable terrain or great stretches of water, combined with their tendency to nest in remote, northern locations, have made it difficult to track their movements. Leg bands, wing tags, and neck bands have assisted in answering questions pertaining to repeated use of nesting or wintering grounds and longevity; however, data yield from band re-sightings is low relative to the number deployed and they do not provide good details on routes, critical stopovers and resting points, or repeated use of remote nesting areas. In decades past, some questions, such as nesting locations of black scoters that wintered in southeast Alaska, could not be answered without years of field research and thousands of man-hours of meticulous exploration (Jim King, USFWS pilot, retired, oral communication, 2011). Radio telemetry promised to help answer these questions more efficiently but the first radios, applied externally, proved to be too disruptive to birds that dive or spend most of their time in the water. Also, the first VHF transmitters had short line-of-sight ranges. Frequent flights of aircraft were required to find the signals and determine location, a process that was both expensive and potentially dangerous as well as severely limited by weather. With the advent of satellite technology, combined with refined transmitter design and implantation techniques that came about in the late 1980s and early 1990s, water birds could be tracked anywhere on the globe at a significantly lower cost and far more safely for the researchers. The value of this modality was proven when at-sea distribution of spectacled eiders were identified after more than a century of traditional efforts were unsuccessful.[1]

Many of basic avian biology questions are still answered today by satellite telemetry including where a species nests, where it winters, and where it prefers to stopover during migration. These are important questions to continue exploring as anthropogenic activities continue to alter the environment; however, with satellite tracking, more specific questions can be answered such as where birds from a specific area tend to go in the winter or summer. Or whether birds from one region tend to share nesting areas with birds from another region (a question which has implications for the spread of pathogens). Also, by developing better models of bird movements, population trends can be more easily estimated based upon ongoing point counts at identified key migratory locations.

Impacts on survival and activity

There always has been concern as to whether any tagging appliance significantly alters the function and behavior of study subjects. It's of great concern, not only for the well-being of the birds but also for the integrity of collected movement data. A few papers published over the years do suggest that even relatively minimal devices, such as flipper-bands on penguins, may have measurable impacts on their behavior.[2] Implanting transmitters is no different in this regard. One study compared behavioral impacts of leg bands versus transmitter implants in Canada geese and found little difference;[3] however, Canada geese are not considered an obligate aquatic species, they do not dive to forage, and the transmitters used were relatively small compared to body mass. Further, these birds were not required to fly long distances during the period of the study. It is generally assumed that the implants, even under perfect circumstances, do create a drain on resources by adding weight, creating a tail-heavy balance moment (which affects flight), changing hydrodynamics during diving, interfering with airsac and organ expansion, and interfering with copulation and egg-laying.

Captive research performed in the late 1980s and early 1990s did identify that implantation of transmitters afforded less impact on behavior than application of external transmitters in mallard ducks.[4] This is because of less disruption of the plumage which is critical to preventing water contact with the skin and, subsequently, interfering with thermoregulation. Further studies with captive scaup helped identify potential impacts of implanted transmitters on breeding and oviposition.[5] Another captive study, utilizing a dive tank, identified hydrodynamic drag created by the vertical whip antenna which led to subsequent changes in antenna design;[6] however, no further related studies have been performed to see if implemented design changes have diminished these impacts. Captive research has contributed positively to the science of deploying implanted transmitters but more is needed to continue the process of refining techniques.

In addition to the impact of the physical presence of the implant, the processes of capture, transport, surgery, and post-operative healing produce changes that have been incompletely investigated. Some specific concerns include stress, bruising of wings or other contusions, damage to feathers, the effects of skin oils, surfactants, and adhesives on feather integrity, and the effects of drugs, pain, and inflammation during the post-operative period on a bird's ability to evade predators, thermoregulate, and forage successfully. Only a few papers documenting some of these concerns have been published to date. Investigation is hampered by 1) an inability to follow controls (eg, untagged birds) in the wild, 2) a lack of captive facilities to perform controlled studies, and 3) funding priorities that focus on deploying more transmitters and not on controlled captive studies. Adding to the investigational difficulties, veterinarians and research teams tend to work independently such that capture, transport, surgical, and medical care protocols vary and basic data collection is not standardized. There are efforts ongoing to accomplish standardization of protocols and data collection to aid multi-variate analyses; however, the author suspects that controlled captive studies must also be performed to optimize medical and captive care procedures specifically.

Techniques

Capture and transport

Capture techniques are typically the prevue of the investigating biologists. In wintertime situations, the typical method used is spotlighting. This involves night operations in a boat, drifting slowly downwind using a strong spotlight to illuminate the birds. If the light is held correctly and the crew practices good noise and light discipline, the bird cannot see beyond the light beam and so does not know that humans are approaching. Once close to the bird, a large pole net is used to retrieve them from the water. On breeding grounds, it's more typical to use bow net traps to capture brooding adults on the nest or they are driven into floating mist nets while they are molting and flightless.

Transport protocols can be influenced by the attending veterinarian. The author prefers windowless carriers with small ventilation holes at top and bottom rims and an interior hammock of knotless netting. Hammocks of knotless netting are utilized extensively by oil spill responders and are believed to inflict less damage to feathers while also sparing focal pressure on the keel and feet.

Pre-surgical assessment

Before surgery, each patient receives a physical examination at the very least. In some cases, pre-operative blood work is possible but in remote locations, instant results are not usually practical. The most important parts of the exam include checking that eyes, wings, and legs are intact and functional. Subjective assessments of fitness and cardiopulmonary function are made based on body mass, conformation, perfusion of mucus membranes, and thoracic auscultation. If anything is in doubt, the bird is rejected and returned to the wild.

Sedation, analgesia, and anesthesia

Midazolam is a commonly used sedative (see below for dosing). It is normally given immediately after capture. As with other species, it's recommended that birds be rested in a quiet, calm area in a covered kennel until the drug takes effect (10 minutes). Butorphanol (1-2 mg/kg IM) and NSAIDs such as ketoprofen or meloxicam have been used for general analgesia and are given just prior to surgery. Ketoprofen use may have led to increased mortality in at least one specific situation involving implanted drake eiders.[7] Lidocaine and bupivacaine (separate or together at total of 2 mg/kg) have been used for local line-block or splash-block analgesia.

Implant subjects may benefit from sedation in an effort to calm them during transport to the surgical facility and possibly to decrease the amount of required anesthesia. Analgesics should decrease pain and, therefore, reduce stress; however, there could be potential problems too. Since the birds must be released back to the rigors of sea life soon after implantation, drugs that impair alertness or falsely elevate a sense of well-being may hamper predator avoidance, self-maintenance behaviors, or foraging. Although we have no pharmacokinetic data on analgesics and sedatives in sea ducks and loons, it can be extrapolated from data in other birds (including some studies with waterfowl) that these substances do not reside long at therapeutic levels in the blood stream. Given the variability of observed responses to midazolam, it may be that dosages used (typically 2 mg/kg IM or IN) are too low. Further research is recommended to determine correct dosages for representative diving bird species.

There are two primary anesthetic protocols in use for PTT implants: total propofol anesthesia (TPA) and isoflurane gas. TPA is used by some veterinarians due to added complexities of transporting oxygen to remote locations. Propofol is administered IV through a 24-gauge catheter inserted in the medial metatarsal vein using an induction bolus of 5-10 mg/kg over 1 minute followed by periodic smaller boluses (2-4 mg/kg) to effect. Positive-pressure ventilation (PPV) is usually required since TPA depresses respiratory function. TPA offers advantages such as simpler logistics, lighter cargo, and the presence of a catheter provides a route for intraoperative fluids or drugs. Disadvantages include poorer oxygen delivery (assuming no compressed oxygen is being delivered) and greater depression of cardiac and respiratory function.

Isoflurane anesthesia is delivered via compressed oxygen in an open circuit. Induction is accomplished by mask and the author finds that butorphanol given at a rate of 1 mg/kg IM and/or midazolam at 2 mg/kg IM 10-15 minutes prior to induction tends to help hasten the process. Once sufficiently anesthetized, a snug-fitting, uncuffed, silicone endotracheal tube is inserted and secured to the bill. One veterinarian utilizes an oxygen generator where electricity is available and reports good results (Glenn Olsen, oral communication, December 2016). The author has always used compressed medical-grade oxygen and has been able to get this delivered to even the most remote locations. It requires significant logistical planning and propofol is still brought as a back-up plan, but so far it has never proven necessary to utilize TPA. The advantages of isoflurane in oxygen are superior oxygen perfusion, less depression of cardiac and respiratory function, and smooth, fast recoveries. Disadvantages include more complicated logistics (due to safety regulations of air transport of oxygen and isoflurane) and more gear to pack (vaporizer and circuit). Catheters are sometimes used in conjunction with isoflurane anesthesia if venous access is desired.

Anesthesia monitoring

Often, no technician or veterinary assistant is available to assist with anesthesia. Instead, biologists are trained on-the-spot to assist. For this reason, monitors are generally kept simple and easily audible by both veterinarian and anesthetist. Also, the constraints of travel to remote locals and the lack of electricity dictate that monitoring equipment be simple, small, and robust.

The most common monitoring performed consists of cloacal or esophageal temperature, visual assessment of breathing rate, and assessment of cardiac rate and rhythm using an esophageal stethoscope. The stethoscope is amplified so that both doctor and anesthetist can hear it. In addition, some doctors utilize a portable pulse oximeter with a reflectance probe directed at the dorsal surface of the oropharynx. More recently, portable in-line end-tidal CO_2 monitors have been utilized to track both ventilation rate and exhaled carbon dioxide.

Surgical technique

The following is based primarily upon the author's technique. After induction and intubation, birds are placed in dorsal recumbency on an insulated pad. The abdomen is wiped clean with sterile water until the sponges come off the feathers clean. The feathers are parted along the ventral midline of the abdomen and sterile water-based bacteriostatic surgical lubricant (Surgilube Foilpac, Savage Laboratories, Melville, NY, USA) is applied to the skin. This gel contains a small amount of chlorhexidine gluconate. Note that no feathers are removed and no sterilizing solutions applied. This is to avoid impairment of waterproofing and thermoregulation immediately after release. Paper tape may be utilized to help retract feathers further followed by a sterile transparent drape (Avian Drape, Jorgensen Laboratories, Loveland, CO, USA).

The entry is via a 3- to 4-cm ventral midline incision taking care to avoid puncturing the intestinal coelomic or air sac membranes. Avoiding air sac disruption is not always possible and is deemed by most veterinarians to be of little consequence if they do rupture. Keeping the membranes intact seems more easily accomplished in individuals with more coelomic fat present. The author believes, based on removal of PTTs from various birds post surgery, that keeping the fat and membranes intact may prevent impression trauma and adhesions to organs. Membranes are separated from the body wall to create a pocket along the right side of the coelomic cavity.

Landmarks are identified for antenna passage. The exact location varies from species to species but is generally along the caudal edge of the ischium, adjacent to the spinal column. The right ureter passes through this general area and must be avoided. A suture is placed in the body wall lateral to this location and the ends are left untied. A trocar is passed down through the site, directed from lateral to medial and a sleeve is slipped over the trocar. The suture needle is then passed through the bioactive collar on the PTT but left untied. The trocar is removed and the PTT antenna is then fed through the sleeve to the outside. The sleeve is then removed and the PTT is inserted completely into the body cavity. The loose suture ends are tied to attach the PTT collar to the body wall. After this, 1 or more sutures are placed in the PTT's mesh covering to further secure it to the bird's body and prevent oscillation while in flight. Body wall and skin are closed in separate layers, usually in simple continuous fashion.

Instead of this sterile-to-dirty, one-site prep technique, some veterinarians prefer to prepare the antenna exit site separately and will insert a cannula or sharp alligator forceps from the outside to assist in passing the antenna through the body wall.

Recovery and release

During recovery, intravenous or subcutaneous crystalloid fluids may be administered, usually at a dose of about 3-5% of body mass. Because most of the birds implanted are highly-obligate water species, they are returned to their habitat as soon as practically possible after surgery. In most cases this will be 2-3 hours post surgery. In some cases they are boated back to their capture site while others may simply be released from a nearby shore. Considerations have to be made as to the presence of predatory birds (eagles and gulls by day, owls by night). In some cases, birds have been held for up to 3 days post surgery but survival was poor and several were observed to have severely impaired waterproofing after release. If birds must be held, preliminary data suggests that regular gavage feeding reduces corticosterone, possibly indicating an amelioration of stress (Scott Ford, unpublished data from 2010 captive study).

Special Considerations and Observations

Multivariate survival analysis

There are many variables that likely affect post-surgical outcome. These include both extrinsic variables and those that are intrinsic to the process. Some external variables, of which there is minimal control possible, include environmental conditions (eg, weather and sea conditions) at the release site, the quality of local foraging opportunities, and the presence of predators. Intrinsic variables, of which we do have control, include the use of sedatives and analgesics, the materials and methods involved in capture, transport and post-operative care, surgical prep agents, anesthesia and surgical protocols, and pre-operative evaluation methods.

An analysis that attempts to collectively compare the effects of each of these variables may help us identify which variables are the most critical to determining positive or negative outcomes. A negative outcome is generally defined as a mortality that occurs within 2 weeks of deployment. Previous attempts at this across multiple studies have been marginally successful and the authors concluded that there were too many uncontrolled variables (Timothy Bowman, oral communication, February 2017). So, the first challenge to overcome is to standardize care protocols and, perhaps more importantly, standardize the data we collect, how it's collected, and create a means of sharing that data. At present, the author and others are attempting to do this and it should facilitate multivariate analyses in the future.

Midazolam

Midazolam sedation has been found to be a very useful tool in pet bird medicine. For the past few years it has seen increased use for satellite transmitter implant surgeries. Generally a dose of 2 mg/kg IN or IM has been utilized but results have been highly variable: no apparent effect in some patients up to mild to moderate sedation with other individuals of the same cohort. Pharmacokinetic data is lacking and there may be unknown changes in drug metabolism dependent upon physiologic conditions. In some species, midazolam appears to have extremely prolonged and profound effects. Most notably, northern gannets were found to experience profound sedation for up to 12 hours post injection. Reversal with flumazenil was not performed in that particular project so it is unknown whether this would have been of assistance or not. At this time, the author recommends the use of midazolam to manage stress but vigilance should be exercised with new species or doses.

Anatomic variations (dive muscle in gannets)

Before any clinician undertakes implantation of transmitters of unfamiliar design or into unfamiliar species, they should conduct practice surgeries with carcasses of the same or similar species. The author has found these exercises abundantly helpful even after performing more than 200 similar surgeries. One case where such benefits was illustrated involved the identification of the prominent caudofemoralis muscle in the northern gannet. Its prominence suggested an important role but the muscle lies directly in the path of the usual antenna exit route. By identifying landmarks pertaining to this muscle, the legs could be positioned caudally and the muscle "strummed" out of the way to safely preserve it during trocarization.

PTT and antenna morphology

The size of PTT electronics has decreased significantly since their introduction. The primary factor limiting size now is the power supply. The shape of batteries also tends to dictate the overall shape of the PTT; however, some features, such as antenna shape and length, can be more easily modified. The author recommends that transmitters be designed to allow placement as far cranial as possible without impinging the right lung or air sac ostia. This can be accomplished by configuring the antenna to exit caudally for 1-2 cm before it bends dorsally. Also the lateral surface of the PTT should ideally be curved convexly to match the round shape of the body wall. Acute corners likely produce greater pressure and discomfort. Finally, the caudal end of the PTT should never be thicker than the cranial end including when PTT collars are attached. This is because the body cavity tapers both vertically and horizontally towards the tail. The main impingements at the caudal abdomen include the cloaca, the caudal division of the kidney, and freedom of movement of the right leg.

References

1. Petersen MR, Douglas DC. Use of implanted satellite transmitters to located spectcled eiders at-sea. *Condor.* 1995;97:276-278.

2. Culik B, Wilson R, Bannasch R. Flipper-bands on penguins: what is the cost of a life-long commitment? *Mar Ecol Prog Ser.* 1993;98(3):209-214.

3. Hupp JW, Ruhl G, Pearce JM, et al. Effects of implanted radio transmitters with percutaneous antennas on the behavior of Canada geese. *J Field Ornithol.* 2003;74(3):250-256.

4. Korschgen CE, Kenow KP, Gendron-Fitzpatrick A, et al.. Implanting intra-abdominal radiotransmitters with external whip antennas in ducks. *J Wildl Manage.* 1996;60(1):132-137.

5. Perry MC, Olsen GH, Wells-Berlin AM. Annual Project Summary for Endorsed Projects. *Sea Duck Joint Venture*. 2008.

6. Latty C, Hollmen T, Petersen M, et al. Abdominally implanted transmitters with percutaneous antennas affect the dive performance of common eiders. *Condor*. 2010;112(2):314-322.

7. Mulcahy DM, Tuomi P, Larsen RS. Differential mortality of male spectacled eiders (*Somateria fischeri*) and king eiders (*Somateria spectabilis*) subsequent to anesthesia with propofol, bupivacaine, and ketoprofen. *J Avian Med Surg*. 2003;17(3):117-123.

Critical Care of Raptors

Scott Ford, DVM, DABVP (Avian Practice)

Session #2105

Affiliation: From Avian Specialty Veterinary Services, 519 S 68th St, Milwaukee, WI 53214, USA.

Abstract: Injured or ill birds of prey often present in a chronically debilitated state resulting in severe emaciation and dehydration. Fluid resuscitation and re-alimentation need to be performed carefully for them to survive. This class will focus on the special needs of these species for optimum stabilization.

Introduction

Injured birds of prey are often admitted with moderate to severe levels of dehydration, malnourishment, and/or physiologic shock from blood loss and trauma. They have very different physiology and nutritional requirements from psittacine birds and other common pet species. Therefore they require specialized medical techniques for recovery from these conditions. This master class will discuss these techniques and provide the practitioner with basic protocols for patient stabilization.

Metabolism

Birds of prey eat protein almost exclusively and so their metabolism is specially adapted for extraction of energy from protein. Non-carnivorous birds typically utilize dietary carbohydrates (glucose) for energy in the fed state. The diet of carnivorous birds contains remarkably little carbohydrate (< 2%) and so amino acids, released from digested protein, are primarily used for gluconeogenesis.[1] In the fasted state they generate glucose from glycerol, liberated by lipolysis, and amino acids liberated from protein stores. As a result, carnivorous birds utilize very little glucose, even when present in their diet. Instead, they continually use amino acids for gluconeogenesis—from their diet when fed and from their own body tissues during periods of fasting (Fig 1).

When administered glucose solutions, several species of raptors have been found to exhibit hyperglycemia with very slow clearing times.[1,2] This is due to the lack of glucokinase, an enzyme responsible for hepatic clearance of glucose. It is also believed that while in this hyperglycemic state, gluconeogenesis from amino acids continues unabated further exacerbating the condition.

In addition to poor clearance, faunivorous birds have poor intestinal disaccharidase activity and glucose-transport capacity. Sugars that remain unutilized in the gut may create an osmotic gradient that decreases GI retention times thus impairing digestion and absorption of water from the gut and cloaca.

Raptors are highly efficient at digesting animal tissues and absorbing their constituents but it comes at some cost. 20% of their potential energy is lost as uric acid during metabolism.[1] Also, great amounts of moisture are required for the production of pepsin, concentrated HCl, mucus, and other digestive secretions.

Soft tissues generally have a poor calcium to phosphorus ratio. Therefore, bones are an essential part of a raptor's diet, even if only partially digested. Periods of impaired GI function may require the addition of other more digestible forms of calcium to the diet or the use of parenteral calcium administration.

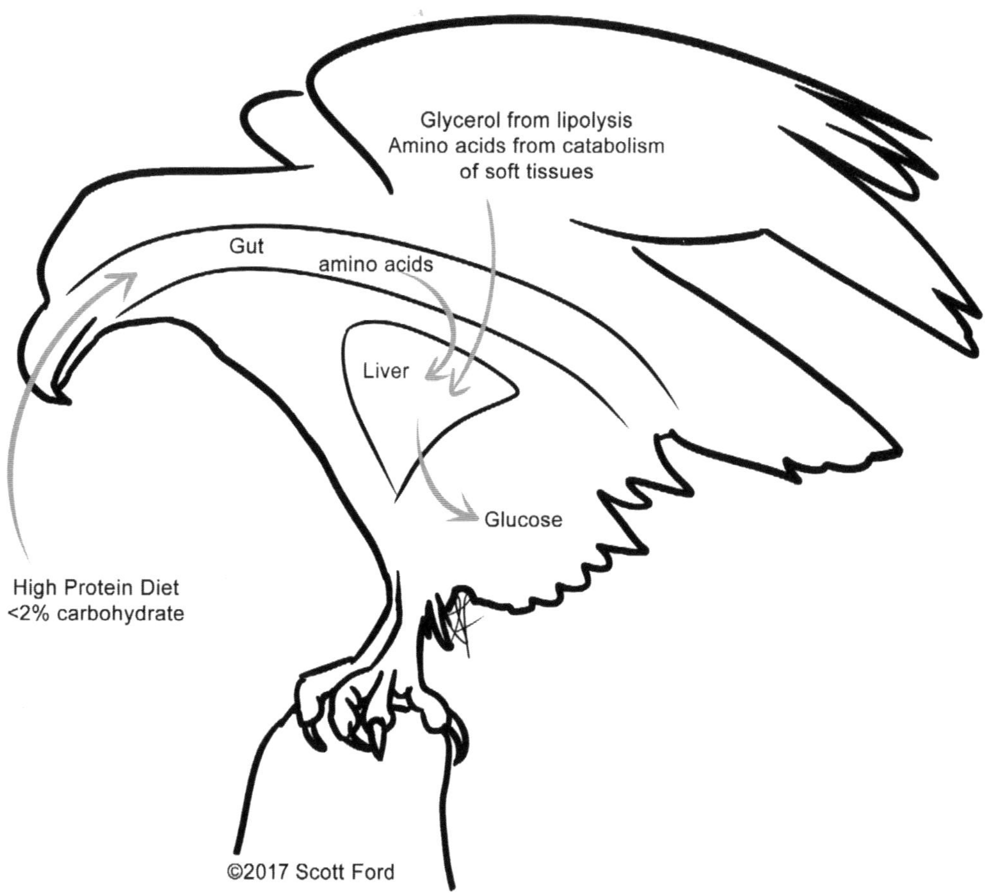

Figure 1. Glucogenesis in birds of prey. Whether fasting or not, raptors generate glucose from amino acids and fats, not from ingested carbohydrates, which make up very little of the natural diet.

Assessment

Fluid deficit

Assessment of fluid deficit is important during admission of a raptor for medical care. Items to check include the eyes, oral mucus membranes, heart sounds, wing tips, basilic vein, cloaca, droppings, and feet. In addition, PCV, total serum solids, and other electrolyte analyses may also be utilized.

The eyes are assessed for roundness, clarity, and brightness. A significantly dehydrated raptor will have a dull look to the eyes and may exhibit slower pupillary reactions or menace response than usual. Bear in mind that some species, such as many owls, may not normally, even when well hydrated, present a marked menace response. Most notably, as the globe loses pressure, it will collapse sufficiently for the lids to close partially creating an almond-shape from what normally should be a round shape to the eye. In extreme cases, the cornea will become tacky in appearance and wrinkle or collapse.

Oral and cloacal mucus membranes should be bright pink, warm, and moist. If they are tacky, pale, or cool to the touch, consider hypovolemia or shock as likely contributors. In addition, in chronic cases of severe dehydration, uroliths may develop within the cloaca. These can be detected by inserting a thermometer and their presence is transmitted as a gritty or stony sensation. Low body temperature (< 103 °F, 39 °C) can also coincide with shock or hypovolemia. Additionally, the droppings of raptors should demonstrate pasty to creamy, formed feces, generally dark in color, along with white mucoid urates, and variable quantities of clear urine. If the bird has not been eating for over 18 hours, there should be small amounts of dark green bile present instead of feces. The urates should never be yellow or green, even when bile is present. Yellowish and/or gritty urates may indicate dehydration as the material is retained in the cloaca and rectum for prolonged periods to recover water.

A low-grade cardiac murmur, best described as a "sloshy" heart sound, may be present in cases of hypovolemia. It occurs as a result of cavitation when blood is under insufficient pressure to fill the heart efficiently. It should diminish as fluid resuscitation proceeds.

The wingtips and feet should be warm and supple. With the foot clenched, the skin on the top should not tent easily and should snap back within a couple of seconds. It should not be dry, flaky, or cracked.

Packed cell volume will increase with hypovolemia and total serum solids, as measured by a refractometer, will increase. Photometric assessment of total protein and albumin is not validated for use in birds so refractometry is recommended. Electrolyte panels may be of some use although normal ranges are sparsely available for specific species and specific point-of-care analyzers.

As a general rule, an assumption of 10% dehydration is safe for any admitted raptor. If you were to replenish fluids at this assumption, even in a normally hydrated patient, there should be no danger of hypertension. However, it would be insufficient for critically dehydrated patients. For less experienced avian clinicians, the author recommends that if 2 or more of the abnormalities described above are discovered, a higher degree of dehydration (eg, 15%) should be assumed and treatment dosages adjusted accordingly.

Degree of emaciation

Emaciation, defined here as the state of being thin, is primarily determined by assessing pectoral muscle mass and abdominal conformation. Radiographs or other imaging may provide additional clues.

The pectoral muscle mass should normally present a slightly convex profile on each side of the keel in birds of prey (other species may vary from this). The more the keel projects above the surrounding breast muscles, the more severe the state of emaciation. The abdomen should present a concave profile along the ventral midline from tip of sternum to the interpubic region. The pubic bones in a healthy individual will be readily palpable at their tips, just cranial of the vent. As they are followed laterally, they should have little palpable tissue covering them until they meet the synsacrum at which point they are lost under the caudal thigh muscles. Additionally, the muscles of the dorsal aspect of the hips should normally make it difficult to detect the wings of the ilium.

One can attempt to assess the percentage of emaciation by expressing it as patient mass divided by ideal weight. However, this method is fraught with inaccuracy due to our inability to differentiate fluid weight from tissue weight. Also, most clinicians do not have access to local average body weights for wild species, which vary widely from region to region or by time of year and stage in circannual rhythm. Therefore, degree of emaciation is usually a subjective determination.

GI function

Ileus is a common sequela to hypovolemia as well as occurring secondary to shock or stress. As mentioned earlier, the GIT requires abundant amounts of water and energy for enzyme and acid-rich secretions that are necessary for breaking down meals of protein. Not only that, but if blood volume is low, so is the perfusion of the GIT so nutrients cannot be efficiently absorbed and transported. For these reasons, it is absolutely critical that fluid resuscitation commences several hours before any attempt at feeding.

Paradoxically, stress may cause an increase in defecation but this is only emptying of the lower GIT and cloaca (which, coincidentally impairs the bird's ability to recover water). Motility of the upper GIT slows in response to stress and, therefore, so does digestion. Stress must be managed judiciously to limit its contribution to ileus. The author recommends placing wild birds in dimly lit quarters out of the normal traffic flow and away from the sounds or potential predators, including people.

GI function is assessed by monitoring and recording, in subjective terms, the quality and quantity of droppings. Typically I recommend that staff record droppings in terms of components. For example, a normal fasted eagle dropping may be described as "Small amounts of green bile in feces, normal quantity of white liquid urates, scant clear urine." Immediately after initiating gavage, many birds will exhibit large amounts of clear fluid to their droppings which is probably a combination of fecal fluid and urine. It seems to occur with chronic inanition particularly and is probably reflective of gut hypermotility. It's logical to assume that if the gut has been inactive for several days or weeks, there's been a loss of smooth muscle mass, villi, enzymatic secretory tissues, and other essential mechanisms for absorption.

Renal function

The presence of white semi-solid, mucoid-encapsulated urates indicates that the kidneys are performing reasonably well. However, serum uric acid determination may be additionally helpful. Be aware that uric acid levels are typically higher (>20 mg/dl) in faunivores than in birds of other diet strategies. Uric acid should be measured during fasting (ideally from a blood sample taken on admission in critical emaciated birds).

Electrolytes and acid-base imbalances

It has been suspected that injured and ill birds of prey may commonly experience metabolic acidosis. This may be the result of hypovolemia decreasing oxygen perfusion.[3] As the acidosis worsens, ventilation rate increases. If this and other efforts fail to correct the acidosis, CNS activity is gradually suppressed resulting in respiratory and cardiac arrest.

In mammalian and human medicine, acid-base or electrolyte imbalances can be more readily identified and corrected. There's a dearth of published information regarding interpretation and appropriate response to electrolyte and acid-base disorders in birds of prey. However, it is suspected that, just as in mammal and human medicine, addressing these disorders could be beneficial during critical stages of care. In recent years, point-of-care analyzers, such as the i-Stat (Abbott Laboratories, Abbott Park, IL, USA) have proven valuable for identifying acid-base disorders in mammal medicine. Practical use in birds has been hampered by the technical challenge of obtaining arterial samples in birds, correcting venous sample results, and interpreting results in light of species variations.[4] Raghav, et al, demonstrated that venous samples can be analyzed by the i-Stat and corrected to closely approximate arterial values.[5] However, built-in temperature correction is not accurate for falcons (their study species) and so correction factors must be determined and applied manually for each species. Base excess is thought to be the most useful diagnostic tool as it can be used for determining appropriate doses of buffer to add to parenteral fluids. As research continues, our ability to respond to acid-base and electrolyte abnormalities will be significantly improved.

Fluid Resuscitation

Treatment Plan

Fluid support should replenish deficit gradually while providing daily maintenance requirements (~5% of the body mass) and ongoing losses from stress or disease.[6] Replacement of half of the deficit is recommended in the initial 12-24 hours, the remainder over the following 2 days. In combination with maintenance requirements and assuming a moderate to severe fluid deficit of 10% (halve these numbers if patient shows only mild dehydration), a fluid resuscitation plan can be created as described below.

Calculate primary dose

Assuming a fluid deficit of 10%, the primary dose would be 5% of body weight in grams (eg, 3500-gram bald eagle's dose would be 175 ml).

Determine route of fluid administration

In most mild to moderate cases of dehydration, a combination of gavage and subcutaneous routes is sufficient. Initially, most of the volume (75%) should be delivered parenterally since the GI tract may not be efficient at absorption. This is gradually shifted to mostly orally as the GI tract is found to be more and more functional. Patients who are in shock or severely hypovolemic may benefit from initial large boluses (10 ml/kg slowly) of fluids delivered IV or IO. Common access sites include the basilic vein, the medial metatarsal vein, the jugular vein, or the distal ulna for IO access. Be sure to include any water used in mixing gavage diets as part of the primary dose.

Determine number of doses per day

As you will see below, the total amount administered per day is controlled by how frequently the primary dose is administered. For patients that are more severely dehydrated, increase the number of doses per day. A good rule of thumb is to increase by 1 dose per day for every 3-5% dehydrated above 10%.

Day 1. Administer dose every 8 hours. Increase frequency to every 6 hours for more severe dehydration. No food for initial 12 hours. If after 12 hours patient is alert and responsive, shows no signs of crop stasis, and produces mutes (droppings), gavage feeding of small amounts of a carnivore critical care diet may be considered.

Day 2. Administer dose every 12 hours. If patient is alert and responsive, shows no signs of crop stasis, and produces mutes, gavage feed gradually increasing amounts of a *carnivore* critical care diet. If mutes demonstrate digested material, a small amount of soft meat (no integument or bones) may be fed in the last half of the day in conjunction with gavage feeding.

Day 3. Administer dose every 12 hours. If patient is stable and demonstrates good GI function, soft meat is increased in conjunction with gavage feeding.

Day 4+. Continue to decrease fluids administration and increase volumes of soft meat while carefully assessing digestive function and attitude. When the bird is self-feeding and gaining weight, it may be transitioned to chunked whole prey (with integument and bones).

Fluid selection

Isotonic crystalloid fluids (eg, Normosol-R, Lactated Ringer's solution) commonly are used in raptors for parenteral administration but can be used orally as well. Commercially available non-flavored electrolyte solutions are acceptable for gavage (Pedialyte, Abbott, Chicago, IL, USA). 6% Hetastarch (10-15 ml/kg IV or IO) in conjunction with crystalloids (reduce dose by 50%) can be utilized in cases of severely depressed total serum solids (eg, < 2 mg/dl).

Re-alimentation

Gavage feeding

Clear fluids are recommended for at least the first dose to promote and assess movement of the GI tract while also allowing time for the process of fluid resuscitation to begin. In cases of severe dehydration, initial doses of oral fluids are poorly assimilated resulting in very loose, clear mutes. The first substances requiring digestion should be easy to digest and assimilate since enzyme systems may not have functioned for a protracted period of time and blunting of intestinal villi, important for absorption, has likely occurred. An elemental diet (Emeraid Carnivore, Lafeber Company, Cornell, IL, USA) may offer advantages in this regard since it should require little or no digestion. Volumes should be kept cautiously small initially and gradually increased.

Ball-tipped gavage needles of different sizes are appropriate for small raptors up to about 1000 grams. For larger birds, a catheter-tipped syringe and an 18 French red rubber catheter are recommended. The tube can be marked with tape or a marker to estimate the distance to the bottom of the crop. In birds lacking a crop, or if you intend to fill the stomach first, the mark should approximate the distance to the tip of the keel. The needle or tube is attached to the syringe and pre-lubricated with a water-soluble jelly (KY, Reckitt Benckiser, Slough, Berkshire, UK). The bird should be held upright on the restrainer's front such that the bird is well below the chest level of the veterinarian. The head is cradled in one hand with the neck lightly stretched upwards and the beak open (a thumb in the corner of the mouth works well). It's recommended to have a good view of the back of the throat during gavage feeding so that you can see reflux of fluid and respond accordingly. The neck must be extended upwards fully to allow good filling of the crop without reflux. The tube is slid past the glottis and down the esophagus, typically from the bird's left to right although in birds of prey, the exact direction of passage is not critical as they have extremely expandable throats. When the established mark is at the oral commissure, the position of the tube is verified. This is done by feeling the neck and crop but also direct visualization of the glottis at the base of the tongue. The fluid dose is then introduced smoothly and slowly, watching the gullet for any reflux. Usually this is due to not having the tube deeply enough or allowing the neck to relax too much. If reflux occurs, pull the neck straight and lift the tube slightly to encourage drainage into the crop. If fluids rise to the point of covering the glottis then withdraw your tube and tip the head sideways and down (while keeping the neck stretched) to allow fluids to drain from the throat. If performing gavage in conjunction with feeding, it should be performed prior to offering food. After feeding, to avoid reflux, care should be taken not to apply pressure to the crop after gavage feeding.

Solid food

Initially, only bits of clean meat should be offered. Clean meat is a term used to describe cut bits of meat where integument and bones have been removed. Alternatively, the skin can be removed and small whole prey ground in a meat grinder to create a safe, easily digested burger-like substance. The basic idea is to provide something that will not slow digestion or require more complex or energy-intensive processes such as pellet formation/

egestion. Bits of clean meat can be dipped in water or crystalloid fluids to moisten them but should not be soaked in fluid as this will leech out blood and nutrients. If the patient is bright and alert, the clean meat may be offered in a dish. Starved raptors still have quick reactions and less fear of humans so there is significant risk of injury when feeding them. It's best to place food in an enclosure before returning the bird or to present it in a covered dish and then uncover it quickly once placed on floor. If the bird is recumbent or depressed, the wisdom of offering any food should be questioned as it's a sign that they may not have sufficient hydration or energy reserves to accomplish digestion. With those that have impaired vision or are simply too stressed to feel comfortable eating, it can be helpful to offer bits of meat on long thumb forceps, either in a small enclosure or during restraint for other treatments.

Crop stasis or ileus

This can occur when food is present in the GI tract but the patient is too dehydrated to digest it. Other causes include obstruction or infectious disease (eg, trichomoniasis or candidiasis) of the GI tract. The crop will fail to empty, there may be an acrid, putrid odor from the mouth, and the bird may regurgitate fetid ingesta. Droppings will exhibit scant or absent feces. A resulting bacterial overgrowth can ultimately cause crop edema and endotoxemic shock.[7] Treatment for ingluvial stasis includes removal of food, either orally or by ingluviotomy, followed by oral antibiotics, such as trimethoprim sulfadimethoxazole at 48 mg/kg PO every 12 hours or clindamycin at 50-100 mg/kg PO every 12 hours until clinical signs resolve. Some authors also lavage or gavage diluted antiseptics (eg, chlorhexidine) into the upper gastrointestinal tract after removal of fermented crop contents. Radiographic images and other diagnostic tests are recommended to identify and address the underlying cause of ingluvial stasis. Food should slowly be reintroduced in conjunction with fluid therapy.

References

1. Klasing K. *Comparative Avian Nutrition*. New York, NY: CABI;2000.

2. Myers MR, Klasing KC. Low glucokinase activity and high rates of gluconeogenesis contribute to hyperglycemia in barn owls (Tyto alba) after a glucose challenge. *J Nutr*. 1999;129(10):1896-1904.

3. Jenkins J. Avian critical care and emergency medicine. In: Altman R, ed. *Avian Medicine and Surgery*. 1st ed. Philadelphia, PA: Saunders;1997:839-863.

4. Montesinos A, Ardiaca M. Acid-base status in the avian patient using a portable point-of-care analyzer. *Vet Clin North Am Exot Anim Pract*. 2013;16(1):47-69.

5. Raghav R, Middleton R, Ahamed R, et al. Analysis of arterial and venous blood gases in healthy gyr falcons (*Falco rusticolus*) under anesthesia. *J Avian Med Surg*. 2015;29(4):290-297.

6. Ford S, Chitty J, Jones M. Raptor medicine master class (bumblefoot, nutrition, stabilization). *Proc Annu Conf Assoc Avian Vet*. 2008;73-192.

7. Samour JH. Management of raptors. In: Harrison GJ, Lightfoot TL, eds. *Clinical Avian Medicine*. Vol 1. Palm Beach, FL: Spix Publishing;2006:915-956.

Section 9

Wednesday Master Classes

Geoff Olsen, DVM, DABVP (Avian Practice)
Moderator

Clinical Cardiovascular Anatomy and Disease States

Brenna Fitzgerald, DVM, DABVP (Avian Practice)

Session #2018

Affiliation: From The Medical Center for Birds, 3805 Main St, Oakley, CA 94561, USA.

Manuscript adapted from: Fitzgerald BC and Beaufrère H. Cardiology. In: Speer BL, ed. *Current Veterinary Therapy in Avian Medicine and Surgery.* St. Louis, MO: Elsevier Saunders;2016:252-328.

Abstract: Cardiovascular disease is common in companion birds. Recognition requires the practitioner to have a thorough knowledge of avian cardiovascular anatomy and physiology, and of clinical disease states and their risk factors. Of clinical significance are distinct anatomic and physiologic differences between avian and mammalian cardiovascular systems and adjacent respiratory structures. Atherosclerosis is likely an underlying factor in the majority of non-infectious cardiovascular diseases diagnosed in pet birds. Lesions are most frequently recognized in the ascending aorta, brachiocephalic trunks, and pulmonary arteries. Unlike humans, recognizable clinical disease is primarily the product of progressive, flow-limiting arterial stenosis rather than thromboembolism. Patients with clinically significant atherosclerosis most often present for falling or collapse, transient or persistent weakness and dysfunction of one or more limbs, with or without altered mentation. In addition to age, female sex, and species, risk factors for development of atherosclerosis also include high-calorie and fat diets, dyslipidemia (eg, hypercholesterolemia), and limited physical activity. Congestive heart failure is a frequent clinical end-point encountered in companion birds. Heart failure can result from primary myocardial failure, ventricular pressure overload, volume overload, conduction disturbances, or diastolic dysfunction. Dilative and hypertrophic cardiomyopathy have been reported in psittacine birds, as well as myocardial disease of infectious, metabolic, nutritional, neoplastic, or toxic origin. Pericardial effusion can develop due to inflammatory changes of the pericardium (pericarditis), trauma or aneurismal rupture, cardiac or pericardial neoplasia, or as a manifestation of metabolic derangements, toxic insult, or, commonly, right-sided congestive heart failure. Infectious pericarditis and associated effusion can occur as an extension of epicarditis, myocarditis, or inflammatory or infectious conditions affecting other adjacent tissues.

Introduction

Cardiovascular disease has traditionally been thought to be a rare occurrence in companion birds, but a growing body of evidence collected over the last few decades indicates otherwise. It is frequently encountered in practice and poses a serious threat to the quality of life and longevity of many avian species. Successful intervention requires a foundational understanding of relevant anatomy and physiology, heightened awareness of risk factors, and recognition of clinical disease states. This paper reviews avian cardiovascular anatomy, as well as recognized disease states and their pathophysiology, clinical and pathologic features, and prevalence.

Anatomy and Physiology

The high-performance cardiovascular system of birds has evolved to meet the high aerobic demands of flight and other intense activity by providing adequate oxygen transport with maximal efficiency. Although heart rate relative to body mass is lower in birds than in mammals, the avian heart is relatively larger and its stroke volume and cardiac output greater. The budgerigar in flight has a cardiac output approximately 7 times that of mammals during exercise. Mean arterial pressure is higher in birds than in mammals, whereas total peripheral resistance is lower.[1,2]

The avian heart is 4-chambered, with left and right atria and left and right ventricles. The left ventricle perfuses the systemic circulation, and the right ventricle perfuses the pulmonary circulation. As resistance to blood flow is lower in the pulmonary circulation than in the systemic circulation, the right ventricle must produce less systolic pressure; it is smaller and the myocardium is thinner than in the left ventricle. The left ventricular wall is 2 to 3 times thicker than the right ventricular wall and can generate 4-5 times the systolic pressure.[1,2] The cone-shaped left ventricle extends to the cardiac apex, and its right wall forms the interventricular septum. The right ventricle is a crescent-shaped cavity that wraps around at least one half of the left ventricle. Its free wall is continuous with the outer wall of the left ventricle and it does not extend to the cardiac apex.[1-3] Paired cranial venae cavae and the single caudal vena cava empty into the right atrium, while the paired pulmonary veins enter the left atrium.[1,2,4] The left atrioventricular (AV) (mitral) valve is membranous and tricuspid (although the cusps are poorly defined) and is attached to papillary muscles via chordae tendineae. These are lacking in the right AV valve, which consists of only a spiral muscular flap.[1,2,5] The pulmonary and aortic valves are each composed of three semilunar cusps.[1-3] The ascending aorta extends from the left ventricle and unlike in mammals, curves to the right. It gives rise to the coronary arteries and paired brachiocephalic trunks, each dividing into common carotid and subclavian arteries, then continues caudally as the descending aorta, giving off the coeliac and cranial mesenteric arteries as its first major branches. The pulmonary trunk arises from the right ventricle and splits into right and left pulmonary arteries.[1,2,4,5]

The heart is enclosed within a fibrous, relatively non-compliant pericardial sac, which normally contains a very small volume of serous fluid that serves to lubricate the movement of the heart within the pericardial sac. The outer fibrous layer of the pericardium is continuous with the surface of the great vessels and attachments exist between the pericardium and the sternal plate, liver, adjacent air sacs and, indirectly, to the vertebral column. The heart is thereby well anchored within the cranioventral coelom, its apex partially enclosed within the lobes of the liver.[1-3]

Adjacent respiratory structures include the paired, non-expansile lungs, resting dorsal to the heart, and the large, complex clavicular air sac which lies within the thoracic inlet and surrounds the heart base and great vessels.[5,6] The position of these structures relative to the heart has great clinical relevance given the potential for cardiovascular involvement in cases of primary respiratory disease.

The cardiac conduction system of birds bears a close resemblance to that of mammals, consisting of the sinoatrial (SA) node, atrioventricular (AV) node, the bundle of His, and its bundle branches. However, there are 3 bundle branches (left, right, and middle) as well as an AV ring of Purkinje fibers surrounding the right AV opening that is unique to birds. A figure 8 is formed as the middle bundle branch passes around the aorta and joins the AV ring. The SA node lies near the openings of the cranial and caudal venae cavae into the right atrium and its pacemaker cells set the heart rate. The wave of excitation initiated in the SA node propagates through the atrial muscle and is transmitted to the ventricles via the AV node, which gives rise to the Purkinje fibers of the AV ring, bundle of His, and the bundle branches. The location of the AV node is a subject of some debate, but is most likely located at the right base of the interatrial septum. Left and right bundle branches emerge from the interventricular septum to supply the left and right ventricles, following the tracts of the coronary arteries into the myocardium (periarterial Purkinje fibers).[1,2]

Vascular microanatomy

Structurally the arteries can be classified into the elastic arteries, consisting of the aortic arch, thoracic aorta up to the coeliac artery, brachiocephalic trunks, and extrapulmonary portions of the pulmonary arteries, and the muscular arteries comprising the remainder of the arterial system. The walls of elastic arteries are composed of 3 layers, the *tunica intima* which is a single-cell layer of endothelial cells adjacent to the arterial lumen, the *tunica media* which is composed of concentric layers of collagen and elastic fibers composed of smooth muscle cells and elastin, and the *tunica adventitia* which is an outer layer of collagen fibers and connective tissue. Muscular arteries have the same 3 tunicae, but the media is only composed of smooth muscle cells and elastin and the adventitia is much thicker and has less compliance.[2]

Systemic arterial blood pressure is a function of cardiac output and the resistance of the arterial system; the latter is partly determined by afterload. Afterload is determined by 1) impedance of the proximal aorta itself, 2) compliance (distensibility) of the elastic arteries, and 3) peripheral resistance to blood flow afforded primarily by small arteries and arterioles. The characteristic impedance of the proximal aorta increases with increased aortic stiffness and decreased radius, resulting in greater left ventricular afterload. Decreased compliance of elastic arteries, as well as increases in peripheral resistance (occurring with decreases in vessel radius), further contribute to afterload.[2,7]

Vasoconstriction (and venoconstriction) are modulated by sympathetic tone and circulating catecholamines (EPI and NE acting via α-adrenergic receptors). Vasoconstriction is also mediated by activation of the renin-angiotensin-aldosterone system (RAAS), arginine vasotocin (AVT, the homolog of antidiuretic hormone in mammals), and endothelin-1.[2]

Recognized Cardiovascular Disease States

Atherosclerosis and stroke

Atherosclerosis is a chronic inflammatory and degenerative disease of the arterial wall wherein the lumen narrows by progressive accumulation of inflammatory cells, fat, cholesterol, calcium, and cellular debris that forms fibrofatty atheromatous plaques within the intima. Potential complications include stenosis, ischemia, thrombosis, hemorrhage, and aneurysm.[8-10] Atherosclerosis is likely an underlying factor in the majority of non-infectious cardiovascular diseases diagnosed in pet birds and is undoubtedly the most common lesion of the cardiovascular system identified post-mortem in companion psittacine birds.

The etiology and development of atherosclerotic lesions can be broadly explained by the response-to-injury hypothesis which postulates that damage to the endothelium lining the artery sets the stage for atherogenesis. Lesion formation involves endothelial dysfunction, damage, and oxidative stress that promote increased endothelial permeability to lipoproteins and intimal adherence and migration of inflammatory cells (particularly monocytes). Subendothelial accumulation of lipoproteins, particularly oxidized low-density lipoproteins (LDL), further attracts monocytes, which once within the subintimal space, differentiate into macrophages and internalize lipoproteins and store cholesterol, thereby becoming foam cells. These cells eventually die by necrosis or apoptosis, resulting in further accumulation of extracellular lipid and necrotic debris. In this way, the process of intimal lipid accumulation, chemoattraction of macrophages, and formation of foam cells drives the progression of the atherosclerotic lesion and eventual development of a necrotic core. Smooth muscle cells play a significant role in lesion development by proliferating and migrating into the intima and subintimal space and producing extracellular matrix components (including collagen and elastic fibers). This accounts for the formation of fibrous tissue which in advanced lesions further enlarges the lesion and creates a fibrous cap overlying the lipid and necrotic core. Smooth muscle cells also accumulate cholesterol and become foam cells.[8-10]

In psittacine birds, lesions are characterized by accumulation of the components discussed above in the intima and luminal side of the vascular media. Lesions can be classified into 7 types (I-VII); types I and II are considered early lesions, type III intermediate, and types IV-VII advanced. Lesion type VI is distinguished from IV and V by the additional complication of fissure, hematoma, or thrombosis, and type VII are calcific lesions characterized by large calcium plaques and osseous metaplasia. Clinical signs are attributed to advanced lesions, while early and intermediate lesions are generally silent and subclinical.[8-10]

Atherosclerotic lesions are most frequently recognized in the ascending aorta, brachiocephalic trunks, and pulmonary arteries. Although not as often appreciated, lesions also appear in peripheral vasculature, most notably the coronary arteries, carotid arteries, and descending aorta, as well as the subclavian arteries, coeliac artery, and ischiatic arteries. Coronary and aortic aneurysms have been known to arise secondary to atherosclerosis owing to decreased compliance of the affected vessels compared to normal elastic arteries.[9]

Unlike humans, recognizable clinical disease in birds is primarily the product of progressive, flow-limiting arterial stenosis rather than thrombosis and hemorrhage of disrupted plaques that results in thromboembolism and acute arterial obstruction. The difference is likely accounted for by the inability of the avian thrombocyte to form shear-resistant arterial thrombi and 3-dimensional aggregates. Acute myocardial infarction is rendered less likely by the differences in avian coronary vasculature and greater collateral circulation compared to humans. In one study, advanced lesions were commonly characterized by severe arterial stenosis and occlusion, but atherothrombotic lesions, aortic dissection and rupture, and lesional hemorrhage were identified rarely (in 1.9% of lesions).[8-10]

Patients with clinically significant atherosclerosis most often present for falling or collapse, frequently accompanied by transient or persistent weakness and dysfunction of one or more limbs, and with or without altered mentation. Extension and rigidity of one leg (and sometimes the ipsilateral wing), clenching of the toes, ataxia or difficulty perching, and tremors or convulsions may be reported. These signs are considered most consistent with stroke, but rarely is this confirmed diagnostically. These events can be difficult to distinguish from seizure activity, and indeed seizures can occur subsequent to ischemic stroke. Signs may be acute and transient, or there may be persistent neurologic abnormities, including reduced mentation, blindness, seizures, vestibular signs, paresis of one or both pelvic limbs, and ataxia. Alternatively, sudden death may occur as a result of a cerebral infarct, aneurysmal rupture, or a lethal cardiac arrhythmia secondary to myocardial ischemia.

Reported prevalence of atherosclerosis in psittacine birds varies widely (1.9-91.8%), with the highest recently reported among Amazons and African grey parrots (*Amazona* spp. and *Psittacus erithacus*, respectively). Most recently, prevalence and epidemiology of clinically relevant atherosclerotic lesions (types IV-VI) were investigated by review of over 7600 psittacine cases representing 5 genera. An overall prevalence of 6.8% was reported, but a significantly higher prevalence of advanced lesions was found for grey parrots, Amazon parrots, and cockatiels (*Nymphicus hollandicus*); cockatoos (*Cacatua* spp.) and macaws (*Ara* spp.) appeared less susceptible. There was a positive association between advanced atherosclerosis and increasing age, female sex, reproductive disease (predominately female), hepatic disease, and concurrent myocardial fibrosis.[9] These findings are subjectively consistent with the author's experience in practice, where severe atherosclerosis is most often encountered in grey parrots and Amazon parrots over 20 years of age, followed by cockatiels, lovebirds (*Agapornis* spp.), Eclectus parrots (*Eclectus roratus*), cockatoos, and macaws.

In addition to age, female sex, and species, proposed (but unconfirmed) risk factors for development of atherosclerosis also include high-calorie and fat diets, dyslipidemia (eg, hypercholesterolemia), and limited physical activity. The predisposition of female psittacines to development of atherosclerosis likely relates to the profound effects of estrogens on lipid, protein, and calcium metabolism in reproductively active hens. Estrogen promotes increased plasma total cholesterol, triglyceride, calcium, and protein levels, as well as production of specific lipoproteins used to transport lipid from the liver (where lipogenesis predominately occurs in birds) to the developing yolk. These changes promote atherogenesis and may at least partially explain the association between reproductive tract disease and increased prevalence of advanced atherosclerotic lesions.[8,9] Species differences in plasma total cholesterol have been found to correlate with relative prevalence of severe atherosclerosis, with more susceptible species having significantly higher total cholesterol levels. One possible explanation is that some species are poorly adapted to the dietary changes related to captivity and are less equipped to metabolize dietary lipids. Diet-induced atherosclerosis has been demonstrated experimentally in budgerigars (*Melopsittacus undulatus*) and Quaker parrots (*Myiopsitta monachus*). Budgerigars fed a 2% cholesterol diet developed advanced atherosclerotic lesions by 3 months, which further progressed by 6 months; hepatic lipidosis developed concurrently. In a recent study, Quaker parrots fed a 1% cholesterol diet developed severe dyslipidemia and advanced atherosclerotic lesions in the aorta, brachiocephalic trunks, and coronary arteries within 4 months.[8-10] In the author's experience, companion birds ultimately diagnosed with advanced atherosclerosis often have a long-term history of a high-calorie and fat diet, regular provision of animal-based products (such as chicken bones, meat, and cheese), clinical obesity, and/or chronic reproductive activity (female birds).

Heart failure

Heart failure occurs when cardiac output becomes inadequate to maintain arterial blood pressure (manifested clinically by low output signs of lethargy, depression, and weakness) and when the heart is unable to empty the venous reservoirs, manifested by vascular congestion and transudation of fluid within tissues and body cavities (congestive signs). In the case of right-sided congestive heart failure, peripheral venous congestion, hepatic congestion, ascites, and pericardial effusion are often present. Pulmonary edema and congestion of the pulmonary veins occur with left-sided congestive heart failure, and a combination of signs may be seen with biventricular failure.[7] Compared to mammals, birds are thought to have a greater propensity for developing pulmonary hypertension and right-sided, rather than left-sided congestive heart failure owing to the morphology of the right AV valve, less deformable nucleated erythrocytes, and the rigid, non-distensible lungs which limit the ability of the blood capillaries to expand and accommodate greater blood flow.

Congestive heart failure is not a primary disease in itself, but an ultimate consequence of structural or functional abnormalities of the cardiovascular or pulmonary systems, compounded by the chronic effects of compensatory mechanisms.[7] Not all cardiovascular disease necessarily leads to congestive heart failure, but it is a frequent clinical end-point encountered in companion birds. The pathophysiology of congestive heart failure has been studied primarily in mammals, but is likely similar across vertebrate taxons owing to shared neuroendocrine regulatory pathways of circulation and hemodynamic constraints.

Heart failure can result from primary myocardial failure, ventricular pressure overload, volume overload, conduction disturbances, or diastolic dysfunction. Myocardial failure can occur with either primary disease of the myocardium or secondary to chronic pressure or volume overload. Pressure overload results from outflow obstruction (eg, from congenital or acquired valvular stenosis, narrowing of the aorta). The affected ventricle(s) ultimately undergoes concentric hypertrophy, in which the ventricular wall thickens with a corresponding decrease in chamber volume. With time, myocardial fibrosis develops and impairs both systolic and diastolic function. As myocardial failure progresses, the ventricle ultimately dilates. Volume overload results from valvular insufficiency (eg, endocardiosis, vegetative endocarditis) and abnormal communications between chambers (eg, ventricular septal defect). The result is eccentric hypertrophy of the affected ventricle(s), characterized by chamber enlargement and increased myocardial mass with little increase in wall thickness. With progression of hypertrophy, eventual myocardial failure ensues, with further chamber dilatation. Diastolic failure results from decreased ventricular compliance, abnormal myocardial relaxation, and cardiac tamponade or constrictive pericardial disease.[7]

Dilative and hypertrophic cardiomyopathy have been reported in psittacine birds, with dilative cardiomyopathy generally considered idiopathic.[2] Hypertrophic cardiomyopathy most often develops secondary to the arterial luminal stenosis and decreased compliance characteristic of atherosclerosis, or to pulmonary hypertension related to atherosclerosis, chronic pulmonary disease, or left-sided congestive heart failure.

Myocardial changes consistent with ischemic injury, including myocardial necrosis, inflammation, degeneration, and fibrosis, have been associated with atherosclerosis of the coronary arteries in psittacine birds. Hypertrophic cardiomyopathy develops secondary to chronic increases in afterload, as occur with the arterial luminal stenosis and decreased compliance characteristic of atherosclerosis. Myocardial ischemia, combined with increased workload and myocardial oxygen demand, can result in reduced contractility, arrhythmias, and reduced cardiac output that may ultimately culminate in congestive heart failure.[7,9]

Myocardial disease may also be of infectious, metabolic, nutritional, neoplastic, or toxic origin. Myocarditis can be caused by a variety of infectious agents, including bacterial, fungal, and viral pathogens, as well as protozoal and helminth parasites.[3] Bacterial infection may involve the epicardium, myocardium, or endocardium and occur by hematogenous spread or by extension from infected adjacent tissues. Fungal infection, most commonly with *Aspergillus* spp., can spread to the heart from the adjacent lungs and air sacs and disseminated infection with *Aspergillus* spp., *Candida* spp., and Zygomycetes can have cardiac involvement by angioinvasion and hematogenous spread. Proventricular dilatation disease (PDD), for which avian bornavirus (ABV) has been recently identified as the etiologic agent, is characterized histologically as lymphoplasmacytic infiltration of nerve ganglia throughout the body. Inflammatory infiltrates are frequently found in the epicardium and myocardium particularly in proximity to Purkinje fibers, potentially precipitating arrhythmias and sudden death. Other pathogens that can produce myocarditis are avian polyomavirus (APV), West Nile virus, *Sarcocystis falcatula*, and *Toxoplasma gondii*.

There are few published studies examining the prevalence and etiology of cardiac disease in psittacines and a relatively small number of case reports. Cardiac disease was recognized in 26 (9.7%) of 269 psittacine birds submitted for necropsy to Ohio State University, College of Veterinary Medicine between 1991 and 1995. In 15 (58%) of these, congestive heart failure was ruled the cause of death, while in the remaining 11 (42%) cardiac lesions were considered incidental and secondary to systemic infectious diseases including pulmonary aspergillosis, bacterial septicemia, APV infection, and PDD. In cases of right-sided congestive heart failure (10 birds), hepatic congestion, capsular fibrin, and ascites were consistently found. Pericardial effusion was identified in 3 of these cases. Pulmonary congestion and edema were observed in cases of left-sided congestive heart failure (5 birds), and birds with biventricular failure exhibited both hepatic and pulmonary lesions.[11]

In another study, 36% of 107 psittacine birds submitted for necropsy exhibited grossly visible abnormalities of the heart, great vessels, or both, while histologic changes in these tissues were found in 99% of birds; 15% had either myocardial hypertrophy or dilatation and 60% had inflammatory infiltrates into the myocardium. Myocardial hemorrhage (19%), aortic chondroid metaplasia (21%), and myocardial fibrosis (10%) were also identified. Atherosclerosis of the great vessels was found in 13% of cases and half of these also had calcification of the coronary arteries. Atherosclerotic disease and congestive heart failure accounted for the majority of non-infectious cardiovascular disease, while a similar spectrum of infectious disease states were identified as in the Ohio State study.[12]

In the author's practice, cardiac disease most consistently takes the form of congestive heart failure subsequent to advanced atherosclerotic disease, and as such, the represented age and species groups mirror those with higher prevalence of atherosclerosis. Cases of valvular insufficiency (including vegetative endocarditis), chronic pulmonary disease with probable hypertension, neoplasms or fungal granulomas encroaching on the heart base and great vessels, infectious myocarditis, cardiac arrhythmias, and congenital defects are encountered less frequently.

Pericardial disease and effusion

Pericardial effusion is characterized by an inappropriate accumulation of fluid within the pericardial sac. It can develop due to inflammatory changes of the pericardium (pericarditis), trauma or aneurismal rupture, cardiac or pericardial neoplasia, or as a manifestation of metabolic derangements, toxic insult, or, commonly, right-sided congestive heart failure. In some cases, an underlying cause cannot be identified and the condition is ruled idiopathic. Depending on the etiology, the effusion may be a pure transudate, modified transudate, or an exudate. Pure transudates accumulating within the pericardial sac (also termed hydropericardium) typically are the result of increased vascular permeability or low vascular oncotic pressure such as occurs with vasculitis or hypoproteinemia, respectively. Modified transudates accumulate with irritation or inflammation of the pericardium (pericarditis), hemorrhage into the pericardial sac (hemopericardium), or increased hydrostatic pressure, as occurs with central venous congestion subsequent to right-sided congestive heart failure.[3]

Infectious pericarditis and associated effusion can be a manifestation of systemic disease or occur as an extension of epicarditis, myocarditis, or inflammatory or infectious conditions affecting other adjacent tissues, including the air sacs, lungs, and liver. Bacterial, fungal, viral, and parasitic agents have been implicated. Visceral gout is a common non-infectious cause of pericarditis and pericardial effusion.[3]

Cardiac tamponade is defined as compression of the heart by a restrictive pericardium or accumulation of pericardial contents, resulting in impairment of ventricular filling (diastolic dysfunction) and subsequent decrease in stroke volume and cardiac output. Chamber filling requires positive intramural pressure (calculated as the difference of intracardiac pressure and intrapericardial pressure), but as intrapericardial pressure increases, intramural (filling) pressure progressively declines. Furthermore, given the Frank-Starling relationship between myocardial stretching and force of contraction, the systolic function of the underfilled cardiac chambers is diminished. The intramural pressure of the thinner-walled right ventricle is overcome more rapidly than that of the left. Therefore, cardiac tamponade results in right-sided congestive heart failure in advance of left-sided failure.

Cardiac tamponade may arise by constriction with a fibrotic pericardium subsequent to chronic, fibrinous pericarditis and adhesion to the epicardium. When pericardial effusion is the cause of compression, the clinical severity of cardiac tamponade depends in large part on the rate at which the effusion develops. As the pericardium is relatively non-compliant, acute accumulations of even small fluid volumes can have profound hemodynamic effects. Conversely, a more chronic onset of pericardial effusion allows time for expansion of the pericardial sac and compensatory physiologic mechanisms, including increased blood volume and heart rate via activation of the renin-angiotensin-aldosterone system and sympathetic stimulation.

Pericardial disease in psittacine birds has been examined in a handful of studies and case reports. In one study, pericardial effusion, accompanied by moderate to marked hepatomegaly, was identified in 6% of 107 psittacine birds submitted for necropsy. Among these birds, pericardial effusion occurred with infectious disease, visceral gout, or was accompanied by right ventricular dilatation and ascites consistent with right-sided congestive heart failure. In 15% of birds with pericardial effusion, fibrinous pericarditis was identified in association with chlamydiosis, psittacid herpesvirus infection, respiratory tract mycosis, PDD, or bacterial septicemia.[12]

In a survey of 60 psittacines and birds of other taxonomic groups presented with varying degrees of illness over a year period, 5 (8.3 %) were diagnosed with pericardial effusion via echocardiography. Four of these birds, 3 grey parrots and 1 blue-fronted Amazon parrot, were presented with lethargy, weakness, and pectoral muscle wasting. In the grey parrots, right ventricular dilatation, hepatic congestion, and ascites were also evident, as consistent with right-sided congestive heart failure. Severe lower respiratory mycosis was diagnosed in all 3 birds. Low-grade pericardial effusion in the Amazon had arisen secondary to right-sided congestive heart failure owing to incompetency of the right AV valve.[13]

In the author's practice, 43 cases of pericardial disease and/or pericardial effusion were reviewed. Infectious pericarditis was identified in 14 cases (32%), including 3 cases of APV infection, 5 cases of aspergillosis with extension from the respiratory tract to the pericardium, 1 case of suspected chlamydiosis, 3 cases of bacterial pneumonia with extension to the pericardium, and 2 cases of bacterial sepsis. Visceral gout occurred in 12 cases (28%) and hypoproteinemia with transudation into the pericardial sac in 2 cases of hepatic failure. Pericardial effusion related to congestive heart failure (either right-sided or biventricular) occurred in 9 cases (20%). One cockatiel developed acute hemopericardium and died as a result of cardiac tamponade. Pericardial thickening was noted in a grey parrot with multicentric lymphoma. In 5 birds (11%), pericardial effusion was ruled idiopathic.

References

1. King AS, McLelland J. Cardiovascular system. In: King AS, McLelland J. *Birds - Their Structure and Function.* 2nd ed. London, UK: BaillièreTindall;1984:214-228.

2. Smith FM, West NH, Jones DR. The cardiovascular system. In: Whittow GC, ed. *Sturkie's Avian Physiology.* 5th ed. San Diego, CA: Academic Press;2000:141-231.

3. Schmidt RE, Reavill DR, Phalen DN. Cardiovascular system. In: Schmidt RE, Reavill DR, Phalen DN, eds. *Pathology of Pet and Aviary Birds.* Ames, IA: Iowa State Press;2003:3-16.

4. Baumel J. Systema cardiovasculare. In: Baumel J, King A, Breazile J, Evans H, Vanden Berge J, eds. *Handbook of avian anatomy: Nomina Anatomica Avium.* 2nd ed. Cambridge, MA: Nuttall Ornithological Club;1993:407-476.

5. Nickel R, Schummer A, Seiferle E, Siller W, Wight P. Circulatory system. In: Nickel R, Schummer A, Seiferle E, Siller W, Wight P, eds. *Anatomy of the Domestic Birds.* Berlin: Verlag Paul Parey - Springer Verlag;1977:85-107.

6. King AS, McLelland J. Respiratory system. In: King AS, McLelland J. *Birds - Their Structure and Function.* 2nd ed. London, UK: BaillièreTindall;1984:110-144.

7. de Morais HA, Schwartz DS. Pathophysiology of heart failure. In: Ettinger SJ, Feldman EC, eds. *Textbook of Veterinary Internal Medicine.* 6th ed. St. Louis, MO: Elsevier Saunders;2005:914-940.

8. Beaufrere H: Atherosclerosis: comparative pathogenesis, lipoprotein metabolism, and avian and exotic companion mammal models. *J Exot Pet Med.* 2013;22(4):320-335.

9. Beaufrere H: Avian atherosclerosis: parrots and beyond. *J Exot Pet Med.* 2013;22(4):336-347.

10. Beaufrere H, Nevarez JG, Holder K, et al.: Characterization and classification of psittacine atherosclerotic lesions by histopathology, digital image analysis, transmission and scanning electron microscopy. *Avian Pathol.* 2011;40(5):531-544.

11. Oglesbee BL, Oglesbee MJ: Results of postmortem examination of psittacine birds with cardiac disease: 26 cases (1991-1995). *J Am Vet Med Assoc.* 1998;212:1737-1742.

12. Krautwald-Junghanns ME, Braun S, Pees M, et al. Research on the anatomy and pathology of the psittacine heart. *J Avian Med Surg.* 2004;18:2-11.

13. Straub J, Pees M, Krautwald-Junghanns ME: Diagnosis of pericardial effusion in birds by ultrasound. *Vet Rec.* 2001;149:86-88.

Diagnosis and Management of Cardiovascular Diseases in Birds

Brenna Fitzgerald, DVM, DABVP (Avian Practice)

Session #2020

Affiliation: From The Medical Center for Birds, 3805 Main St, Oakley, CA 94561, USA.

Manuscript adapted from: Fitzgerald BC and Beaufrère H. Cardiology. In: Speer BL, ed. *Current Veterinary Therapy in Avian Medicine and Surgery.* St. Louis, MO: Elsevier Saunders;2016:252-328.

Abstract: Cardiovascular disease is common in companion birds, but poses a particular diagnostic challenge, in part due to the limited sensitivity of available investigative methods. Successful intervention requires accurate and timely diagnosis, and the practitioner must be familiar with diagnostic options and their strengths and weaknesses. Key elements of a baseline diagnostic work-up include comprehensive serum biochemistry panel, complete blood count (CBC), radiographic imaging (potentially enhanced with contrast), ultrasonography, electrocardiography (ECG), and in some clinical settings, endoscopy and advanced imaging (computed tomography [CT] and magnetic resonance imaging [MRI]). Diagnosis begins with careful history collection and thorough physical examination. Clinical pathology, overall, has a low sensitivity for specific cardiovascular disease. ECG is an appropriate diagnostic step for characterization of arrhythmias, but must be used in conjunction with imaging techniques. Radiographic imaging allows assessment of the size, contour, and radiodensity of the cardiac silhouette and great vessels, while coelomic ultrasound allows identification of specific cardiac abnormalities, pericardial effusion, ascites, and hepatic parenchymal changes that cannot be distinguished radiographically. Angiography can supplement echocardiographic findings and identify vascular abnormalities. Endoscopy allows both direct visual inspection of cardiovascular structures and targeted collection of diagnostic samples. CT can be used to image the vasculature and to diagnose cardiomegaly, ventricular dilatation, pericardial effusion, ascites, pulmonary edema, and venous congestion. Successful intervention in avian cardiovascular disease requires innovative treatment approaches, incorporating both available information specific to birds and components of therapy in human and small animal cardiovascular medicine. Pharmacologic data in avian species is limited, but treatment approaches that have been reported, utilized by the author, or show promise for future application are discussed.

Introduction: Diagnosis

Cardiovascular disease has traditionally been thought to be a rare occurrence in companion birds, but a growing body of evidence collected over the last few decades indicates otherwise. It is frequently encountered in practice and poses a serious threat to the quality of life and longevity of many avian species. Successful intervention requires accurate and timely diagnosis, but cardiovascular disease poses a particular diagnostic challenge given the potential variability of its clinical manifestations (to include subclinical, slowly progressive disease), and the limited sensitivity of available investigative methods. These proceedings review current diagnostic options that can facilitate detection of pathologic changes. A baseline diagnostic work-up is appropriate in any patient presenting with clinical signs consistent with cardiovascular disease, in order to begin to assess overall systemic health status. Key elements may include comprehensive serum biochemistry panel, complete blood count (CBC), and radiographic imaging (which may include imaging enhanced with contrast), ultrasonography, electrocardiography (ECG) and blood pressure measurement to further characterize the nature and extent of cardiovascular disease. Endoscopy and advanced imaging, including computed tomography (CT) and magnetic resonance imaging (MRI), also have application in diagnosis and characterization of cardiovascular disease in some settings.

History, Clinical Signs, and Physical Examination Findings

Diagnosis of cardiovascular disease begins with careful history collection and thorough physical examination, to be performed in a manner in keeping with patient stability. Examination findings, as well as relative predisposition to cardiovascular disease given the species, age, gender, lifestyle, and diet should be taken under consideration when formulating initial differential diagnoses and prioritizing diagnostic maneuvers. Clinical presentations of cardiovascular disease share many commonalities, particularly as more than one condition can exist concurrently, but there are distinguishing features of some disease states that warrant mention. Many cases are likely subclinical and go unrecognized for many years until well advanced. Clinical signs can have a subtle, insidious onset, with owners reporting progressively declining activity level, reduced appetite, and waning interest in household activities, toys, and vocalizations that were historically of interest to the bird. In some cases, no premonitory signs are appreciated prior to sudden death. Presenting complaints common to most forms of cardiovascular disease are lethargy, weakness, reduced appetite, respiratory distress, and exercise intolerance. Physical examination findings may include cachexia, tachypnea, dyspnea, harsh lung sounds, cyanosis, tachycardia, arrhythmia, systolic murmur, coelomic distension and ascites, peripheral edema, and altered mentation.

Atherosclerosis

Patients with clinically significant atherosclerosis most often present for falling or collapse, frequently accompanied by transient or persistent weakness and dysfunction of one or more limbs, and with or without altered mentation. Extension and rigidity of one leg (and sometimes the ipsilateral wing), clenching of the toes, ataxia or difficulty perching, and tremors or convulsions may be reported. These signs are considered most consistent with stroke, but rarely is this confirmed diagnostically. These events can be difficult to distinguish from seizure activity, and indeed seizures can occur subsequent to ischemic stroke.[1] Signs may be acute and transient, or there may be persistent neurologic abnormities, including reduced mentation, blindness, seizures, vestibular signs, paresis of one or both pelvic limbs, and ataxia. In cases with concurrent congestive heart failure, dyspnea and exercise intolerance may be part of the history and appreciable upon physical examination along with other congestive signs.

Congestive heart failure

Birds with congestive heart failure present most consistently with respiratory compromise owing to either intracoelomic air sac compression caused by ascites and organomegaly, pulmonary edema, or a combination of the two. Coelomic distention is among the congestive signs appreciable upon physical examination in cases of right-sided or biventricular failure. Peripheral venous congestion and edema may also be noted. In the case of cardiogenic pulmonary edema, increased respiratory noise is typically heard upon auscultation of the lung fields in a calm bird, but characteristic "crackles and wheezes" are generally absent except in severe cases, and cough is exceptionally rare.

Clinical Pathology

Clinical pathology, overall, has a low sensitivity for specific cardiovascular disease. Abnormal findings are more often attributable to secondary effects on other organ systems, underlying primary disease processes, or to concurrent, unrelated disease. In principle, disease affecting the myocardium may cause elevations in aspartate aminotransferase (AST), creatine kinase (CK), and lactate dehydrogenase (LDH).[2] Other biochemical abnormalities may be seen depending on the systems secondarily or concurrently involved, particularly elevations in uric acid and bile acids reflecting renal and hepatic functional impairment, respectively. These may occur as a consequence of renal hypoperfusion or hepatic congestion secondary to heart failure, or as concurrent manifestations of an underlying disease process affecting multiple systems. Alternatively, hyperuricemia can be causally associated with cardiac disease (ie, visceral gout), as can hypoproteinemia (which may result in pericardial effusion) and electrolyte disturbances which can also produce cardiac arrhythmia. Hematologic changes, including elevated total white blood cell (WBC) count and shifts in the leukocyte differential are seen with many forms of inflammatory or infectious disease, as well as with stress, and can be seen in patients with cardiovascular disease.

Electrocardiography and Blood Pressure

Electrocardiography (ECG) is an appropriate diagnostic step for characterization of arrhythmias, and can provide information to suggest certain cardiac abnormalities, including chamber enlargement. However, it must be used in conjunction with other diagnostic modalities, namely imaging techniques, to diagnose specific cardiac disease. Severe pathology and mechanical dysfunction can exist in the absence of ECG changes.[2]

Accurate measurement of arterial blood pressure requires the direct method, for which an arterial catheter is placed; given that this technique is challenging and often impractical in the clinical setting, indirect methods are most commonly used in veterinary practice. However, it has been consistently demonstrated that values obtained with this method does not agree with direct systolic blood pressure measurements and may therefore be of low clinical value as a diagnostic tool.[3] In general, hypotension is defined as a systolic blood pressure lower than 90 mmHg and a mean below 60 mmHg. Values for hypertension have been poorly defined in birds, but are expected to be higher than in mammals owing to their greater blood pressure. Systolic values over 200 mmHg have been proposed as hypertensive.[4]

Radiography

Radiographic imaging is generally an insensitive diagnostic method to identify cardiovascular disease, but allows assessment of the size, contour, and radiodensity of the cardiac silhouette and great vessels, including the aorta, brachiocephalic trunks, and pulmonary arteries and veins.[5] In large psittacines, using high-resolution digital imaging, the common carotid and axillary arteries may be visible on the ventrodorsal view, while some major branches of the descending aorta, namely the coeliac and cranial mesenteric arteries, may be seen on the lateral view.

Enlargement of the cardiac silhouette can be seen with pericardial effusion, myocardial hypertrophy or inflammation, chamber dilatation, cardiac or pericardial masses, and aneurismal dilatation. A rounded, globoid cardiac silhouette is suggestive of pericardial effusion, but accurate confirmation and differentiation between possible etiologies is impossible without echocardiography. The size of the cardiac silhouette on the ventrodorsal view can be quantified by taking standardized measurements and calculating ratios relative to skeletal landmarks. In addition to enlargement of the cardiac silhouette, radiographic abnormalities that may be seen in cases of congestive heart failure include an enlarged hepatic silhouette and caudodorsal displacement of the adjacent gastrointestinal tract, increased radiodensity of the pulmonary parenchyma, intracoelomic air sac compression, or loss of coelomic detail due to ascites.[5] The hepatic silhouette may appear enlarged due to hepatomegaly, perihepatic effusion (fluid accumulation in the hepato-peritoneal cavities), or cardiomegaly or pericardial effusion, considering that the cardiac apex is positioned between the liver lobes. Increased pulmonary density, particularly of the reticular pattern, may be seen with cardiogenic pulmonary edema or with primary disease of the pulmonary parenchyma (inflammatory or neoplastic infiltration, fibrosis).[5] Varying degrees of compression of the cranial and caudal thoracic and abdominal air sacs are typically seen with ascites and/or organomegaly.

The clavicular air sac normally provides the contrast necessary to visualize the heart base and great vessels. When this space is lost or diminished, as occurs with space-occupying masses or accumulation of exudate within the air sac, visibility of the great vessels is obscured.

Increased radiodensity and tortuosity of the great vessels is suggestive, but not conclusive, of atherosclerosis.[5] Assessment of increased opacity is highly subjective and subject to variability in radiographic technique, though focal and linear mineralization associated with one or more vessels is highly suggestive of advanced, calcified atherosclerotic lesions. In psittacines, these appear most often along the ascending and descending aorta, brachiocephalic trunks, and occasionally along smaller arteries, including the coeliac artery.

Ultrasonography

Coelomic ultrasound allows identification of cardiac abnormalities, pericardial effusion, ascites, and hepatic parenchymal changes that cannot be distinguished radiographically beyond enlargements of the cardiac and hepatic silhouettes.[5] It can most often be performed in the awake or sedated patient, and represents a good option for those too compromised to withstand extended restraint or general anesthesia. Birds can be allowed to stand or be held in an upright position to reduce positional respiratory or circulatory compromise[5] and should ideally by fasted prior to examination to reduce interference from the gastrointestinal tract. Even so, the presence of echogenic material (eg, grit) within the ventriculus can hamper visibility.

Instrumentation and technique, anatomic limitations and approaches

Two-dimensional (B-mode) echocardiography is a well-established technique to assess morphologic and functional status in birds, but M-mode is not useful because only longitudinal and semitransverse views are attainable in most birds.[5-7] M-mode echocardiography is possible in pigeons and chickens. Minimal technical requirements of the ultrasound system include a frame rate of 100 frames per second, high-frequency probe ($\geq$ 7.5 MHz) with a small coupling surface (microconvex array, phased array, or one end of a linear array), Doppler function, and capability to record still images and video clips for later review.[5,7] Simultaneous ECG facilitates interpretation of images in relation to the cardiac cycle, but is not essential as end-diastolic and end-systolic frames can be visually selected.

Ultrasonographic examination is complicated by unique anatomic features of birds, most notably the presence of a large sternal plate and air sacs, that substantially limit available acoustic windows.[6,7] Cardiac views in particular are constrained by the situation of the heart in an indentation of the sternum and surrounded by the clavicular and cranial thoracic air sacs. Given these limitations, two approaches are possible for transcoelomic ultrasound: ventromedian and parasternal. The ventromedian approach is the most commonly used for psittacine and raptorial birds. The probe is placed just caudal to the sternum where a normally featherless area (apterium) lies along the ventral midline between adjacent feather tracts in most birds (excluding waterfowl). The beam plane is directed craniodorsally and the liver serves as an acoustic window, avoiding the air sacs laterally and the sternum ventrally.[5-7]

The heart can be visualized in two longitudinal views, horizontal (4-chamber view) and vertical (two-chamber view), the latter obtained by rotating the probe 90°. The horizontal view allows visualization of the left and right ventricles, interventricular septum, left and right atria, left atrioventricular (AV) valve, muscular right AV valve, and aortic valve, though it is difficult to distinguish the borders of the atria from the surrounding tissue. The right ventricle appears triangular in shape, tapering toward the cardiac apex. The vertical view permits visualization of the left atrium, left ventricle, and left AV valve; a marginal sliver of the right ventricle may also be visible adjacent to the sternum. Reflection of the ultrasound beam off the sternum may produce a mirror artifact in which a duplicated partial image of the heart appears on the opposite side of the sternum; this can create a false impression of chamber dilatation or aneurysm.[5-7]

The hepatic parenchyma should appear homogenous and of moderate echogenicity, with a fine, granular texture. Hepatic vessels appear as anechoic channels passing through the parenchyma. The parasternal approach can be utilized in birds whose ribs do not extend as far caudally, namely pigeons and gallinaceous birds, allowing sufficient space to accommodate the probe on the lateral flank between the last rib and the pelvis. The right lateral approach is generally chosen to avoid the ventriculus which often contains echogenic material that obstructs visibility of the heart. The leg is extended either cranially or caudally and the probe placed caudal to the last rib and dorsal to the lateral margin of the sternum, with the beam plane directed craniomedially. This approach provides longitudinal and semitransverse views of the heart.[5-7] Although views are limited for transcoelomic echocardiography, acoustic windows are greatly improved and visibility facilitated by cardiomegaly, hepatomegaly, ascitic fluid, and pericardial effusion, all frequently present in cases of congestive heart failure.

Subjective assessments

Subjective assessments of chamber size, wall thickness, ventricular contractility, diastolic function, and valvular morphology and function can be made.[7] Chamber dilatation, myocardial hypertrophy, valvular insufficiency and anomalies (to include vegetations), cardiac masses, septal defects, some aneurysms, pericardial effusion, ascites, and hepatic venous congestion can be identified by ultrasound assisted by color Doppler.[5-7] Ascitic fluid can be collected via ultrasound-guided coelomocentesis and pericardial fluid by pericardiocentesis.[5]

Hyperechogenicity of the walls of the ascending aorta is suggestive of mineralization that can accompany advanced atherosclerosis. In cases of right-sided congestive heart failure, the right ventricle is often demonstrably dilated such that it is of nearly equal size as the left. Thickening of the myocardium, to include the muscular right AV valve, is often appreciated. The right atrium may be dilated such that it is readily visible and pericardial effusion, when present, appears as an anechoic area between the heart and pericardium. Dilatation of the hepatic veins and uniform hyperechogenicity of the hepatic parenchyma are consistent with passive congestion.[7] Similarly, ventricular dilatation, myocardial thickening, and reduced myocardial contractility can be appreciated in cases of left-sided congestive heart failure and a combination of these findings is seen with biventricular congestive heart failure.

Cardiac measurements and calculation of functional parameters

An adequate morphologic and functional assessment can often be made qualitatively in cases where disease is advanced and pathologic changes are readily apparent. However, quantitative evaluation may allow the practitioner greater sensitivity in identifying more subtle changes and monitoring trends in a given patient over time. Ventricular and atrial dimensions, aortic diameter, and interventricular septum width can be measured in both systole and diastole, and some reference values are available for pigeons as well as for a few psittacine and raptorial species.[5-7] The probe is positioned such that the cardiac chambers appear maximally expanded and measurements are taken at their widest points using the inner edge method. It should be noted that accurate measurement of the atria (particularly the right atrium) can be hindered by indistinct borders and that of the right ventricle by its complex three-dimensional configuration and motion of the right AV valve. Measurements cannot be successfully taken in all individuals due to variation in image quality from one bird to the next. Ventricular widths can be used to calculate fractional shortening

$$\left[(diastolic\ diameter - systolic\ diameter)\times\frac{100}{diastolic}diameter\right]$$

for evaluation of ventricular contractility (reference values are available for some psittacine species).[5,6] Accuracy and precision of echocardiographic measurements have been questioned by some authors considering the small size of the structures imaged, rapid heart rates of avian species, current equipment resolution, and the fact that different observers can add up to 30% variability. Measurements should be taken from several cardiac cycles and averaged to obtain representative values. Follow-up measurements should be obtained by the same operator using the same equipment and technique, and only changes in measurements larger than 20% considered genuine. Transcoelomic echocardiography is also thought to underestimate the true fractional shortening of the highly efficient avian heart, as values obtained through transesophageal echocardiography were much higher.[8]

Angiography

Angiography, which can be performed using either fluoroscopy or computed tomography (CT), can provide an additional diagnostic modality to supplement echocardiographic findings, and to identify vascular abnormalities.[5,9] Fluoroscopic angiography allows visualization of the heart and vascular tree in real time. The brachiocephalic trunks, aorta, pulmonary arteries, pulmonary veins, and caudal vena cava can be visualized on the ventrodorsal view. Ventricular hypertrophy, chamber dilatation, valvular or vascular stenosis, and aneurysms can be identified using fluoroscopic angiography.[5] CT angiography (CTA) shows great promise for identification of vascular disease, including arterial calcification and luminal stenosis related to atherosclerosis, aneurysms, and congenital vascular anomalies.[9] Unfortunately, these techniques are not yet readily accessible to most veterinary practices.

Endoscopy and Advanced Imaging

Endoscopy allows both direct visual inspection of cardiovascular structures and targeted collection of diagnostic samples as appropriate. Endoscopy can only detect gross abnormalities, including cardiomegaly, pericardial effusion, pericardial thickening or exudate, arterial discoloration (to suggest atherosclerotic disease) or gross structural changes, and neighboring masses or granulomas. Standard lateral approaches permit visualization of the heart, great vessels, lungs, and liver via the cranial and caudal thoracic air sacs and of the descending aorta and its major branches, caudal vena cava, and ischiatic veins via the abdominal air sacs. The interclavicular approach (via the clavicular air sac) is the best means of visualizing the heart base, ascending aorta, brachiocephalic trunks, carotid and subclavian arteries, pulmonary arteries, and jugular veins. The ventral midline approach, with entry into the ventral hepato-peritoneal cavities, permits evaluation of the liver, heart, and pericardium.[2,9]

Computed tomography and magnetic resonance imaging (MRI) are seldom used to image the avian heart as scans cannot be gated to the fast cardiac cycle in birds to reduce motion artifacts and improve diagnostic value. However, CT can be used to image the vasculature and can readily diagnose arterial calcification associated with advanced atherosclerosis, as well as cardiomegaly, ventricular dilatation, pericardial effusion, ascites, pulmonary edema, and venous congestion. Cerebral complications such as ischemic and hemorrhagic strokes can be diagnosed using CT or MRI, though concurrent atherosclerosis cannot be detected unless lesion calcification is adequately severe.[1,9]

Ancillary Diagnostics

Fluid analysis, cytology, culture, histopathology, and pathogen-specific testing of antemortem diagnostic samples can assist in determining the etiopathogenesis of cardiovascular disease, particularly when it involves infectious disease or neoplasia. Certainly myocardial samples cannot realistically be obtained in the living bird, but pericardial and ascitic fluid, caseous exudates, and biopsies of the pericardium, other affected organs (such as liver, lung, and air sac), or tumors can be collected for analysis.

Introduction: Management

Successful intervention in avian cardiovascular disease requires innovative treatment approaches, which should incorporate both available information specific to birds and components of therapy in human and small animal cardiovascular medicine. Although limited pharmacologic data in avian species complicates effective medical management of cardiovascular disease, treatment approaches that have been reported, utilized by the author, or show promise for future application are discussed. Medications that have been used empirically, and those for which pharmacokinetic and pharmacodynamic data are available are presented in Table 1.

Table 1. Selected agents for treatment of cardiovascular disease in birds.

Drug	Species	Dose / Frequency	Basis
Diuretics			
Furosemide	Parrots, raptors, mynah birds	0.1-2.2 mg/kg IM, PO q6-24h	EU[a]
	Parrots	1-5 mg/kg IM q2-12h (acute treatment)	
		1-10 mg/kg PO q8-12h (maintenance)	EU
		2.5 mg/kg IM	(author)
		5 mg/kg PO	EU
	Chickens		(author)
	(*Gallus gallus domesticus*)		PD[b]
			PD
Spironolactone	Chickens	1 mg/kg PO	PD
	Grey parrot (*Psittacus erithacus*)	1 mg/kg PO q12h	EU
Vasodilators			
Isoxsuprine	Amazon parrot	10 mg/kg PO q24h	EU
	(*Amazona auropalliata*)		
	Parrots	10 mg/kg PO q12h	EU
			(author)
Enalapril	Pigeons (*Columba livia*)	1.25 mg/kg PO q8-12h	PK[c]
	Amazon parrots		
	Parrots	1.25 mg/kg PO q8-12h	PK
		2.5-5 mg/kg PO q12h	EU
			(author)
Benazepril	Grey parrot	0.5 mg/kg PO q24h	EU
Positive inotropes			
Digoxin	Budgerigars	0.02 mg/kg PO q24h	PK
	(*Melopsittacus undulatus*)		
	Quaker parrots		
	(*Myiopsitta monachus*)	0.05 mg/kg PO q24h	PK
	Sparrows		
	Indian ringneck parakeet		
	(*Psittacula krameri*)	0.02 mg/kg PO q24h	PK
	Indian hill mynah	0.025 mg/kg PO q24h	EU
	(*Gracula religiosa*)		
	Chickens		
		0.01 mg/kg PO q24h	EU
		0.01 mg/kg PO q24h	EU

Table 1 Continued. Selected agents for treatment of cardiovascular disease in birds.

Positive inotropes (cont'd)			
Pimobendan	Hispaniolan Amazon parrots (*Amazona ventralis*)	10 mg/kg PO q12h	PK
	Parrots	6-10 mg/kg PO q12h	EU (author)
	Harris' hawk (*Parabuteo unicinctus*)	0.25 mg/kg PO q12h	PK,EU
	Parrots	0.25 mg/kg PO q12h	EU
Dobutamine	Hispaniolan Amazon parrots (*Amazona ventralis*)	5-15 µg/kg/min (CRI)	PD
Dopamine	Hispaniolan Amazon parrots	5-10 µg/kg/min (CRI)	PD
Negative inotropes			
Carvedilol	Turkeys (*Meleagris gallopavo*)[d]	1-20[e] mg/kg PO q24h	PD
	Turkeys[d]		
Propranolol	Turkeys[d]	10 mg/kg PO q8h	PD
		10-30 mg/kg PO q8h	PD
Atenolol	Most species	0.2 mg/kg IM, 0.04 mg/kg IV	EU
	Turkeys[d]	10-30 mg/kg PO q24h	PD
Carteolol	Turkeys[d]	0.01-10 mg/kg PO q12h	PD
Nifedipine	Turkeys[d]	10-50 mg/kg PO q8h	PD
Diltiazem	Most species	1-2 mg/kg PO q8-12h	EU
Parasympatholytics			
Atropine	Most species	0.01-0.1 mg/kg IM,IV	EU
	Moluccan cockatoo (*Cacatua moluccensis*)	0.05 mg/kg IM	EU
Glycopyrrolate	Most species	0.01-0.02 mg/kg IM	EU
Other Antiarrhythmics			
Lidocaine (preservative-free)	Chickens	6 mg/kg IV	PD
	Chickens	2.5 mg/kg IV	PK,PD
Propantheline	Moluccan cockatoo	0.3 mg/kg PO q8h	EU

[a]Empirical use.
[b]Pharmacodynamic study.
[c]Pharmacokinetic study.
[d]Broad-breasted white turkey poults.
[e]Some birds receiving carvedilol at 20 mg/kg had significantly decreased heart rate and blood pressure for up to 8 hours.

Atherosclerotic Disease

Treatment of atherosclerosis involves both controlling risk factors and managing sequellae, including peripheral hypoperfusion, ischemic stroke, and congestive heart failure. Atherosclerotic lesions cannot be resolved, but diet, husbandry, and lifestyle changes may help to prevent, slow progression, or decrease the size of lesions. Peripheral vasodilation may produce clinical improvement by promoting improved peripheral perfusion and reducing afterload by decreasing vascular resistance. Isoxsuprine is a peripheral vasodilator that causes vascular smooth muscle relaxation predominately through α-adrenoceptor blockade. Its empirical use in birds with clinical atherosclerotic disease has resulted in apparent clinical improvement, including reduced frequency and severity of stroke-like events and neurologic signs.[10] Angiotensin-converting enzyme (ACE) inhibitors, including enalapril and benazepril, effect vasodilation by blocking the formation of angiotensin II (AII). Angiotensin II promotes vasoconstriction and venoconstriction by mediating release of catecholamines, which act on vascular smooth muscle via α-adrenergic receptors. Although the relative vasodilatory effect of an ACE inhibitor compared to isoxsuprine in birds is not known, it is conceivable that the two used in combination would have synergistic effects: an ACE inhibitor by limiting α-adrenoreceptor stimulation and isoxsuprine by α-adrenoreceptor antagonism. In mammals, pentoxifylline promotes passage of erythrocytes through damaged microvasculature by increasing their flexibility. It may have value in improving peripheral perfusion in birds with atherosclerotic disease. It has been used in birds as an adjunct therapy for frostbite, at a dose of 15 mg/kg PO q8-12h for 2-6 weeks with no apparent adverse effects.

Patients with signs of stroke may have marked neurologic deficits to include reduced mentation, limb paresis, and ataxia that prevent normal eating and drinking and impair mobility. They may experience seizures or suffer injuries from falls. Supportive care measures include fluid and nutritional support, analgesia, anticonvulsant therapy when needed, and management of secondary conditions such as trauma and aspiration pneumonia. Exercise restriction should be part of the longer-term treatment plan, as well as appropriate housing modifications to accommodate and protect birds with persistent deficits. Operant learning methods should be employed to facilitate low-stress medication administration. Managements that may also have preventative value, particularly for at-risk species, include moderation of dietary calories and fat to prevent and resolve obesity, avoidance of dietary sources of cholesterol, and control of female reproductive activity. It is possible that increasing opportunities for exercise beginning early in life, generally lacking for captive birds, may have preventative value.

Congestive Heart Failure

Appropriate treatment requires that known or hypothesized underlying cause(s) be addressed, if possible. The mainstays of treatment of congestive heart failure in small animal medicine, namely diuretics, ACE inhibitors, and positive inotropes, can be applied to treatment of the condition in birds. Beta-blockers, considered part of standard treatment in humans, may also have potential application. The goal of initial treatment is to stabilize the patient presenting with acute or decompensated failure, followed by design of a longer-term management strategy. In the author's experience, this has met with highly variable success, but some patients have been maintained in stable condition for up to 3 years.

Diuretics

Marked reduction of excessive circulating plasma volume (hypervolemia), edema, and effusion is an immediate treatment priority accomplished through the use of diuretics, principally furosemide. In the author's experience, a dose range of 1-5 mg/kg IM has been most efficacious when stabilizing psittacine patients with severely decompensated cardiovascular disease. An initial frequency of 2 hours is usually followed by a shift to 6- to 12-hour intervals. Once the patient has been stabilized, furosemide can be administered orally, with the dose increased at least 2-fold (presuming oral bioavailability is 60-75% as in humans) at a frequency of 8-12 hours. The ultimate goal for long-term management is to identify the lowest dose that controls congestive signs. Spironolactone may be used concurrently with furosemide when congestive signs cannot be controlled with furosemide and an ACE inhibitor alone.

ACE inhibitors

ACE inhibitors are an essential component of long-term medical management of congestive heart failure by blunting the effects of the renin-angiotensin-aldosterone system (RAAS). Enalapril has been the most commonly used ACE inhibitor in birds and empirical evidence suggests it is both safe and efficacious.[5] With inclusion of enalapril in the treatment regime, increased quality of life and longevity have been reported in birds with severe cardiac pathology.

Positive inotropes

Positive inotropes are used to enhance myocardial contractility and are appropriate for treatment of heart failure due to systolic dysfunction. They are contraindicated in cases of hypertrophic cardiomyopathy (where diastolic dysfunction is the primary problem), and of outflow obstruction (eg, aortic stenosis).[5] Likewise, it should be questioned whether their use is appropriate in birds with heart failure secondary to atherosclerotic disease and luminal stenosis of major arteries. In those cases with severe systolic dysfunction, however, it is fair to consider their inclusion in the treatment regime, at least in the short term, in an effort to stabilize the patient. Digoxin and pimobendan have been used or proposed for treatment of heart failure in avian species, but pharmacodynamic data is lacking and pharmacokinetic data and information as to their efficacy and margin of safety is extremely limited.[5] Digoxin, a digitalis glycoside, may have value in certain heart failure cases when used in combination with diuretics and ACE inhibitors, particularly when supraventricular tachyarrhythmia is a feature.[2] However, the narrow therapeutic index is a concern and chronic, maintenance administration may prove problematic.[5] Pimobendan enhances myocardial contractility without increasing myocardial oxygen consumption. Pharmacokinetic data suggests higher doses may be needed in psittacine species, relative to humans and dogs, to achieve therapeutic levels. However, extrapolation of this data to other species groups should be done with caution. Based on a case report of congestive heart failure in a Harris' hawk (*Parabuteo unicinctus*), the dose for pimobendan appears to be much lower for this raptor species.[11] Dopamine and doputamine, rapid-acting, intravenous sympathomimetics, increased heart rate and arterial blood pressure when administered by CRI to Hispaniolan Amazon parrots, but their effect on systolic function and cardiac output is unknown.

Negative Inotropes

Beta-blockers are part of core therapy for heart failure in humans, both in early and advanced stages. Treatment is generally initiated at a very low dose and gradually titrated upwards. The beneficial effects of β-blockers have also been documented in furazolidone-induced model of dilated cardiomyopathy (DCM) of turkey poults. Cardioprotective efficacy of β-blockers is dose-dependent and greatest with non-selective β-blockers; efficacy is ranked in descending order as: carteolol > propranolol > atenolol ≥ metoprolol. β-blockers were found to improve systolic performance and decrease mortality in turkeys with advanced furazolidone-induced heart failure. The cardioprotective effects of calcium channel blockers (nifedipine > verapamil) were found to rival those of β-blockers in furazolidone-induced DCM of turkey poults.[12] The author suggests that β-blockers may have application as adjunctive treatment of congestive heart failure, provided the condition is first stabilized using conventional management strategies. Hypothetically, β-blockers may also have merit for treatment of HCM in birds.

Antiarrhythmic agents

Prior to antiarrhythmic therapy, factors underlying or contributing to the arrhythmia (eg, hypoxia, electrolyte disturbances, intoxications) must be identified and addressed. It is only appropriate to treat arrhythmias that are symptomatic, causing hemodynamic instability, and complicating or precipitating heart failure. Aside from digoxin, β-blockers and CCBs are utilized in small animal cardiology for treatment of supraventricular tachyarrhythmias, including atrial fibrillation, in order to slow the ventricular response rate. Lidocaine is a parenteral antiarrhythmic agent used to control life-threatening ventricular tachyarrhythmias. Antimuscarinic agents atropine and glycopyrrolate are used to treat bradycardias, including SA arrest and first- and second-degree AV block, and to determine whether these arrhythmias are related to increased vagal tone or other cause.[5] Another indication is as an antidote for organophosphate and carbamate intoxication. Propantheline is an oral antimuscarinic agent used in the treatment of anticholinergic-responsive bradycardias.

Supportive care and husbandry considerations

General supportive principles apply to the avian heart failure patient, namely rest, minimizing stress with judicious, limited handling and restraint, and provision of supplemental oxygen when appropriate. Oxygen supplementation is indicated for patients with pulmonary edema, while physical fluid removal is more efficacious to stabilize those with air sac compression and cardiac tamponade. Patients may be presented in a severely debilitated state. In-patient supportive care measures may include fluid and nutritional support, but parenteral fluid administration must be carefully questioned. Maintaining the delicate balance between management of hypervolemia and concurrent dehydration and renal dysfunction is a profound clinical challenge, requiring close monitoring for changes in clinical status and adjustment of the treatment plan accordingly. Longer-term husbandry considerations, including dietary changes (which should also entail sodium restriction), and lifestyle and housing modifications, mirror those discussed for atherosclerosis. Given that congestive heart failure often arises secondary to atherosclerotic disease in psittacine birds, preventative considerations may be applicable to both conditions.

Pericardial Effusion and Cardiac Tamponade

Treatment is based first on removal of the fluid, and second on treatment of the underlying etiology of fluid accumulation. Fluid removal can be accomplished either by ultrasound-guided or endoscopic pericardiocentesis, or by endoscopic or surgical fenestration of the pericardium. Diuretics are contraindicated in cardiac tamponade because they decrease the preload necessary to maintain ventricular filling pressure and cardiac output; however, furosemide with or without an ACE inhibitor may be useful to reduce pericardial fluid of low volume related to congestive heart failure.

References

1. Beaufrere H, Nevarez J, Gaschen L, et al. Diagnosis of presumed acute ischemic stroke and associated seizure management in a Congo African grey parrot. *J Am Vet Med Assoc.* 2011;239(1):122-128.

2. Lumeij J, Ritchie B. Cardiology. In: Ritchie BW, Harrison GJ, Harrison LR, eds. *Avian Medicine: Principles and Applications*. Lake Worth, FL: Wingers Publishing Inc;1994:695-722.

3. Acierno MJ, de Cunha A, Smith J, et al. Agreement between direct and indirect blood pressure measurements obtained from anesthetized Hispaniolan Amazon parrots. *J Am Vet Med Assoc.* 2008;233:1587-1590.

4. Lichtenberger M, Ko J. Critical care monitoring. *Vet Clin North Am Exot Anim Pract.* 2007;10(2):317-344.

5. Pees M, Krautwald-Junghanns ME, Straub J. Evaluating and treating the cardiovascular system. In: Harrison GJ, Lightfoot TL, eds. *Clinical Avian Medicine.* Vol 1. Palm Beach, FL: Spix Publishing;2006:379-394.

6. Krautwald-Junghanns ME, Schulz M, Hagner D, et al. Transcoelomic two-dimensional echocardiography in the avian patient. *J Avian Med Surg.* 1995;9:19-31.

7. Pees M, Krautwald-Junghanns ME. Avian echocardiography. *Semin Avian Exot Pet Med.* 2005;14:14-21.

8. Beaufrere H, Pariaut R, Rodriguez D, et al.: Comparison of transcoelomic, contrast transcoelomic, and transesophageal echocardiography in anesthetized red-tailed hawks (*Buteo jamaicensis*), *Am J Vet Res.* 2012;73(10):1560-1568.

9. Beaufrère H, Pariaut R, Rodriguez D, et al. Avian vascular imaging: a review. *J Avian Med Surg.* 2010;24: 174-184.

10. Simone-Freilicher E. Use of isoxsuprine for treatment of clinical signs associated with presumptive atherosclerosis in a yellow-naped Amazon parrot (*Amazona ochrocephala auropalliata*). *J Avian Med Surg.* 2007;21(3):215-219.

11. Brandão J, Reynolds CA, Beaufrère H, et al. Cardiomyopathy in a Harris hawk (*Parabuteo unicinctus*). *J Am Vet Med Assoc.* 2016;249(2):221-227.

12. Glass MG, Fuleihan F, Liao R, et al. Differences in cardioprotective efficacy of adrenergic receptor antagonists and Ca^{2+} channel antagonists in an animal model of dilated cardiomyopathy. *Circ Res.* 1993;73(6): 1077-1089.

Section 10

Wildlife

Julia Ponder, DVM, MPH;
Maureen Murray, DVM, DABVP
(Avian Practice)
Moderators

On the Quest to Prevent Extinctions: Do the Costs Really Outweigh the Benefits?

Julia B. Ponder, DVM, MPH

Session #2080

Affiliation: From The Raptor Center, College of Veterinary Medicine, University of Minnesota, 1920 Fitch Avenue, St. Paul, MN 55108, USA.

Eradication of invasive rodents from islands is an increasingly used conservation tool to protect endemic island species and seabird colonies.[1] While successful eradications are possible, they present a dilemma. Broadly applied brodifacoum, the most commonly used eradication method, poses potential risks to native and endemic species.[2,3] Efforts to avoid non-target impacts must be integrated throughout the planning and implementation process. These efforts begin with *a priori* risk assessments to identify exposure pathways and potential harm to non-targets.[4,5] Potential risk management plans are developed and evaluated for feasibility. While risk assessments are based on the best available knowledge, uncertainty exists in predicting all possible outcomes and can lead to unexpected negative impacts. This uncertainty needs to be accounted for in planning, particularly when endemic, threatened or endangered species are involved.

Strategies used for risk management include avoidance of exposure, minimization of risk to non-target poisoning and mitigation of impacts after they have occurred.[6] Tactics used to protect native avian species include (1) timing of rodenticide application, (2) captive holding of at-risk species, (3) translocation, and (4) optimization of bait choice and presentation.[5] Despite implementation of these strategies, non-target impacts to native and endemic birds have occurred as a result of island rodent eradication projects.[7]

In this presentation, case studies will be used to demonstrate the benefits and risks of island rodent eradication to avian species. Removal of invasive mammals is a critical tool for conservation. There are penalties to be paid in the pursuit of success, however. An ethical approach is needed to balancing risk, benefit, cost and uncertainty.

References

1. Russell JC, Holmes ND. Tropical Island Conservation: Rat eradication for species recovery. *Biol Conserv.* 2015;185:1-7.

2. DIISE. The Database of Island Invasive Species Eradications. Available at: http://diise.islandconservation.org. Accessed April 4, 2017.

3. Eason CT, Murphy EC, Wright GRG, Spurr EB. Assessment of risks of brodifacoum to non-target birds and mammals in New Zealand. *Ecotoxicology.* 2002;11:35-48.

4. Broome K, Cox A, Golding C, et al. Rat eradication using aerial baiting: current agreed best practice used in New Zealand (v3.0). Wellington, New Zealand: Department of Conservation Internal Document. 2014:24.

5. Keitt B, Griffiths R, Boudjelas S, et al. Best practice guidelines for rat eradication on tropical islands. *Biol Conserv.* 2015;185:17-26.

6. U.S. Department of Agriculture Natural Resource Conservation Service. Pest Management Environmental Risk Analysis. Available at: http://directives.sc.egov.usda.gov/ViewRollUp.aspx?hid=17015&sf=1. Accessed April 4, 2017.

7. Hoare JM, Hare KM. The impact of brodifacoum on non-target wildlife: gaps in knowledge. *NZ J Ecol.* 2006;157-167.

Successful Decontamination of Polyisobutylene Exposure with Margarine in Avian Wildlife

Cameron M. Ratliff, DVM,
Sharon Schmalz,
Rachel Rech, DVM,
J. Jill Heatley, DVM, DABVP (Avian Practice), DABVP (Reptile and Amphibian Practice), Dipl ACZM

Session #2081

Affiliation: From the Department of Veterinary Clinical Sciences, Texas A&M College of Veterinary Medicine, 4474 TAMU, College Station, TX 77843, USA (Ratliff, Rech, Heatley), The Wildlife Center of Texas, 7007 Katy Rd, Houston, TX 77024, USA (Schmalz).

Two great blue herons (*Ardea Herodias*) and an anhinga (*Anhinga anhinga)* were presented to the Wildlife Center of Texas covered in a black sticky substance. Based on physical characteristics and historical evidence of a nearby chemical spill, the substance was determined to be polyisobutylene (PIB), a synthetic rubber polymer used in manufacturing. To address clinical signs and hematologic abnormalities, the birds were first provided supportive care upon triage, including fluid therapy, antibiotics, antiparasiticides and pain management as indicated. Margarine was recommended to emulsify the PIB, as detergents and oils were ineffective. Topical application of margarine successfully removed PIB from the plumage. One of the great blue herons died 1 day after removal of PIB. Necropsy findings included a friable liver and PIB contents within the gastrointestinal tract. Histopathology on select tissues is pending. The use of margarine for decontamination of PIB is unreported, and could prove useful in future decontamination events, especially in removing the substance from the plumage of avian species.

Anticoagulant Rodenticides in Birds of Prey in Massachusetts, USA: Recent Findings

Maureen Murray, DVM, DABVP (Avian Practice)

Session #2069

Affiliation: From Tufts Wildlife Clinic, Cummings School of Veterinary Medicine at Tufts University, North Grafton, MA 01536, USA.

Exposures to and mortalities from from anticoagulant rodenticides (ARs) in free living birds of prey are a well recognized problem in multiple countries.[1-4] Prior research in 4 species in Massachusetts, USA from 2006-2010 identified that 86% of 161 birds were positive for residues of second-generation ARs (SGARs), primarily brodifacoum.[1] The Environmental Protection Agency (EPA) issued restrictions on sales and packaging of ARs, which went partially into effect in 2011. These restrictions prohibit the sale of the more potent and longer acting second-generation ARs (SGARs) to general consumers, while allowing use by pest management professionals (PMPs) and agricultural users.[5] The current study examined the same 4 species in the same geographic area for exposure to and mortality caused by ARs from 2012-2016. Species included are red-tailed hawks (*Buteo jamaicensis*), barred owls (*Strix varia*), great horned owls (*Bubo virginianus*), and Eastern screech owls (*Megascops asio*). Sixteen birds diagnosed with AR toxicosis as the cause of death and 78 asymptomatic birds were tested. Ante-mortem and post-mortem signs of AR toxicosis were documented. Ninety-six percent of all birds were positive (100% of AR toxicosis diagnoses and 95% of asymptomatic birds). The majority of birds were exposed to more than one SGAR. Common presenting signs and post-mortem lesions include profound anemia, subcutaneous and/or intramuscular hemorrhage, excessive bleeding from a minor wound, and pallor of internal organs. These findings demonstrate that exposure to SGARs remains an ongoing risk to birds of prey and will assist clinicians in recognizing signs of toxicosis in these species.

References

1. Murray M. Anticoagulant rodenticide exposure and toxicosis in four species of birds of prey presented to a wildlife clinic in Massachusetts, 2006-2010. *J Zoo Wildl Med.* 2011;42:88-97.

2. Stone WB, Okoniewski JC, Stedelin JR. Anticoagulant rodenticides and raptors: recent findings from New York, 1998-2001. *Bull Environ Contam Toxicol.* 2003;70:34-40.

3. Huang AC, Elliott JE, Hindmarch S, et al. Increased rodenticide exposure rate and risk of toxicosis in barn owls (*Tyto alba*) from southwestern Canada and linkage with demographic but not genetic factors. *Ecotoxicology.* 2016;25:1061-1071.

4. Langford KH, Reid M, Thomas KV. The occurrence of second generation anticoagulant rodenticides in non-target raptor species in Norway. *Sci Total Environ.* 2013;450-451:205-208.

5. US Environmental Protection Agency, EPA. Risk mitigation decision for ten rodenticides. USEPA Office of Prevention, Pesticides and Toxic Substances. 2008. Available at: http://www.regulations.gov/#!documentDetail;D=EPA-HQ-OPP-2006-0955-0764 Accessed February 17, 2017.

Medical Rescue and Applications of Falconry Techniques during Rehabilitation of a Wild Bonelli's Eagle (*Hieraaetus fasciatus*) with a Chronic Tibiotarsal Fracture and a Cloacal Prolapse

Victor Daniel Mateuta Sirbu, DVM, and
Melodiya Nyela Fampo Magno, DVM

Session #2028

Affiliation: From the Wildlife Division, Wrsan, P.O. Box 77338, Abu Dhabi, United Arab Emirates.

Bonelli's eagles (*Hieraaetus fasciatus*), commonly sighted in the Middle East preying on captive falcons during the hunting season, were not favored and this impacted their welfare. This case describes an injured Bonelli's eagle after being trapped and kept in inhumane conditions, and the surgical and medical management of its pre-existing chronic cloacal prolapse and tibiotarsal fracture, as well as the role of falconry during its rehabilitation. The cloacal prolapse was infected, with myiasis and it partially occluded the cloacal opening. It was surgically corrected to invert the remaining healthy tissue into the cloaca. Moreover, radiographs revealed a closed, simple transverse fracture that resulted in an angular deformity of the distal tibotarsus and was enclosed by a fibrous callus. Surgery was performed using an external skeletal fixator and intramedullary pin, tie-in technique with 2 transverse pins fixed together with screwed clamps on an external bar. Two weeks post-surgery, the foot was fully functional for perching and gripping food. At 4 weeks, osteosynthesis incorporated the tibiotarsal fragments and callus. Fixators were removed as endosteal reconstruction was achieved. At 6 weeks, there was proper remodelling of callus. Complete endosteal union was achieved in the tenth week when molting commenced. Flight and stamina were restored using falconry techniques including manning, conditioning with flights towards the glove, free-soaring in the local wilderness with a transmitter, recovering by positive operant conditioning. After 3 months of training, molting commenced, which necessitated delayed release into the wild. Training will be resumed after completion of the molt, with the plan to release the eagle during the next migration season.

References

1. Bennett RA. Orthopedic surgery. In: Altman RB, Clubb SL, Dorrestein GM, Quesenberry K, eds. *Avian Medicine and Surgery.* Philadelphia, PA: Saunders;1997.

2. Bueno I, Redig PT, Rendahl AK. External skeletal fixator intramedullary pin tie-in for the repair of tibiotarsal fractures in raptors: 37 cases (1995-2011). *J Am Vet Med Assoc.* 2015;247(10):1154-1160.

3. Egger EL. External skeletal fixation—general principles. In: Slatter DH, ed. *Textbook of Small Animal Surgery.* 2nd ed. Philadelphia, PA: Saunders;1993.

4. Harcourt-Brown NH. Tendon repair in the pelvic limb of birds of prey. In: Lumeij JT, Remple JD, Redig PT, et al, eds. *Raptor Biomedicine III.* Lake Worth, FL: Zoological Education Network;2000.

5. Harcourt-Brown NH. Foot and leg problems. In: Beynon PH, Forbes NA, Harcourt-Brown NH, eds. *Manual of Raptors, Pigeons and Waterfowl.* Ames, IA: Iowa State University Press; 1996.

6. Martinez JE, Zuberogoitia I, Gomez G, et al. Attack success in Bonelli's eagle *Aquila fasciata*. Spain. *Ornis Fennica.* 2014;91:67-78.

7. Ontiveros D, Pleguezuelos JM. Infuence of prey densities in the distribution and breeding success of Bonelli's eagle (*Hieraaetus fasciatus*): management implications. *Biol Conserv.* 2000;93:19-25.

8. Redig PT, Ponder J. Management of orthopedic issues in birds. In: Samour J, ed. *Avian Medicine.* 3rd ed. St. Louis, MO: Elsevier Mosby Ltd;2016:506-511.

Development of an Eagle Detection System for Use on Wind Farms

Seth Oster, DVM

Session #2029

Affiliation: From Auburn University College of Veterinary Medicine, 1350 Pratt-Carden Rd, Auburn University, Auburn, AL 36849, USA.

Introduction

Wind energy is a growing form of energy production throughout the globe. In the United States, it accounts for 5% of energy production annually. At the state level, it accounts for 66% of energy production in Colorado, during windy seasons.[1] There are currently over 50,000 operating wind turbines in the United States.[2]

Bird collision with wind turbines has been widely documented. These collisions are fatal in most cases. Large birds, such as raptors, appear to have a higher relative rate of fatal collisions than smaller birds, like passerines.[3] Studies have been performed on various aspects of turbines, such as tower height and blade construction to determine relevance to bird collisions.[3]

Legal action was taken against wind farms by the U.S. Fish and Wildlife Service (USFWS). This resulted in changes to the construction and placement of wind turbines to lower the incidents of fatalities as well large fines against the wind farms. In addition the USFWS has mandated the wind energy industry to develop and deploy detection technology on wind farms to further reduce the incidents of collisions.[4]

Current methods of preventing collisions consist of trained human spotters working to identify protected species, such as eagles, moving in the area of wind farms. Spotters report to farm staff the locations of eagles and shut down turbines in their flight path. Spotters also do carcass surveys on site and report moralities to the USWFS.

Development of Eagle Detection Systems

Multiple private companies have sought to meet the demand for automated eagle detection equipment. Methods used include stereoscopic video and radar. These methods would allow for much larger observation areas than human observers can cover. Furthermore, software designed for these systems could send alerts to wind farm staff sooner than human observers could, allowing for more time to shut down turbines.

Two private companies (Lauder Wind and RES America) began work on detection systems at the National Wind Technology Center in Boulder, Colorado in early 2015. This wind lab is designed for testing various systems in the wind energy sector and is run by the National Renewable Energy Laboratory (NREL), a division of the U.S. Department of Energy.

In order to learn what an eagle looks like on stereoscopic video and radar, GPS coordinates of an eagle's flight path are needed to correlate with the data each of these systems collect. Due to the nature of the GPS data needed and the manpower required to operate these new systems, observation of wild eagles with these systems was ineffective in gathering the data required.

Use of Trained Eagles to Obtain Data

Auburn University maintains an educational collection of nonreleasable raptors. These animals are housed and trained at the Southeastern Raptor Center (SRC), which is fully permitted by the USFWS. Among the birds in the collection are a bald eagle and golden eagle. These eagles are both trained for free flights over large distances.

The SRC was contacted in October of 2015 by staff at NREL to discuss free flights on site in Boulder, Colorado. The goal of these flights would be to collect data for the radar and stereoscopic detection systems. SRC staff consulted with USFWS, after which they agreed to participate in this research.

In order to perform these trained flights on site in Boulder, a number of permits and documents were required. They included 2 National Environmental Policy Act worksheets by NREL, 2 safe work permits by NREL, 1 scientific collecting permit by USFWS, 1 scientific collecting permit by Colorado, 1 raptor importation permit from Colorado, 1 set of import numbers by USDA, 1 eagle exhibit permit by USFWS, 1 falconry permit by USFWS, 2 health certificates by USDA, and 1 Rocky Flats Wildlife Refuge special use permit by USFWS.

In addition, SRC outlined protocols for the trip. This included two trainers and one veterinarian to travel with the birds to the site. First aid equipment was kept on hand, and the staff at Colorado State University Veterinary Teaching Hospital agreed to provide assistance if more extensive care was required. The eagles were to be assessed visually during each flight and in-between each flight for any signs of injuries or abnormal behavior. Furthermore, each eagle received a physical exam at the end of each day to monitor for any abnormalities. These protocols were approved by Auburn University's Institutional Animal Care and Use Committee.

Traditionally, the flighted birds in the SRC collection are trained using the same methods as falconers. Birds are rewarded with food when they strike a presented lure, or leather target. These same methods would be used to call the eagles from their release site to the trainers in Boulder. These eagles were flown free during programs from July to December. The flights in Boulder were scheduled for February 2016. The eagles were maintained at a flight weight and flown to the lure 5 times per week during January to keep them in condition for February.

On site at the National Wind Technology Center, NREL staff stopped all other onsite projects, including any functioning wind turbines. NREL staff were also responsible for the safety of all personnel on site. This included having first aid on hand, distribution and use of safety equipment, and monitoring weather conditions. If wind speeds exceeded 15 mph, personnel would not be allowed to work in any elevated structures. If wind speeds exceeded 25 mph, personnel would need to shelter indoors. Up-to-the-second weather data was available onsite.

The two detection systems onsite were set up and calibrated prior to the arrival of the eagles. These systems were kept in fixed locations. Multiple sites were selected to release the eagles from. These multiple locations would provide the detection systems with different angles to view the eagles from. This allowed for a composite image of an eagle from all angles to be developed. Flight paths also were designed to allow SRC staff multiple options. Ideally, SRC staff wanted to fly the eagle with a tail wind, to prevent them from being blown off course.

During each flight, a GPS logger was attached to the tail of the eagle. These loggers would allow the researchers to correlate their data with the eagle's position on the site. Eagles were assessed based on body weight and behavior to determine if they were ready to fly. Behavior was assessed by offering a small amount of food to the bird and observing its response. If the eagle quickly left the glove and aggressively hit the leather training target (which had its food attached to it), this was considered a positive response and the eagle was deemed ready for flight. The eagle was then loaded into a commercial plastic crate (dog kennel) that was secured to a JLG lift (JLG Industries, Inc).

One SRC staff member was placed in the lift with NREL staff, while the other two SRC staff members were placed in the field to achieve the desired flight path. NREL staff was position around the SRC staff in the field to act as spotters if the bird did not land on target. The JLG lift was raised to a height of 100 feet. Once in position, the detection systems were calibrated. SRC staff on the ground signaled to SRC staff in the lift when they were ready. SRC staff in the lift made the final decision to release based on wind speed and direction. The bird then flew from the lift to the ground SRC using the training methods described earlier. Once the bird was secured on the ground, the JLG lift was lowered, the GPS logger collected, and the process repeated.

Overall 26 flights were performed over a course of 4 days. One of these days resulted in zero flights, as weather conditions did not permit flights that day. At the end of the first day of flights, both eagles were found to be very mildly dehydrated and were given 50 ml/kg SC Lactated Ringers Solution. Fluid loss was attributed to the relatively lower humidity of Colorado resulting in more evaporative losses. Soaked food items were fed for the remainder of the trip, and both birds maintained a normal hydration after this. No other health issues were encountered during this trip.

Both systems collected a sufficient amount of radar data to begin to build a model of an eagle in these systems. The next phase is to build a drone that mimics the radar signature of an eagle. An algorithm also must be developed to automatically detect the eagle's signature. This algorithm can be refined through further testing with the drone. The system then will be re-validated with live eagles again. These flights are tentatively planned for December of 2018.

References

1. U.S. Department of Energy. Frequently Asked Questions About Wind Energy. Available at: https://energy.gov/eere/wind/frequently-asked-questions-about-wind-energy Accessed May 30, 2017.

2. American Wind Energy Association. American Wind Power Breezes Past 70-gigawatt Milestone. Available at: http://www.awea.org/MediaCenter/pressrelease.aspx?ItemNumber=8255 Accessed May 30, 2017.

3. U.S. Department of Energy. AWWI Wind Wildlife Interactions Factsheet. Available at: https://energy.gov/sites/prod/files/2015/03/f20/AWWI-Wind-Wildlife-Interactions-Factsheet.pdf Accessed May 30, 2017.

4. U.S. Fish and Wildlife Service. Land-Based Wind Energy Guidelines. Available at: https://www.fws.gov/ecological-services/es-library/pdfs/WEG_final.pdf Accessed May 30, 2017.

Neurologic Disease in Raptors: Pesticide Toxicosis, A Cause That is Talked About but Rarely Diagnosed

David E. Scott, DVM

Session #2016

Affiliation: From the Carolina Raptor Center, 6000 Sample Road, Huntersville, NC 28078, USA.

Abstract: Neurologic disease in raptors is very common and there are many etiologies including trauma, infectious disease as well as various toxins. The treatment and prognosis are very much dependent on a quick and accurate diagnosis. Clinical disease due to pesticide toxicosis is often referred to but little actual clinical data is available. This study will summarize the findings of cases involving disease the central nervous system in wild raptors. The findings indicate that pesticide intoxication is a common cause and should be included in any differential list for neurologic disease.

Materials and Methods

All red-shouldered hawks (*Buteo lineautus*) that presented with signs of disease of the central nervous system (ie, tremors, seizures, extreme lethargy) in a 12-month period were included in this study. All were subjected to various diagnostic tests based on the clinical course. A definitive diagnosis was made in the majority of the cases and the results were tabulated.

Results

Sixteen cases were included. The most common identifiable causes of neurologic disease were determined to be pesticide toxicosis (6, 37%), trauma (5, 31%) and West Nile virus (1, 6%). Toxicosis with pesticides was diagnosed post-mortem using GC/MS on samples of liver or brain. The most commonly identified pesticides were dieldrin and chlordane. All cases of pesticide toxicosis where fatal within 24 hours of admission.

Conclusions

Toxicosis from pesticides should always be included in the list of differentials for any raptor admitted with neurologic disease. In this study, pesticide toxicosis was more common than trauma as a cause of admission.

References

1. Mineau P, Fletcher MR, Glaser LC, et al. Poisoning of raptors with organophosphorus and carbamate pesticides with emphasis on Canada, US and UK. *J Raptor Res.* 1999;33(1):1-37.
2. Mineau P, Tucker KR. Improving detection of pesticide poisoning in birds. *J Wildl Rehab.* 2002;25(2):4-13.

3. Redig P. Lead from spent ammunition: a source of poisoning in bald eagles. *Proc Annu Conf Assoc Avian Vet.* 2009;427-428.

4. La Douceur E, Kagan R, Scanlan M, Viner T. Chronically embedded lead projectiles in wildlife: a case series investigating the potential for lead toxicosis. *J Zoo Wildl Med.* 2015;46:438-442.

5. Behmke S, Fallon J, Duerr A, et al. Chronic lead exposure is epidemic in obligate scavenger populations in eastern North America. *Environ Int.* 2015;79:51-55.

6. Jackson A, Evers D, Collin A, et al. Mercury risk to avian piscivores across western United States and Canada. *Sci Total Environ.* 2016;568:685-696.

7. Pimentel D, Acquay H, Biltonen P, et al. Environmental and economic costs of pesticide use. *BioScience.* 1992;42(10):750-760.

8. Redig P, Arent L. Raptor toxicology. *Vet Clin North Am Exot Anim Pract.* 2008;11(2):261-282.

9. Pennsylvania Animal Diagnostic Laboratory. *Toxicology Report.* Kennett Square, PA: New Bolton Center;2016.

Diagnostic Use of Fibrinogen Levels in Brown Pelicans (*Pelecanus occidentalis*)

J. Jill Heatley, DVM, MS, DABVP (Avian Practice), DABVP (Reptile and Amphibian Practice) Dipl ACZM, Jordan Gentry, DVM, Amelia Gessner-Knepel, DVM, Johanna Heseltine DVM, MS, DACVIM, Karen E. Russell, DVM, PhD, DACVP, Sharon Schmalz, BS

Session #2061

Affiliation: From the Department of Small Animal Clinical Sciences (Heatley, Gentry, Gessner-Knepel, Heseltine) and the Department of Pathobiology (Russell), College of Veterinary Medicine and Biomedical Sciences, Texas A&M University, College Station, TX 77843, USA and from the Wildlife Center of Texas, Houston, TX 77024 USA (Schmalz).

Brown pelicans (*Pelecanus occidentalis*) often are affected by natural and manmade environmental disturbances such as hurricanes and oil spills and fishing and may require medical care and rehabilitation.[1] Reference intervals for this species are limited, dated, and typically from small sample sizes.[2,3] Seventy-one presumed healthy brown pelicans were sampled as a part of their pre-release examination from rehabilitation at the Wildlife Center of Texas after a large-volume stranding in December 2014 to January 2015. Select venous analytes were measured to establish reference data for brown pelicans. Fibrinogen also was measured using the well-established common, patient side test, the heat precipitation method and the Abaxis VSPro Equine Fibrinogen cartridge in an attempt to validate the VSPro for use in avian species. Abaxis VS2 Avian/Reptile Chemistry panel, iSTAT CG4+, and iSTAT Chem8+ results in addition to body condition score, spun PCV, cloacal temperature and fibrinogen were measured. Fibrinogen results were not comparable between the gold standard method and the VSPro, indicating that the VSPro is not appropriate for clinical use in brown pelicans. After initial scientific assessment of fibrinogen the Zoological Medicine service used Fibrinogen for pre-endoscopic assessment of health in another 8 brown pelicans presented for fishhook removal. Most pelicans survived. Based on student T test of the variables examined, only total solids was close to predicting outcome with $P = 0.196$ with dead birds having lower total solids.

References

1. Selman W, Hess TJ, Salyers B, Salyers C. Short-term response of brown pelicans (*Pelecanus occidentalis*) to oil spill rehabilitation and translocation. *Southeastern Naturalist*. 2012;11(1):G1-G16.

2. Zais J, Fox WP, Cray C, Altman NH. Hematologic, plasma protein and biochemical profiles of brown pelicans (*Pelecanus occidentalis*). *Am J Vet Res*. 2000;61(7):771-774.

3. Ferguson LM, Norton TM, Cray C, Oliva M, Jodice PGR. Health assessments of brown pelican (*Pelecanus occidentalis*) nestlings from colonies in South Carolina and Georgia, USA. *J Zoo Wildl Med.* 2014;45(4): 802-812.

4. Hawkey C, Hart MG. An analysis of the incidence of hyperfibrinogenaemia in birds with bacterial infections. *Avian Pathol.* 1988;17(2):427-432.

5. Miesbach W. et al. Comparison of the fibrinogen Clauss assay and the fibrinogen PT derived method in patients with dysfibrinogenemia. *Thromb Res*. 2010;126(6):e428-433.

6. Pindyck J; et al. The structural characteristics of chicken fibrinogen. *Biochim Biophys Acta.* 1977;492: 377-386.

Addendum. Texas A&M University case # 224413 Galveston pelican flock.

Band #	Problem	WBC	Heterophils (%)	Fibrinogen (mg/dl)	Plasma total solids (mg/dl)	Outcome
105	Single J hook at the isthmus to the ventriculus	26.7	56		5.4	Endoscopic removal successful
197	Hooks, myopathy sepsis	37.8	73	300	2.8	Necropsy: hematoma left neck, lung hemorrhage, necrosis and hemorrhage ventricular mucosa
120	Single J hook at the isthmus to the ventriculus	3.4	73	<100	5.0	Pelican #120 found deceased at 6:30 AM on morning rounds, endoscopy resulted in tear of tissue during retraction of the hook, post hook removal proventriculus had moderate to severe inflammation and trauma to the lumen walls. Necropsy: hemoceolom, airsacculitis, necrotizing myositis
230	Metallic treble fish hook in the entrance of the ventriculus	16.6	62	<100	4.3	Endoscopic removal successful
133	Single J hook at the isthmus to the ventriculus	20.9	71	300	4.8	Endoscopic removal successful
296	Hurt leg	26.1	73	500	3.8	1 hook removed from proventriculus, 2 from wings
297	Ingested fish hook, hooked wing	19.9	71	600	5.5	2 hooks removed from ventriculus
135	Metallic treble fish hook in the ventriculus, one of the prongs of the hook extends beyond the soft tissue contour of the ventriculus	42	73	1000	5.0	A tear of tissue during hook retraction hook and exam of proventriculus post-removal of hook revealed moderate to severe inflammation and trauma to the lumen walls, a large piece of tissue was removed with the hook and submitted for histopath for evaluation of perforation/full thickness tear, biopsy revealed nonfull thickness tissue, contrast study iohexol revealed intact GI tract however patient regurgitated iohexal into the lungs
119	Treble fish hook in ventriculus	11.4	73	300	4.9	Tear of tissue during hook retraction of the hook and examination of the proventriculus post-removal of the hook revealed mild inflammation and trauma to the lumen walls, a small barb of the treble hook was not removed with the other pieces and could not be appreciated on endoscopic exam
139	Sinker weight in the thoracic esophagus at the level of the coracoid with 2 treble hooks in the ventriculus	13	73	<100	4.8	Bird regurgitated weight and two treble hooks
191	Treble fish hook at entrance to ventriculus and metal sinker	18.4	59	<100	4.1	Endoscopic removal successful
249	Single J metallic hook in the proventriculus	9.1	69	<100	4.2	Endoscopic removal successful

Treatment for Bilateral Ulcerative Pododermatitis in a Hooded Vulture (*Necrosyrtes monarchus*)

Melodiya Nyela Fampo Magno, DVM

Session #2027

Affiliation: From the Falcon Research Center, Khattm Al Shiklah street, Al ain, United Arab Emirates P.O. Box 1200, Jurong Bird Park, Wildlife Reserve Singapore, 2 Jurong Hill Singapore 628925.

Captivity allows birds to thrive in safer environments, increasing their chance for potential propagation but one medical challenge commonly encountered is pododermatitis. This case describes the treatment of bilateral ulcerative pododermatitis in a hooded vulture (*Necrosyrtes monarchus*), a critically endangered species. Initial approaches to the severe pedal disorder consisted of moving the patient into an aviary with several perch types lined with grass carpet, and implementing empirical therapy of enrofloxacin and meloxicam to control severe inflammation and bacterial infection of *Escherichia coli* and *Proteus* sp. Wound contamination was managed by surgical debridement in conjunction with antibiotic therapy and donut bandages to relieve the pressure from the surgical site, and to promote blood perfusion. Although weekly wound dressing, application of topical anti-bacterial and granulation medications, and use of NSAID were implemented within 2 months upon initial presentation, these resulted in minimal impact to the edematous condition and the granulation of tissues. Because this kind of medical and surgical management did not eliminate the recurrence of pododermatitis, alternatively, the perch substrate was modified from grass carpet to fine, granular sand contained in a round container placed on top of a vertical log perch. In 10 days, the ulcerative pedal lesions revealed noticeable reduction of edema. In 2-3 weeks, marked degree of granulation was noted and complete healing was achieved in 1.5 months, post-modification. The significance of this study is that it demonstrates the impact of simple and practical conservative treatment such as changing the type of perch substrate in minimizing the post-surgery recovery time of ulcerative pododermatitis especially for critically endangered species.

References

1. Burke HF, Swaim SF, Amalsadvala T. Review of wound management in raptors. *J Avian Med Surg*. 2002;16(3):180-191.

2. Reidarson TH, McBain J, Burch L. A novel approach to the treatment of bumblefoot in penguins. *J Avian Med Surg*. 1991;13(2):124-127.

3. Gentz EJ. *Fusobacterium necrophorum* associated with bumblefoot in a wild great horned owl. *J Avian Med Surg*. 1996;10(4):258-261.

4. Sander S, Whittington JK, Bennet A, et al. Advancement flap as a novel treatment for a pododermatitis lesion in a red-tailed hawk (*Buteo jamaicensis*). *J Avian Med Surg*. 2013;27(4):294-300.

5. Cooper JE. Foot conditions. In: Cooper JE, ed. *Birds of Prey: Health and Disease*. 3rd ed. Hoboken, NJ: Wiley-Blackwell;2002:345.

6. Harcourt-Brown NH. Bumblefoot. In: Samour J, ed. *Avian Medicine*. 3rd ed. Mobsy Elsevier;2008:209-215.

7. Legler M, Seiler C, Siewert C, et al. Influence of lure flying and vertical jumping training on the skin temperature and radiation heat loss of the feet of falcons. *Wiener Tierärztliche Monatsschift*. 2016;103 (1-2):9-16.

8. Wyss F, Wenker C, Hoby S, et al. The effect of fine granular sand on pododermatitis in captive greater flamingos (*Phoenicopterus roseus*). *Anim Welf.* 2014;23(1):57-61.

9. Jacob FG, Baracho MDS, Nääs IDA, et al. The use of infrared thermography in the identification of pododermatitis in broilers. *Engenharia Agricola*. 2016;36(2):253-259.

Effects of Atorvastatin on Plasma Lipid Profile in Hispaniolan Amazon Parrots (*Amazona ventralis*)

Jessica A. Robertson, DVM,
David Sanchez-Migallon Guzman, LV, MS DipECZM (Avian), DipECZM (Small Mammal), Dipl ACZM,
James L. Graham, BS,
Jamie Douglas, DVM, MS,
Peter J. Havel, DVM, PhD,
Hugues Beaufrère, DVM, PhD, Dipl ACZM, DABVP (Avian Practice), DipECZM (Avian),
Valerie J. Wiebe, PharmD,
Heather Knych, DVM, PhD,
Joanne R. Paul-Murphy, DVM Dipl ACZM, Dipl ACAW

Session #2056

Affiliation: From the William R. Pritchard Veterinary Medical Teaching Hospital, (Robertson, Wiebe), Department of Medicine and Epidemiology (Sanchez-Migallon Guzman, Paul-Murphy, Douglas), Department of Molecular Biosciences (Havel, Knych) Department of Nutrition (Graham, Havel), School of Veterinary Medicine, University of California, Davis, One Shields Avenue, Davis, CA 95616 USA, Health Sciences Centre, Ontario Veterinary College, University of Guelph, Guelph, ON, N1G 2W1, Canada (Beaufrere).

Atherosclerosis and dyslipidemia commonly affect companion psittacine birds. As in primates and laboratory mammals, hypercholesterolemia is thought to be a risk factor for the development of atherosclerotic lesions in psittacines. Statin drugs are prescribed for treatment and prevention of dyslipidemia and atherosclerotic diseases in humans, as statins have been found to lower blood cholesterol, reduce atherosclerotic burden, and reduce risk of death from atherosclerotic diseases in humans. To investigate the plasma lipid-lowering effects and safety of atorvastatin use in parrots, 16 hypercholesterolemic Hispaniolan Amazon parrots were randomly assigned to receive atorvastatin or placebo suspension over 30 days. Birds were deemed otherwise healthy prior to study based on physical examination, packed cell volume (PCV), total solids (TS) and plasma biochemistry. Compounded atorvastatin suspension (10 mg/mL) was administered via gavage tube into the crop q24h at 10 mg/kg. Plasma lipid profiles were determined on day 0 (before drug administration), 14 and 30, which included determination of total cholesterol (TC), high-density lipoprotein cholesterol (HDL-C), non-high density lipoproteins (nHDL-C) and triglycerides. Following venipuncture, samples collected were centrifuged and supernatant HDL fraction was removed. Plasma samples and HDL fractions were then assayed for cholesterol and triglyceride concentrations via enzymatic assays. Subtraction of HDL values from total cholesterol yielded the non-HDL level for each assay. The results were analyzed using linear mixed models. There was no significant difference in TC, HDL-C, nHDL-C or triglycerides between treatment and control groups during at any time point. Further studies using different dosages and frequencies of atorvastatin are needed in Psittaciformes.

Acknowledgments: Support from Richard M. Schubot Parrot Wellness and Welfare Program and Louisiana State University School of Veterinary Medicine. Funding was provided by the Center for Companion Animal Health (CCAH) at the School of Veterinary Medicine, UC Davis.

Macrorhabdus ornithogaster in a Wild Yellow-rumped Warbler (*Setophago coronata*)

Cheryl B. Greenacre, DVM, DABVP (Avian Practice), DABVP (Exotic Companion Mammal Practice)

Session #2071

Affiliation: From the College of Veterinary Medicine, Department of Small Animal Clinical Sciences, 2407 River Drive, C-247, University of Tennessee, Knoxville, TN 37996, USA.

Macrorhabdus ornithogaster, also known as avian gastric yeast (AGY) and previously known as megabacterium, is an ascomycetous yeast that can cause proventriculitis and weight loss in a wide variety of species. The prevalence and clinical significance of AGY in wild birds is not well known, but a few cases have been reported including wild Eurasian siskins (*Carduelis spinus*) in Germany, and various species of finches in Britain. To the author's knowledge, this is the first report of AGY in a wild bird, in the United States. In 2009, necropsies were performed on three yellow-rumped warblers that were found dead near the same back yard feeder over a one week period. All were found to have salmonellosis with mild multifocal heterophilic and granulomatous hepatitis. One bird also had a mild chronic lymphoplasmacytic proventriculitis with fungi (*M. ornithogaster*), as well as a mild acute focal heterophilic ventriculitis, a mild enteritis with rare intraepithelial coccidia, and a fractured humerus. It is suspected concurrent salmonellosis precipitated clinical disease in this case as it did with the finch cases described in Britain.

References

1. Lanzarot P, Blanco JL, Alvarez-Perez S, et al. Prolonged fecal shedding of 'megabacteria' (*Macrorhabdus ornithogaster*) by clinically healthy canaries (*Serinus canaria*). *Med Mycol.* 2013;51(8):888-891.

2. Legler M, Stelter R, Jung A, et al. First detection of *Macrorhabdus ornithogaster* in wild Eurasian siskins (*Carduelis spinus*) in Germany. A case study. *Tierarztl Prax Ausg K Kleintiere Heimtiere.* 2015;43(3): 161-165.

3. Tomaszewski EK, Logan KS, Snowden KF, et al. Phylogenetic analysis identifies the 'megabacterium' of birds as a novel anamorphic ascomycetous yeast, *Macrorhabdus ornithogaster* gen. nov., sp. nov. *Int J Syst Evol Microbiol.* 2003;53(4):1201-1205.

4. Pennycott TW, Ross HM, McLaren IM, et al. Causes of death of wild birds of the family Fringillidae in Britain. *Vet Rec.* 1998;155-158.

Section 11

Posters

Management of Multiple Diseases in an Aged Cockatiel (*Nymphicus hollandicus*)

Stephanie Lamb, DVM, DABVP (Avian Practice)

Session #2046

Affiliation: From Arizona Exotic Animal Hospital, 744 N Center St Suite #101, Mesa, AZ 85201, USA.

Abstract: A suspected geriatric female cockatiel presented to the clinic after being found outside by a good Samaritan. A mass was present and identified as a squamous cell carcinoma. The patient was also found to be in congestive heart failure. Treatments included surgical debulking of the tumor, radiation, and supportive care for congestive heart failure. Although the patient did succumb to its diseases this case report describes its management.

Introduction

As animals age, they can be presented for veterinary care with a combination of ailments. Neoplasia, cardiac disease, atherosclerosis, cataracts, chronic malnutrition, and obesity are amongst the more common problems that can be seen.[1] Managing these disorders when they present in tandem can be more demanding than when they present as singular events.

Squamous cell carcinoma has been reported in various avian species and is often found within the alimentary tract or on the skin.[1-4] Diagnosis is made by biopsy and various treatments have been tried with different success. Surgical debulking, cisplatin, carboplatin, strontium-90 radiation, cobalt radiation, and cryotherapy have all been described.[2-5] Palliative care with debulking using a CO_2 surgical laser has also shown efficacy and resolution of the disease in some cases.[6]

Cardiac disease and atherosclerosis are being identified more and more in pet birds. Birds with atherosclerosis can present with signs of falling, transient of persistent weakness, ataxia, altered mentation, convulsions or sudden death. Clinical signs of congestive heart failure signs can vary based on if it is left or right sided but can include ascites, hepatomegaly, distended veins, edema and increased respiratory sounds.[7] Congestive heart failure can occur secondary to atherosclerosis. Diagnosis of these disorders often includes radiographic imaging and echocardiography. Additional diagnostics options including angiography, CT, MRI and endoscopy may also prove to be useful.[1,7] Treatment of congestive heart failure often includes medications including, but not limited to, furosemide, enalapril, and pimobendan.[1,7]

Atherosclerosis treatment often includes the use of peripheral vasodilation.[7] Additional supportive care measures may be necessary based on how the patient is presenting and can include oxygen supplementation, coelomocentesis, fluid or nutritional support.[7] Patients need to be closely observed and have their medications adjusted accordingly for the individual's signs, and any concurrent diseases. Certain medications may require close monitoring of the patients electrolytes and renal function.[7]

Reproductive disease is common in birds and can manifest in various ways. Soft shelled eggs, abnormal shaped eggs, egg binding and dystocia are just some of the problems that can be seen in female birds. Malnutrition from vitamins A, D, E, and selenium or calcium deficiencies have been known to result in these issues.[8,9] Treatment of disorders such as these mentioned involves correcting the underlying nutritional disorder if present and addressing the reproductive problem through medical or surgical management.

Clinical Report

A suspected geriatric female cockatiel presented to the hospital with an ulcerated wound on her preen gland. She had been found outside by a good Samaritan; therefore, it was unknown how long the lesion had been present. At physical examination, the ulcerated lesion was found to be present in an area that was about 13 × 10 mm and the skin was slightly raised. There also was a grade II/VI systolic heart murmur. Initial diagnostics included radiographs, a CBC and a biochemistry panel. The radiographic heart size measured within reference intervals but a calcified structure was found in the caudal left coelom. This structure was postulated to be an old, compressed egg. There were several opacities seen in the medullary cavity of numerous long bones. The biochemistry panel was interpreted to be within normal reference intervals. The patient was started on amoxicillin trihydrate/clavulanate potassium (125 mg/kg PO q12h for 14 days; Clavamox, Zoetis, Kalamazoo, MI, USA), meloxicam (1 mg/kg PO q12h for 14 days; Metacam, Boehringer-Ingelheim, St. Joseph, MO, USA) and silver sulfadiazine (apply to lesions q24h for 14 days; Silver Sulfadiazine Cream, USP 1%, Crow Laboratories, Johnson City, TN, USA). Although the presence of the egg in the coelom was considered to be abnormal, it was suspected to be incidental at that time and further treatment was not explored, outside of ensuring appropriate nutritional needs of the bird.

After 2 weeks the patient was rechecked and found to have no changes in the appearance of the preen gland lesion however the tissue was thicker and more mass like. A biopsy of the ulcerated lesion was performed with the use of conscious sedation including butorphanol (2 mg/kg IM once; Torbugesic, Pfizer, New York, NY, USA) and midazolam (1.5 mg/kg IM once; Midazolam Injection USP, Akorn Inc, Lake Forest, IL, USA) and local anesthesia. It was submitted for histopathology and squamous cell carcinoma was confirmed.

To further treat the squamous cell carcinoma, radiation therapy with strontium 90 was elected. The consulting oncologist requested that the mass be further debulked in order to have a better chance at a successful outcome with the strontium 90 treatment. Therefore, 1 week after the biopsy, the patient was placed under sevoflurane anesthesia (3-5% in 1.5 L O_2, Sevoflo, Zoetis, Kalamazoo, MI, USA) to obtain wider surgical margins. During the procedure the patient maintained a stable heart rate and rhythm and was breathing well on her own, though intermittent positive pressure ventilations were being administered by the technician. Due to the extensive nature of the mass the edges were able to be sutured but the center of the mass was left open. At the end of the anesthetic episode a bandage was being placed to cover the open portion of the mass and the patient's heart rate suddenly dropped and apnea occurred. Intermittent positive pressure ventilations were used every 6 seconds with continued oxygen flow at a rate of 1.5 L/min. A single dose of atropine (0.2 mg/kg IM once; Atropine sulfate, Med-Pharmex Incorporated, Pomona, CA, USA) was given and this increased the heart rate back to normal levels; however, the heart murmur became more audible. A 22-gauge needle was placed as an intraosseous catheter in the proximal right tibiotarsus, and a 1-ml bolus of fluids (Lactated Ringers Solution, Hospira, Lake Forest, IL, USA) was given slowly. A pocket of air was found to be forming on the left side of the body. This was suspected to be a result of an undiagnosed or new air sac rupture and it was aspirated. The patient slowly regained consciousness and the heart murmur intensity also reduced to its original intensity. The patient was continued on meloxicam that was previously prescribed for pain relief but also had tramadol (15 mg/kg PO q12 h; Tramadol, Amneal Pharmaceuticals of NY, Hauppauge, NY, USA) added in and doxycycline (25 mg/kg PO q24h for 21 days; Vibramycin, Pfizer Labs, New York, NY, USA) for antibiotic coverage since a significant portion of the surgery site was left open.

Three weeks later (6 weeks after initial presentation) radiation therapy with strontium 90 was initiated. The patient was heavily sedated with butorphanol (2 mg/kg IM once; Torbugesic, Pfizer, New York, NY, USA) and midazolam (1.5 mg/kg IM once; Midazolam Injection USP, Akorn Inc, Lake Forest, IL, USA) for this procedure due to concerns about the prior anesthetic episode. She received 6 treatments totaling 100 Gy. The probe was positioned so that there were 4 treatments on the dorsal surface and 2 on the ventral surface.

Two weeks after radiation therapy (8 weeks after initial presentation) the patient was found to be experiencing a mild increase in respiratory effort. On auscultation no rales were heard, but the lung sounds subjectively seemed louder. Repeat radiographs, CBC and serum biochemistry were performed. Although the CBC data remained within reference intervals, the uric acid was increased at 28.1 mg/dl. The bile acid level was elevated at 248 µmol/L. The total calcium was low at 6.8 mg/dl. Radiographs revealed the cardiac silhouette width to thoracic width measurement as 63%. The liver margins appeared enlarged and the caudal air sac space was reduced with a fluid or soft tissue opacity. Air sac lines were present at the lung and caudal thoracic air sac junction. The calcified structure in the caudal left coelom remained the same as previous radiographs. Treatments were altered, adding milk thistle (600 mg/kg PO q24h; Milk Thistle, Nature's Answer, Hauppauge, NY, USA) and lactulose (0.5 ml/kg PO q24h; Constulose Lactulose Solution USP, Actavis Inc, Morristown, NJ, USA). Meloxicam use was discontinued due to possible side effects on the kidneys. Tramadol (15 mg/kg PO q12h) and doxycycline (25 mg/kg PO q24h) were continued. Allopurinol was considered but withheld until the patient could be evaluated by a cardiologist.

One week later (9 weeks after initial presentation) the patient was seen by a cardiologist for an echocardiogram. This evaluation revealed regurgitant blood flow in the heart at the left atrioventricular valve. The liver appeared normal; however, there was scant ascites within the caudal coelom. Treatments recommended included furosemide (2 mg/kg PO q12h; Furosemide Oral Solution USP, Morton Grove Pharmaceuticals Inc, Morton Grove, IL, USA), enalapril (0.5 mg/kg PO q12h; Enalapril, Diamondback Drugs, Scottsdale, AZ, USA), and pimobendan (10 mg/kg PO q12h; Vetmedin, Boehringer Ingelheim, St. Joseph, MO, USA). Doxycycline was discontinued. Atherosclerosis was considered a differential in this case but treatment with isoxsuprine was decided against at the time of the cardiology consult. Approximately 48 hours after starting this treatment regime, the patient was breathing more comfortably and subjectively was more energetic.

One week after the evaluation by the cardiologist (10 weeks after presentation) the patient represented for a recheck examination. The patient was continuing to breathe well and was showing normal behaviors. A recheck chemistry panel showed the bile acids at 207 µmol/L and uric acid was 14.9 mg/dl. The calcium was within reference ranges. Medications were kept the same.

The patient's preen gland mass was being monitored on a weekly basis and at 11 weeks after the initial presentation the mass appeared to be increasing in size. Therefore, a second surgery was elected for further debulking. Heavy sedation was not enough this time to keep the patient comfortable and immobile, and so anesthesia with sevoflurane was performed. A radiocautery unit was used to shave off pieces of the mass and reduce its overall size. The mass was shaved down to what appeared to be normal tendon on the top of the caudal vertebrae and debulked at the edges to what appeared to be normal skin. The patient's recovery from anesthesia was slow, but she woke up without intervention. The patient was continued on furosemide, enalapril, pimobendan, tramadol, and milk thistle. Topical ointment (Heal X Soother Plus Topical Cream; HBD International Inc, Brentwood, TN, USA) was applied to help prevent infection at the site.

Four weeks after starting cardiac medications (13 weeks after presentation) the patient was again presented for evaluation. The bird was reported to be stable at home but had a mild increase in respiratory effort noted intermittently. This was confirmed on physical examination. The CBC remained in normal limits. The chemistry panel showed an improvement in bile acids level at 71 µmol/L. Uric acid level was increased to 50.0 mg/dl. Radiographically, there was reduced fluid opacity in the caudal coelom and more air sac space. The mineralized structure was still present in the caudal coelom and unchanged. Since the patient was breathing with increased effort intermittently there was concern that cardiac medications needed to be increased; however, with the increase in the uric acid level there was concern that increasing cardiac medications could result in dehydration and further impairment of renal function. It was elected to increase the enalapril dose to 1.25 mg/kg PO q12h but leave other cardiac medications the same.

Two days later the patient was found to be lethargic, dehydrated and in respiratory distress. Morning doses of furosemide and enalapril were skipped though other medications were continued. Lactated Ringer's solution (25 ml/kg SC once; Hospira Inc, Lake Forest, IL, USA) was given to help rehydrate but the bird's attitude declined gradually over the next several hours. Euthanasia was elected.

At gross necropsy, the patient had scant ascites in the caudal coelom, pulmonary edema and cardiomegaly. Erythema was noted on the ventricular serosal surface. The liver appeared small and the kidneys were pale. The cranial and middle right kidney divisions appeared markedly reduced in size. A small, compressed, mineralized egg with an irregular surface was present within the distal oviduct which was inactive. The mass and ulceration near the uropygial region remained the same as it had been two days prior.

Histopathology again confirmed squamous cell carcinoma at the uropygial gland region. The kidneys had mild to moderate interstitial fibrosis with renal tubular epithelial necrosis and intratubular mineralizations. The lung had diffuse severe pulmonary congestion and collapse with focal lipid pneumonia. The heart was found to have severe atherosclerosis and the liver was congested with mild Kuepfer cell proliferation and erythrophagocytosis.

References

1. Greenacre CB. Birds of a certain age-overview of geriatric disease. *Proc Annu Conf Assoc Avian Vet.* 2014;149-153.

2. Zehnder A, Graham J, Reavill DR, McLaughlin A. Neoplastic diseases in avian species. In: Speer BL, ed. *Current Therapy in Avian Medicine and Surgery.* St. Louis, MO: Elsevier Saunders;2016:107-141.

3. Nemetz LP, Broome M. Strontium-90 therapy for uropygial neoplasia. *Proc Annu Conf Assoc Avian Vet.* 2004;15-19.

4. Abu J, Wunschmann A, Redig PT, Feeney D. Management of a cutaneous squamous cell carcinoma in an American flamingo (*Phoenicopterus ruber*). *J Avian Med Surg.* 2009;23(1):44-48.

5. Pignon C, Mayer J. Radiation therapy of uropygial gland carcinoma in psittacine species. *Proc Annu Conf Assoc Avian Vet.* 2011;263.

6. Van Sant F, Munir K. Surgical laser treatment of squamous cell carcinoma. *Proc Annu Conf Assoc Avian Vet.*2015;137-139.

7. Fitzgerald BC, Beaufrere H. Cardiology. In: Speer BL, ed. *Current Therapy in Avian Medicine and Surgery.* St. Louis, MO: Elsevier Saunders;2016:252-328.

8. Schmidt RE, Reavill DR, Phalen DN. Reproductive system. In: Schmidt, RE, Reavill DR, Phalen DN, eds. *Pathology of Pet and Aviary Birds.* 2nd ed. Ames, IA: John Wiley and Sons, Inc;2015:145-160.

9. Speer BL. Diseases of the urogenital system. In: Altman RB, Clubb SL, Dorrestein GM, Quesenberry K, eds. *Avian Medicine and Surgery.* Philadelphia, PA: Saunders;1997:625-644.

Low-level Laser Therapy in Backyard Chickens: Something to Crow About (And You Can Eat the Eggs)!

Janet Gordon Palm, DVM, CVCP

Session #2072

Affiliation: From Animobility Integrative Veterinary Services, PO Box 749, Fruita, CO 81521, USA.

Abstract: In the last 7 years, low-level laser therapy (LLLT) has become an accepted modality in the areas of rehabilitation, wound healing, and pain management. Ongoing studies and numerous case responses are paving the way for veterinary confidence necessary to invest in this seemingly non-invasive treatment modality. Successful safe management of multiple inflammatory and systemic conditions in multiple species was performed using a non-heat-producing, frequency-specific, low-power, visible-light wavelength laser with line-generated beams. Another application exclusive and unique to this laser system is the ability to affect the autonomic nervous system as an aid in calming fractious animals, as well as increase in gut motility and appetite. Immune enhancement applications have also been experienced with supportive references in the literature. As with other species, the poultry industry could see benefit from this non-invasive modality, as there is no withdrawal time required for eating the eggs produced. It is now well documented that light energy stimulates the innate pharmacy within the body to advance healing and reproductive efficiency.

Introduction

The appropriate tool is vital to the ability to provide optimum results. Along with the surge of interest has been an equal amount of confusion, misinformation, and sometimes less than stellar results. Laser devices vary in power, wavelength, frequencies, and whether they are in fact a true laser, or an LED. There is a lack of clarity in what should be required knowledge of vital information regarding the device specifications prior to purchase. Knowledge of the benefits and limitations of the device allows appropriate and safe use as well as optimum results. Improper or cavalier use can result in poor results and occasionally harm. Although there can be no question that high power Class 3b and 4 units can have beneficial results in areas of joint and muscular rehabilitation, they have potential to cause thermal injury to tissues intolerant of this temperature variation. Using a heat-producing device on an acute injury or incision is encouraged by marketing representatives, justifying this on the basis of the heat-induced vasodilation to speed healing. It is this author's opinion, validated by consequences such as the "smell of burning feathers" and other evidence of thermal injuries, that this is an unacceptable risk, when non-invasive options are available.

A non-heat-producing Class 2 laser is a safer option that exists, having more versatility in accessing tissue and disease conditions that are, or should be, contraindicated for use with higher power Class 4 lasers with 500 mW power. These Class 2 lasers cause vasodilation and biostimulation without generating heat.

An insight into what is possible with the safety and versatility of using a Class 2 frequency-specific low-level laser (5 mW and 7.5 mW) with visible-light wavelengths of 635 nm (red) and 405 nm (violet) is offered. Examples include abilities to expedite major wound healing, manage musculoskeletal issues, arthritis, abscesses, infections, immune stimulation, and endocrine balance. Also included is the ability to affect the sympathetic and parasympathetic nervous system to calm animals and improve gut motility.

Versatility of Class 2 Low-Power Lasers

The now outdated concept that more power is essential to penetrate deep tissues, is challenged by obvious patient response. While this may be true to deliver a calculated therapeutic dose to a measured distance in a mathematic calculation or cadaver leg, it isn't relevant when considering a living biologic system with the properties of biophotons, biologic water, and the intercellular Integrin system of signal transduction. Therapeutic response can occur to deeper structures (6 cm) within a treatment time of 6 minutes, as measured by MRI (Erchonia Corp, McKinney, TX 75069, USA).[1] Established laser experts Tuner and Hode have since discussed that the idea of therapeutic dose needs to be re-evaluated.[2]

Albert Einstein proposed this stimulated emission of radiation phenomenon. Absorption of the light energy results in the photons exciting the electrons orbiting the nucleus of the atom to a higher orbit. When these electrons decay back to the ground state, they emit a photon to the neighboring atom, whose electrons excite, also emitting a photon as they retreat back to ground state, and the process continues in a 3 dimensional cascade effect eventually affecting the entire organism. Lending credence to this continuing cascade effect, Mester has shown that laser effects are systemic.[3] Studies on burn victims have shown the laser-treated area healed faster than the control group; however, the opposite non-treated area also healed faster than the controls. Incisions on rats behaved similarly. This concept is important to note, as many of the treatment protocols involving lameness involve treating the alternate legs simultaneously in order to strengthen tissues to aid in prevention of future breakdown. This has been successfully performed in managing canine ACL ruptures in the canine as well.

Communication and Case Response

Therapy laser is not a magic wand. Rehabilitation at the cellular level works best when considered as an adjunct to healing along with rehabilitation at the spinal and connective tissue levels with some form of neuronal modulation such as acupuncture, chiropractic, VOM and myofascial release techniques. Other building blocks for healing include adequate nutrition, hydration, exercise for lymphatic drainage and circulation, as well as mental enrichment to aid in managing the deleterious effects of stress.

Having performed thousands of laser therapies over 7 years on multiple species including birds, dogs, cats, pocket pets, horses, chickens, raptors, and other wildlife, the author is familiar with a myriad of responses. This can range from the "wow" response such as a hen with a fractured leg and questionable nerve sensation of 2 weeks duration, suddenly responding to toe pinch after two 60-second LLLT treatments, to gradual improvement noticed after several treatments as tendons and ligaments heal.

In the event of slight to no response within 6 laser treatments, the rationale includes several possibilities: 1) either the cells have rehabilitated to the extent they can, (prompting a review of nutrition, hydration, nutraceutical status) as healthy cells do not respond to biostimulation as pathologic tissues do, 2) there may be an underlying condition yet to be diagnosed, 3) the wrong diagnosis was made, or 4) the treatment protocol needs to be tweaked and thus the need for close follow up and progress exams.

Communicating these factors to owners is paramount to compliance and expectations. It is not practical to imply a prediction of who will receive a "wow" response, and who may gradually improve in 1-2 weeks. That is why close follow-up calling is imperative within 1-3 days and medical progress exams initiated when appropriate.

The Role of Frequency-specific Low-level Laser Therapy in Healing

There is controversy concerning the validity and importance of frequency (oscillations or pulsations) of the laser light in the overall effects of healing response. In this author's experience, as well as other operators of laser devices with similar specifications and capabilities, the answer is clear. Frequency does make a difference. It is definitely worth pursuing in an evidence base medicine research setting.

Considering that the String Theory of Quantum Physics suggests that the physical universe consists of a series of vibrations, James Oschman's book, *Energy Medicine: The Scientific Basis*, denotes that all tissues have a vibrational state unique to it.[4] Royal Raymond Rife discovered in the 1930s that varying the frequencies of micro electric current, different microorganisms are affected. Laser light, another form of electromagnetic energy, can be pulsed in varying frequencies. This altering of the vibration or oscillation of the beam is measured in Hertz.

When varying the laser frequencies, combined with the tissue-compatible (non-heat-producing) low-power, visible-light wavelength, different tissues can be affected more specifically. This allows the practitioner to practice the "art" of veterinary medicine as individualized treatment protocols can be prescribed.

Treatment Protocols

It is important to note that this is a Class 2 frequency-specific low-power laser. The protocols used with this laser system cannot translate to other lasers with higher power, and non-visible spectrum wavelengths. These unique specifications are found to be tissue friendly, non-heat-generating, and thus have more tissue versatility than other laser devices.

The treatment protocols vary on a case-by-case basis, which is where the true "art" of medicine comes in. Determining factors are as follows:

Physical exam and supporting diagnostics,
Diagnosis or list of differentials,
Chronicity of condition or symptoms,
Owner compliance to schedule and costs,
Response to treatment, and
Thus follow up is imperative!

A. Accelerated plan: severe trauma, pain

Two treatments 90+ minutes apart, 3 days in week 1, then 1 treatment/day for 3 days in week 2, then 1 treatment/day for 2 days/week for a total of 8-13 treatments.

60-120 seconds for each module.

70-90 minutes between treatments allows physiologic changes from biostimulation to occur in order to receive optimal benefit as a second treatment. LLLT is cumulative and the benefits will be seen sooner with the accelerated plan. Noticeable benefits may be witnessed within hours with acute injuries, or several days if healing connective tissues and fractures.

B. Chronic plan: long-term low-grade pain

Two treatments on day 1; 2 treatments once weekly for 2 weeks, then 1-2 treatments/week.

C. Try and see plan: diagnosis open or owner has financial limitations

This involves 2 treatments on day 1, then 1 treatment/week.[5]

Autonomic Nervous System Applications

The autonomic nervous system controls all life functions and is always in a state of balance. When the sympathetic tone is high (fright or flight) the parasympathetic tone is low, (homeostasis, digestion, appetite).

The sympathetic ganglion, reside along the paralumbar muscles between T1-L3-4. The line-generated beams of the laser device can access this entire area. As photonic energy is absorbed, causing the excitation of electrons to higher orbits, then subsequent photonic emissions to neighboring atoms in a cell-to- cell cascade, the sympathetic tone is continually stimulated. When the pre-programmed light stops in 60-120 seconds (180 seconds in mammals), the sympathetic tone plummets, allowing the parasympathetic tone to increase. Within minutes, a fractious, painful, or emotional animal can be calmed to a more harmonious state. Appetite, peristalsis, and focus are restored. Blood pressure has been documented to decrease.[6]

Similarly, the parasympathetic tone can be further enhanced. By following the sympathetic protocol, the calming can be even more stimulated. The areas to receive the laser light input would be cervical and sacral, where the parasympathetic ganglion reside.

It is important to note that the sympathetic stimulation should not be used in shock, anesthetic emergencies, or near death cases. In these cases, the parasympathetic stimulation alone is indicated. Further stimulation of a sympathetic tone in these fragile patients could be detrimental, and send them over the edge.

Prepare the staff to have the laser available, set for parasympathetic settings to have at hand for any shock, severe trauma, anesthetic emergency, collapse, or fragile state. This will save time in the event of an emergency. If there can be a fraction of a second plateau in an otherwise downward spiral that can make the difference between life and death, it is well worth it.

Immune System

Borrowing on the concept of varying the frequencies for specific tissues, allows for access to enhancing the immune system. Accessing the entire chest and abdomen with the line generated beams for the preprogrammed time (60-120-sec), allows stimulation of lymphatics including the thymus (bursa of Fabricius in birds) and spleen.

LLLT has been found to stimulate mesenchymal stem cell proliferation.[7] This involves lasering through the thoracic inlet, as well as over the entire chest and abdomen, especially the spleen. The light energy absorption and eventual response to the thymus, spleen and lymphatics is intended to speed the production of macrophages and killer T cells.[8]

Wounds, Tendons, Ligaments, and Neuropathies

The proposal that laser energy merely promotes cellular energy rather than changes cell function explains why injured tissue cells respond to laser therapy whereas there is little effect on non-injured cells.

Laser energy is more effective in treating pathologic states, therefore, when healthy subjects are used, the outcomes may be subdued. Although tissue healing is accelerated, no hyperplastic effects have been reported. During wound healing, laser wounds had more collagen and a higher tensile strength than the controls, until healing was complete, at which time both lasered and control groups were similar. Laser energy catalyzes normalization rather than creating a super normal effect.

Laser protocol

Varying degrees of wounds would determine mixing and matching protocols to the issues each individual demonstrates. A combination of the immune system enhancement, with treating over the nerve root or entire spine, along with settings to increase circulation and lymphatic drainage over the wound or soft tissue injury would be in order.

The violet wavelength and its documented bactericidal and sporicidal properties allows for benefits against bacteria and fungi. The author has used LLLT successfully to manage cellulitis, aspergillosis, and MRSA infections.

Arthritis, Degenerative Joint Diseases, and Fractures

Depending on the degree of pain, anxiety, the sympathetic overstimulation can be used prior to laying a hand on the animal. This author is an advocate of developing a rapport with the animal and managing his/her primary need before proceeding. Approach and retreat, calm and slow approach for the insecure, versus treat rewards for the more confident.

LLLT is used for pre and post-op surgical applications in soft tissue as well as orthopedics. Studies have shown that when using the bactericidal violet 405 nm 5 mW on osteoblasts in cell culture when studying infected arthroplasties, over 15 mW actually altered osteoblast function. It was concluded that 5 mW was the maximum power recommended to stimulate osteoblast function.[9] As a reminder, popular Class 3b units are generally over 60 mW and Class 4 are 500 mW. Use of the violet/red low-power laser in orthopedics for fracture repair and infection control is exciting.

Laser protocol

Sympathetic-parasympathetic balance should be performed if appropriate. Nervous system settings for the nerve root that innervates the area of involvement, as well as pain/inflammation settings for the area of involvement would be suggested. These can be accessed simultaneously with some Class 2 lasers that offer two probes, or sequentially as with single probes. The line-generated beam allows for entire spine, leg, or multiple legs lasered at the same time.

Endocrine Diseases

Endocrine diseases can also be treated, as there exists no contraindication due to the non-heat-producing qualities of this lasers ability to cause biostimulation. Although a cure is not suggested, there have been cases of reversal of early onset diabetes when pancreatic inflammation has been decreased, and islet cells stimulated to regenerate their insulin producing abilities. Egg production can be increased as reproductive tissues normalize as healing is advanced. This was demonstrated on multiple occasions with hens laying eggs immediately post trauma with the addition of LLLT to their treatment protocol. Low power laser irradiation using 633 nm at a radiation density of 0.1 mW/cm^2 produced a significant increase in the hatchability of layer and broiler eggs.[10]

For example, in the experiment using layer hen eggs that were stored for 20 days prior to incubation, the hatching rate for the control eggs ($n = 38$) declined to 47.22%, while the treatment eggs ($n = 83$) receiving red laser treatment one day prior to incubation was 74.32% ($P < 0.001$).[10] A study by Kusuma in 2013 showed use of a 632.8 nm 5 mW laser stimulated acceleration of growth, development and maturation of gonads, and stimulation of oocytes maturation in catfish. Laser puncture exposure at the reproductive acupoint for 15 seconds increases gonadotropin hormone production, which is the regulator for steroidogenesis, oogenesis, and oocyte maturation in catfish.[11]

Conclusions

The abilities of the frequency-specific low-level laser with its line-generated beam, low-power and visible-light wavelengths, allow safe treatment of a myriad of infectious, inflammatory, and degenerative diseases in these stress-sensitive animals. The ability to affect the autonomic nervous system to calm and stimulate appetite and gut motility, treat endocrine, gonads, and sensitive tissues not amenable to heat, shows promise. Managing arthritis and joint disease, as well as expediting wound, fracture healing, and neuropathies effectively without the additional risks of higher power units, has been proven.[12-17] The safe use and lack of need for eye protection allows for confident use and client and handler safety. In addition, the ability to treat from a safe distance without sedating the animal, with no risk of retinal damage, is invaluable.

References

1. Neira R. Low level laser-assisted lipoplasty appearance of fat demonstrated by MRI on abdominal tissue. *Am J Cosmet Surg.* 2001;18:3.

2. Hode L, Tuner J. *Laser Phototherapy-Clinical Practice and Scientific Background.* Coeymans Hollow, NY: Prima Books;2014.

3. Mester E, Jaszagi-Nagi E. The effect of laser radiation on wound healing and collagen synthesis. *Stud Biophys.* 1973;35:227-230.

4. Oschman J. *Energy Medicine: The Scientific Basis.* London, UK: Churchill Livingstone;2000.

5. Inman W. Advanced laser therapy. *Adv Techn Semin.* 2012.

6. Palm JG. Evidence supporting low power laser therapy: increasing applications increases success. *Erchonia Laser Healthcare Semin.* 2016.

7. Gianni F, Soares DM, Barreto MP, Barboza C. Effect of LLLT on mesenchymal stem cell proliferation: a systemic review. *Laser Med Sci*. 2015;30:2189-2194.

8. Yu W, Chin LH, Naim JO, Lanzafame RJ. Improvement of host response to sepsis by photobiomodulation. *Lasers Surg Med*. 1997;21(3);262-268.

9. McDonald RS, Gupta S. 405nm light exposure of osteoblasts and inactivation of bacterial isolates from arthroplasty patients: potential for new disinfection applications? *Eur Cell Mater*. 2013;25:204-14.

10. Yakimenko V. Effects of low-intensity red laser irradiation level on hatching eggs of chicken and quail. *Int J Poult Sci*. 2002;1(1):6-8.

11. Hariani D. Profile of catfish (*Clarius* sp) ooocyte exposed by laserpuncture. *J Biol Life Sci*. 2014;5:188-195.

12. Jann, H; Equine wound healing: Influence of low level laser therapy on an equine metacarpal wound healing model; Dept Veterinary Clinical Sciences Dept., Oklahoma State University, Stillwater, OK, 74078, USA

13. Gordon J. Cold laser therapy (3LT) in the management of avian inflammatory conditions. *Proc Annu Conf Assoc Avian Vet*. 2011;69-73.

14. Joensen J, Demmink JH, The thermal effects of therapeutic lasers with 810 and 904 nm wavelengths on human skin. *Photomed Laser Surg*. 2011;29(3):145-53.

15. Denegar C, Saliba E, Saliba SF. *Biologic Effects of Laser Therapy; Therapeutic Modalities for Musculo-skeletal Injuries*. 3rd ed. Champaign, IL: Human Kinetics;2010.

16. Karu TI, Afanasyeva NI. Cytochrome oxidase as primary photo acceptor for cultured cells in visible and near IR regions. *Doklady Akademii Nauk*. 995;342:693-695.

17. Karu T. *Ten Lectures on Basic Science of Laser Phototherapy*. Coeymans Hollow, NY: Prima Books;2007.

Tibiotarsal Fracture in a Toco Toucan (*Ramphastos toco*) Secondary to Nutritional Hyperparathyroidism

Jennilee Morrison, DVM,
Andrew C. Cushing, BVSc, Cert AVP (ZooMed), Dipl ACZM,
Cheryl B. Greenacre, DVM, DABVP (Avian Practice),
DABVP (Exotic Companion Mammal Practice)

Session #2084

Affiliation: From the Departments of Small Animal Clinical Sciences, College of Veterinary Medicine, University of Tennessee, 2407 River Drive, C-247, Knoxville, TN 37996, USA.

An 8-week-old, hand-raised female Toco toucan *(Ramphastos toco)* on a psittacine hand-feeding formula diet presented for evaluation of an acute non-weight bearing lameness of the right pelvic limb. Physical examination and radiographs revealed decreased bone density, mild varus deformity of the left tibiotarsus, and 2 fractures of the right tibiotarsus (closed transverse fracture proximal to a mid-diaphyseal greenstick fracture). The bird had low 25-hydroxyvitamin D_3 (5 nmol/L; various avian species reference interval 5-45 nmol/L). The fracture was repaired using an intramedullary pin with Type I external fixator tie-in. The bird was exercise-restricted and the diet changed to a low-iron hand feeding formula. Calcium carbonate (100 mg/kg PO q12h), and ultraviolet light exposure (12 h/day) were prescribed due to the diagnosis of nutritional secondary hyperparathyroidism. This diagnosis was identified as primary and renal secondary forms were excluded and no other causes could be identified. The bird was radiographed at 2, 4, and 6 weeks postoperatively. The fracture healed well. The apparatus was destabilized by removing the external fixator at week 4 and the intramedullary pin at week 6 postoperatively. Repeat vitamin D_3 levels showed improvement but still low levels at week 6 postoperatively (23 nmol/L). The calcium and UV light supplementation were continued. The clutchmate also was diagnosed with hypovitaminosis D_3, low bone density, and a mid-diaphyseal transverse fracture of the right tibiotasus, which responded to the same treatments provided to this female toucan, further supporting the diagnosis of nutritional secondary hyperparathyroidism.

References

1. Bennett RA, Kuzma AB. Fracture management in birds. *J Zoo Wildl Med.* 1992;23(1):5-38.

2. Martin HD, Ritchie BW. Orthopedic surgical techniques. In: Ritchie BW, Harrison GJ, Harrison LR, eds. *Avian Medicine: Principles and Applications.* Lake Worth, FL: Wingers Publishing Inc;1994: 1137-1171.

3. Degernes LA, Roe SC, Abrams CF. Holding power of different pin designs and pin insertion methods in avian cortical bone. *Vet Surg.* 1998;27(4):301-306.

4. Van Wettere AJ, Redig PT, Wallace LJ, et al. Mechanical evaluation of external skeletal fixator-intramedullary pin tie-in configurations applied to cadaveral humeri from red-tailed hawks (*Buteo jamaicensis*). *J Avian Med Surg.* 2009;23(4):277-285.

5. Van Wettere AJ, Wallace LJ, Redig PT, et al. Mechanical evaluation of various external skeletal fixator-intramedullary pin tie-in configurations using a tubular plastic bone model. *J Avian Med Surg.* 2009;23(4):263-276.

6. Stanford M. Calcium metabolism. In: Harrison GJ, Lightfoot TL, eds. *Clinical Avian Medicine*. Vol 1. Palm Beach, FL: Spix Publishing;2006:141-151.

7. Orosz SE, Monks D, de Matos R. Clinical endocrinology of the protein hormones. In: Speer BL, ed. *Current Therapy in Avian Medicine and Surgery*. St. Louis, MO: Elsevier Saunders;2016:379-399.

8. Dacke CG. Parathyroids, calcitonin, and vitamin D. In: Whittow GC, ed. *Sturkie's Avian Physiology*. 5th ed. London, UK: Academic Press;2000:472-485.

9. Stanford M. Effects of UVB radiation on calcium metabolism in psittacine birds. *Vet Rec*. 2006;159:236-241.

10. Lidor C, Dekel S, Edelstein S. The metabolism of vitamin D_3 during fracture healing in chicks. *Endocrinology*. 1987;120:389-393.

11. Bickle DD. Vitamin D metabolism, mechanism of action, and clinical applications. *Chem Biol*. 2014;21(3):319-329.

12. Stanford M. Clinical pathology of hypocalcemia in adult grey parrots (*Psittacus erithacus*). *Vet Rec*. 2007;161:456-457.

13. Tangredi BP, Krook LP. Nutritional secondary hyperparathyroidism in free-living fledgling American crows (*Corvus brachyrhynchos brachyrhynchos*). *J Zoo Wildl Med*. 1999;30:94-99.

14. Stanford, M. Measurement of 25-hydroxycholecalciferol in captive grey parrots (*Psittacus erithacus*). *Vet Rec*. 2003;153:58-59.

15. Phalen DN, Drew ML, Contreras C. Naturally occurring secondary nutritional hyperparathyroidism in cattle egrets (*Bubulcus ibis*) from central Texas. *J Wildl Dis*. 2005;41:401-415.

16. Morgan KJ, Alley MR, Gartrell BD, Perriman, L. Fibrous osteodystrophy in two Northern royal albatross chicks (*Diomedea sanfordi*). *NZ Vet J*. 2011;59(5):248-252.

17. Nichols D, Montali RJ, Pickett C, Bush C. Rickets in double-crested cormorants (*Phalacrocorax auritus*). *J Zoo Anim Med*. 1983;14 (3):115-124.

18. Adkesson MJ, Langan JN. Metabolic bone disease in juvenile Humboldt penguins (*Spheniscus humboldti*): investigation of ionized calcium, parathyroid hormone, and vitamin D_3 as diagnostics parameters. *J Zoo Wildl Med*. 2007;38(1):85-92.

19. Harcourt-Brown NH. Incidence of juvenile osteodystrophy in hand reared parrots (*Psittacus e. psittacus*). *Vet Rec*. 2003;152:438-439.

20. Stanford M. Measurement of ionized calcium in African grey parrots (*Psittacus erithacus*): the effect of diet. *Proc Euro Assoc Avian Vet*. 2003;276-282.

21. Howard LL, Kass PH, Lamberski N, Wack RF. Serum concentrations of ionized calcium, vitamin D_3, and parathyroid hormone in captive thick-billed parrots (*Rhynchopsitta pachyrhyncha*). *J Zoo Wildl Med*. 2004;35(2):147-153.

22. Woodhouse SJ, Rick M. The effect of UVB radiation on serum vitamin D and ionized calcium in the African spoonbill (*Platalea alba*). *J Zoo Wildl Med.* 2016;47(2):447-456.

23. Seibels B, Vince M. *Toucan Husbandry Manual for AZA Piciformes TAG*. Draft. Riverbanks Zoo;2001.

24. St. Leger J, Vince M, Jennings J, et al. Toucan hand feeding and nestling growth. *Vet Clin North Am Exot Anim Pract.* 2012;15:183-193.

25. Koutsos E, Gelis S, Echols MS. Advancements in nutrition and nutritional therapy. In: Speer BL, ed. *Current Therapy in Avian Medicine and Surgery.* St. Louis, MO: Elsevier Saunders;2016:142-176.

26. Teare JA. Ramphastos_toco_No_selection_by_gender__All_ages_combined_Conventional_American_units__2013_CD.html. In: Teare JA, ed. *ISIS Physiological Reference Intervals for Captive Wildlife: A CD-ROM Resource Species360.* Bloomington, MN: International Species Information System;2013.

Involvement of Birds of Prey in the Epidemiology of the Lyme Disease Agent *Borrelia Burgdorferi*: An Update

Markus Büker, Dr med vet, MSc, PhD, MRCVS, and
Klaus Hansen

Session #2085

Affiliation: From the Veterinary Practice for Companion Animals and Exotics "Tierarztpraxis am Hiddeser Berg," 32756 Detmold, North Rhine-Westphalia, Germany (Büker) and Detmold Municipal Zoo "Adlerwarte Berlebeck," 32760 Detmold, North Rhine-Westphalia, Germany (Büker, Hansen).

Abstract: Borrelia burgdorferi, the causative agent of Lyme disease, is the most important zoonosis-pathogen in the Northern Hemisphere.[1] Similar to small mammals, birds – primarily *Passeriformes* but also sea birds – play an important role in the transmission, distribution and maintenance of this disease.[2] Previous studies on birds have focused mainly on the detection of *B. burgdorferi*-positive ticks. The present update analyzes the presence of *B. burgdorferi* antibodies in sera from free-living birds of prey in Eastern Westphalia, Germany. Of 29 samples analyzed, 4 samples were positive: 2 in eagle owls (*Bubo bubo*) and 2 in common buzzards (*Buteo buteo*) by using a modified indirect immunofluorescent test (IFT). The results show that birds of prey are susceptible to *B. burgdorferi* and react immunologically to an infection. In addition, a peregrine falcon (*Falco peregrinus*) was presented with a swollen intertarsal joint, diarrhea and a reduced general condition. Radiographs were inconspicuous. Antibacterial treatment against the bacterium *Escherichia coli* found in the intestine and joint did not lead to success. A serologic testing for *B. burgdorferi* was positive. The bird recovered well after therapy against borreliosis similar to humans and mammals. In the future, it should be taken into account that birds of prey can be vectors for *B. burgdorferi*.

Introduction

The importance of the Lyme borreliosis, a tick-borne disease caused by the bacterium *Borrelia burgdorferi* sensu lato, is constantly increasing. It is the most frequent arthropod-borne disease in the Northern hemisphere today.[1] The annual worldwide number of reported human cases is about 85,000.[3,4] Lyme disease is a multi-systemic infectious disease. It appears that different genospecies have certain organ tropisms. The highest risk of infections results from infected tick nymphs, because they are easily overlooked due to their small size and are widespread. A large number of bird species, primarily ground-foraging passerines but also sea birds, act as competent reservoirs for *B. burgdorferi* and are capable of transmitting this spirochete to host-seeking ticks.[2,5] To date, no studies about *B. burgdorferi* antibodies in raptors or reports about clinical cases in birds exist.

Material and Methods

A total of 29 blood samples were obtained from 10 different species of birds of prey, mainly common buzzard (*Buteo buteo*, 11 samples), kestrel (*Falco tinnunculus*, 4 samples), barn owl (*Tyto alba*, 4 samples), and eagle owl (*Bubo bubo*, 3 samples) between September 2010 and April 2011. The samples were taken at the bird of prey shelter station Adlerwarte Berlebeck, Detmold, county of Lippe, North Rhine-Westphalia, Germany. Indirect immunofluorescence was used for the detection of IgG-antibodies against *B. burgdorferi* sensu lato (MegaScreen®

FLUOBORRELIA, MegaCor Diagnostik GmbH, Hörbranz, Austria). The test kit was modified using anti-chicken antibody-conjugate (Goat-Anti-Chicken IgG, AbD Serotec, Oxford, UK). The appropriate dilution was selected after a previous IFT with positive *Chlamydophila psittaci* reference samples from birds. For negative controls, 20 blood samples from indoor-housed laying hens (*Gallus gallus domesticus*) were taken.

Results

Of 29 samples analyzed, 4 samples (2 in eagle owls and 2 in common buzzards) showed titres between 1:64 and 1:256 (13.8%) and were assessed as positive according to a previous experimental study performed by Burgess.[6]

Case Report

A 3-year-old female peregrine falcon (*Falco peregrinus*) weighing 980 g was presented due to a mild swollen left intertarsal joint. The owner observed diarrhea the previous 5 days. The bird was kept at a weathering of about 5 × 2.5 m and was trained for falconry. On clinical examination, the falcon showed a reduced general condition and a swollen left intertarsal joint. There were slight signs of diarrhea.

Following on these clinical signs, radiographs were taken (latero-lateral and dorso-ventral beam) of the whole bird, which revealed a periarticular soft-tissue swelling without lysis of the articular surfaces of the left intertarsal joint. The walls of the intestinal loops were thickened and showed slight gaseous distention. Blood samples for biochemical and hematologic analysis were taken from the brachial vein (*V. ulnaris*). Interestingly, all parameters were within normal ranges.

The parasitologic fecal examinations (direct smear and flotation process) were negative in both cases. Subsequently, bacteriologic examinations of the feces and the periarticular soft-tissue swelling were performed. *Escherichia coli*, a gram-negative nonspore-forming bacillus could be isolated from the cloacal swap and also from the joint in middle-rate quantity. An antibacterial treatment with enrofloxacine 10 mg/kg PO q24h (Enrotron 25, Levisto, Sendenhorst, NRW, Germany) was performed based on the bacterial culture and sensitivity. Unfortunately, there was no significant clinical improvement during the next 6 days after starting the treatment. A new blood sample was taken and tested for *Borrelia burgdorferi* antibodies using the modified indirect immunofluorescence-test. The bird showed an antibody-titre of ≥ 1:256 (Fig 1). An antibiotic treatment similar to the protocol used in humans and dogs was performed with doxycycline 50 mg/kg PO q24h (Ronaxan, Merial, Hallbergmoos, Bavaria, Germany). The falcon recovered well; after 10 days there were no signs of diarrhea and the joint appeared to be nearly of normal size.

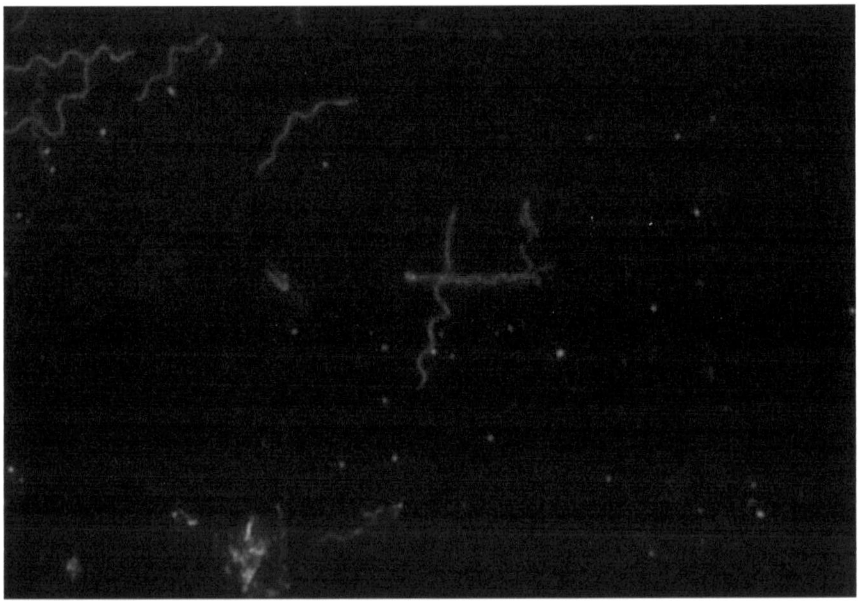

Figure 1. Microscopic photo of modified positive IFA test for *Borrelia burgdorferi* in a peregrine falcon (*Falco peregrinus*).

Discussion

To the authors' knowledge, this is the first study identifying *B. burgdorferi* antibodies in raptors and it seems to be the first case report of borreliosis in a bird of prey. Until now, it appeared that *B. burgdorferi* is asymptomatic in avian species.[7] Infected birds are thought to play a role in the transmission, maintenance and long-distance movement of Lyme disease.[7-9] A large number of bird species, primarily ground-foraging passerines but also sea birds, act as competent reservoirs for *B. burgdorferi*.[7,9] Information about the prevalence of *B. burgdorferi* in different bird species or in birds generally is difficult. Large-scale studies with more than 1000 birds report values of 4.4-19%.[10-12]

Two different enzootic cycles for the widespread of *B. burgdorferi* by birds are described: the terrestrial enzootic cycle and the marine enzootic cycle.[7] In the terrestrial enzootic cycle many birds are associated with the dispersal of vector ticks and therefore the distribution of *B. burgdorferi* across their annual migration routes.[13-17] During these routes, migrating birds use different stopover sites where they feed and rest. At these locations ticks may attach and later detach further along the migration routes or even in breeding and wintering areas. New foci of tick-borne diseases may become established in this way.[7] In the marine enzootic cycle the seabird-associated tick *Ixodes uriae* is the main vector in this cycle. Seabirds often live in large colonies of thousand to millions of individuals, especially during the breeding season; therefore, ticks and also *Borreliae* can easily be spread.[7,18] A global transmission cycle including a transhemispheric exchange also is assumed, because the same *B. garinii* spirochetes were found in seabirds in the Northern and Southern polar region, even on mammal-free islands.[7,19] The relatively low body temperature of seabirds may play a role in the maintenance of spirochetemia.[7,18,20]

Raptors, in particular common buzzard and eagle owl, could have an important role in the terrestrial enzootic cycle of *B. burgdorferi*. The importance of raptors in Lyme disease likely has been underestimated until now. In the future, birds of prey should be considered as potential carriers of Lyme disease.

References

1. Franke J, Moldenhauer A, Hildebrandt A, Dorn, W. Are birds reservoir hosts for *Borrelia afzelii*?, *Ticks Tick Borne Dis*. 2010;1:109-112.

2. Olsen B. Borrelia. In: Thomas NJ, Hunter DB, Atkinson CT, eds. *Infectious Diseases of Wild Birds*. Ames, IA: Blackwell Publishing;2007.

3. Hubálek Z. Epidemiology of Lyme borreliosis. *Curr Probl Dermatol*. 2009;37:31-50.

4. Radolf JD, Salazar JC, Dattwyller RJ. Lyme disease in humans. In: Samuels DS, Radolf JD, eds. *Borrelia Molecular Biology, Host Interaction and Pathogenesis*. Poole, UK: Caister Academic Press;2010:487-533.

5. Humair P-F, Gern L. The wild hidden face of Lyme borreliosis in Europe. *Microbes Infect*. 2000;2:915-922.

6. Burgess EC. Experimental inoculation of mallard ducks (*Anas platyrhynchos*) with *Borrelia burgdorferi*. *J Wildl Dis*. 1989;25:99-102.

7. Olsen B. *Borrelia*. In: Thomas NJ, Hunter DB, Atkinson CT, eds. *Infectious Diseases of Wild Birds*. Ames, IA: Blackwell Publishing;2007:341-351.

8. Reed KD. Birds, migration and emerging zoonoses: West Nile virus, Lyme disease, influenza A and enteropathogens. *Clin Med Res*. 2003;1(1):5-12.

9. Humair P-F, Rais O, Gern, L. Transmission of *Borrelia afzelii* from *Apodemus* mice and *Clethrionomys* voles to *Ixodes ricinus* ticks: differential transmission pattern and overwintering maintenance. *Parasitology*. 1999;118:33-42.

10. Dubska L. Differential role of passerine birds in distribution of *Borrelia* spirochetes, based on data from ticks collected from birds during the postbreeding migration period in central Europe. *Appl Environ Microbiol*. 2009;75(3):596-602.

11. Kjelland V, Stuen S, Skarpaas T, et al. *Borrelia burgdorferi* sensu lato in *Ixodes ricinus* ticks collected from migratory birds in Southern Norway. *Acta Vet Scand*. 2010;52(59):1-6.

12. Comstedt P. Migratory passerine birds as reservoirs of Lyme borreliosis in Europe. *Emerg Infect Dis*. 2006;12(7):1087-1095.

13. Anderson JF. Involvement of birds in the epidemiology of the Lyme disease agent *Borrelia burgdorferi*. *Infect Immun*. 1986;51(2):394-396.

14. Anderson JF, Magnarelli LA. Avian and mammalian hosts for spirochete-infected ticks and insects in a Lyme disease focus in Connecticut. *Yale J Biol Med*. 1984;57:627-641.

15. Ishiguro F. Prevalence of Lyme disease *Borrelia* spp. in ticks from migratory birds on the Japanese mainland. *Appl Environ Microbiol.* 2000;66(3):982-986.

16. Rand PW. Participation of birds (Aves) in the emergence of Lyme disease in Southern Maine. *J Med Entomol.* 1998;35(3):270-276.

17. Smith JR, Rand PW, Lacombe EH, et al. Role of bird migration in the long-distance dispersal of Ixodes dammini, the vector of Lyme disease. *J Infect Dis.* 1996;174:221-224.

18. Olsen B, Jaenson TG, Noppa L, et al. A Lyme borreliosis cycle in seabirds and *Ixodes uriae* ticks. *Nature.* 1993;362:340-342.

19. Olsen B, Duffy DC, Jaenson TG, et al. Transhemispheric exchange of Lyme disease spirochetes by seabirds. *J Clin Microbiol.* 1995;33(12):3270-3274.

20. Olsen B, Jaenson TG, Bergström, S. Prevalence of *Borrelia burgdorferi sensu lato*-infected ticks on migrating birds. *Appl Environ Microbiol.* 1995;61(8):3082-3987.

Idiopathic Splenomegaly in a Female Hyacinth Macaw and a Review of Splenomegaly in Large Psittacine Birds

April Romagnano, PhD, DVM, DABVP (Avian Practice),
Angela Lennox, DVM, DABVP (Avian Practice),
Carolyn Cray, PhD,
Drury Reavill, DVM, DABVP (Avian Practice), DABVP (Reptile and Amphibian Practice), DACVP

Session #2103

Affiliation: From Avian & Exotic Clinic of Palm City, Palm City, FL 34990, USA (Romagnano), Avian & Exotic Animal Clinic of Indianapolis, IN 46268, USA (Lennox), Division of Comparative Pathology, Department of Pathology & Laboratory Medicine, University of Miami, Miami, FL 33101, USA (Cray), Zoo/Exotic Pathology Service, Carmichael, CA 95608, USA (Reavill).

A necropsy on a previously apparently healthy female hyacinth macaw found dead, revealed a very remarkable splenomegaly. On histology there was splenic necrosis, hemorrhage, and histiocytic accumulation of hemosiderin. The remainder of the necropsy was unremarkable and other mild non-specific histologic lesions and special stains and other testing did not lead to identification of an etiology for the splenomegaly nor the bird's death. Months before death this bird had normal labwork. The conspecific male cagemate had an elevated WBC, positive *Sarcocystis*-AB, and a high blood zinc at the time of the female's death. Interestingly, an archived 1.5-year-old female green wing macaw had a remarkable splenomegaly and on histopathology splenic necrosis and hemorrhage also of unknown etiology. This pattern lead us to do a retrospective splenomegaly histopathology review. We noted that infrequent idiopathic splenomegaly in macaws and other large psittacine birds is histologically characterized by extensive necrosis, hemorrhage, and variable extramedullary hematopoiesis. The potential etiologies for these idiopathic splenomegalies include secondary responses to low grade infections (bacteria, fungus, *Chlamydia*, other infectious agents). Toxicity, including lead toxicity, and hemorrhage with or without necrosis from clotting deficiencies also are suspect. Known etiologic causes for splenomegaly in psittacine birds includes: sarcocystis, *Chlamydia*, polyomavirus, bornavirus secondary infections, iron storage disease, altered fat metabolism and accumulation of lipids, lymphoma, *Mycobacteria*, trauma and hematoma, metastatic tumors, and other bacterial, protozoal, fungal or parasitic infections. A summary of 60 cases (33 of unknown etiology) of splenomegaly in macaws and other large psittacine birds will be presented, compared, and discussed.

References

1. Schmidt RE, Reavill DR, Phalen D. *Pathology of Pet and Aviary Birds*. 2nd ed. Ames, IA: John Wiley and Sons Inc;2015.

Index

Q

R

S

T

U

V

W

X

Y

Z

Association of Avian Veterinarians
PO Box 9
Teaneck, NJ 07666
USA

ISBN 978-1-5108-4949-5

CURRAN ASSOCIATES INC.
proceedings.com

26th Annual Meeting of the North American Membrane Society (NAMS 2016)

Bellevue, Washington, USA
21-25 May 2016

North American Membrane Society